Dangerous Fishes of New Zealand

Laith A. Jawad

Dangerous Fishes of New Zealand

Springer

Laith A. Jawad
School of Environmental and Animal Sciences
Unitec Institute of Technology
Auckland, New Zealand

ISBN 978-3-031-69697-8 ISBN 978-3-031-69698-5 (eBook)
https://doi.org/10.1007/978-3-031-69698-5

This Springer imprint is published by the registered company Springer Nature Switzerland AG
The registered company address is: Gewerbestrasse 11, 6330 Cham, Switzerland

To my lovely family…………..I dedicate this book

Preface

More than 15 years have passed, but I can clearly recall the conversation I had with a few colleagues at the Auckland War Memorial Museum regarding potentially harmful fish and whether they exist in New Zealand. At that time, I explained to my coworkers who the dangerous fish were. At the end of the discussion, I informed them about my interest in writing a book about the dangerous fishes of New Zealand. One of them asked me, "Are you serious about writing such a book? I do not think there are dangerous fishes in the waters of New Zealand." Later, I went exploring some literature on dangerous fishes and then checked those species considered dangerous with the checklist of fishes of New Zealand, and it was a surprise when I found several New Zealand fish species falling into the different categories of what is known as dangerous fishes.

Two motives were behind my preference to write a book about the dangerous fishes of New Zealand: these are the high number of species that are allocated to the different groups of dangerous fishes and the unawareness of the New Zealanders to the hazards of the different groups of dangerous fishes living in their environment. Dangerous aquatic organisms may be encountered during recreational or fishing activities in coastal environments. Such organisms differ greatly and are, in general, of local or regional importance. The likelihood and nature of human experience are often determined considerably by the type of activity concerned. Because of the wide variety of organisms that may be confronted, the coverage of these fish in the present book is limited to only the fish species that are known to be dangerous to humans.

In each dangerous group, the information given is a historical background about the fish group, the causative agent, symptoms, treatment, and prevention. The account of the species includes identification, world distribution, habitat and ecological role, biology, economic value, and conservation status. To draw the attention of the readers and make them aware of the dangers of the species dealt with in this book, examples of cases of attack by dangerous fish or incidences of intoxication are included. Thanks to the fully informative species account used in Fishbase, I decided to adopt this format in the present book.

Two kinds of hazards can be marked relative to the dangerous aquatic species: the risk of infectious disease transmitted by species with lifecycles linked to the aquatic environment and injury or intoxication (e.g., ciguatera, histamine poisoning, etc.) rising from direct confrontation with large animals or venomous species. The obvious consequences involving dangerous fish may have critical economic significance in areas that depend greatly on recreational tourism and fisheries as a source of income. Among the aims of the present book is to educate locals about the dangers of this group of fish as part of a public education and awareness program that should be initiated in coastal areas. Consequently, it is imperative to recognize and evaluate the hazards posed by various fish in each region and bring the results to public attention. Moreover, at locations where hazards, including dangerous fish, have been recognized, procedures should be developed for treating any wounds incurred.

I would like to express my thanks to all the amateur and professional photographers who gave their permission to use their photos of various dangerous fish species in the wild and in the laboratory, as well as those who assisted in providing valuable references and information about the poisonous organisms in New Zealand.

Auckland, New Zealand Laith A. Jawad

Acknowledgments

For the following who granted me their permission to use the fish photos from their works, I would like to express my sincere thanks.

Aihartza, J. - Argazki-bilduma, N., Alpha, Melbourne, Australia, Amaral, A., Asman, P., FishBase, Ayling, T., Bailly, N., Philippines, Fishbase, Bañón Díaz, R., Spain, Barria, A., Wikimedia, Bartholomew, A. (1834-1909), Wikipedia, Belouahem, S., Ain Barbar Bay (Gulf of Annaba), Algeria, Bornatowski, H., Brandi, N., NOAA Fishbase, Bray, D. J., Australia, Bryant, P. J., Burkhead, N. Wikipedia, Cada. R. N., FishBase, Caine, C., Cihant, Turkey FishBase, Citron, Wikimedia, Citron, Wikipedia, CSIRO National Fish Collection, De Maddalena, A., De Vroe, J., FishBase, Demelas, M., Sardinia, Italy, Descouens, D., Wikipedia, Duarte, P. N., Portugal FishBase, Dubosc, J., Fishbase, Duffy, Clinton, New Zealand, Ehrensperger, T., Estrada, R., FishBase, Field, R., FishBase, Finucci, B., Giannakenas, C., Fish Base, Gomes, L., Brazil FishBase, Gratwicke, B., Wikipedia, Green. A. J., Reef life Survey, Gullibles travels, Headden, J. A., Wikipedia, Hobgood, N., Wikipedia, Hongming, C., Wikipedia, Hooker, Y., FishBase, Hutson, K., Adelaide, Australis, Iranja, N., Madagascar FishBase, Ishikawa, Y., Yukio Ueta, Izuzuki Wikipedia, Mora, J., Wikipedia, Janke, M., Kenya Fishbase, Johnson, S., FishBase, Justine, J.-L. Wikipedia, Kabasakal, H., Turkey, Fishbase, Karelj, Prague Czech Republic, Khan, M. M., Pakistan, Krajangdara, T. Phuket, Kretzberg, R., Maldives, FishBase, Martínez, J. A. G, Wikipedia, Matsuura, K. Fishbase, McDowall, R.M, New Zealand, Moore, D., Fishbase, Nesnad, Wikimedia, Holtzha, N. Wikipedia, Nobel, B., Fish Base, Oddone, L., Wikipedia, Orlov, A., FishBase, Osmany, H. B., Pakistan, Rault, N., FishBase, Reyes, R. Philippines FishBase, Schneider, F., Fish Base, Serrano López, A., Spain, Fishbase, Shao FishBase, Ian Shaw, Australia, Shebs, S., Wikipedia, Skipworth, I., Wikipedia, Smith, D., NMNH, Steele, Mark A., California, USA, Stef, South Africa FishBase, Tassapon, K., Fishbase, Thaman, R., Kadavau, E. Forika, Tokushima Pref, Japan, Tvskova, E., Fishbase, Vasenin, A., Wikipedia, Wayne, A., London, UK, Wikipedia, Wirtz, P., FishBase, Bernard Yau, Australia, Zsilavecz, G., South Africa.

A special thanks to Joacim Näslund, Department of Aquatic Resources, Institute of Freshwater Research, Swedish University of Agricultural Sciences, Drottningholm, Sweden, for editing the hand-drawn images in Photoshop.

My thanks are also due to Dr Philippa Johnstone, Hunter Corners Medical Centre, Auckland, for providing a valuable information about cases of ciguatera poisoning and to Dr Lesley Rhodes, Cawthron Institute, Nelson, New Zealand, for providing references on fish and shellfish poisoning in New Zealand.

Finally, I would like to thank all staff of Springer Publishing House for their distinguished service and valuable advice and suggestions during the publication process of this book.

Contents

Part II Poisonous

Part III Venomous Fishes

About the Author

Laith A. Jawad obtained a degree (MSc) in fish taxonomy from the Zoology Department, University of Bristol, UK, in 1980. He continued as fish taxonomist at Basrah University, Iraq, where he worked for more than 20 years before he immigrated to New Zealand in 1997. During this time, he started the biochemical taxonomy of fishes of Iraq and published over 550 scientific papers and book reviews in leading scientific journals. He is the author and co-author of several textbooks in biology published in Arabic. In 2016, he contributed five chapters to a book about coastal fishes, *Coastal Fishes: Habitat, Behavior and Conservation*, published by Nova Publishers, Canada, and in 2017 he authored a book *Dangerous Fishes of the Eastern and Southern Arabian Peninsula*, published by Springer. He served as fish biodiversity expert and consultant at the Ministry of Agriculture and Fisheries in Oman for the period 2008–2012 during which he co-authored two papers describing a new fish species from the Omani waters and reported over 80 fish species as a new record to the Omani waters. He authored a guide to the fishes of the southern coasts of Oman published by the Ministry of Agriculture and Fisheries in Oman in 2018. He published three books as an editor. These are *Tigris and Euphrates Rivers: Their Environment from Headwaters to Mouth*; *The Arabian Seas: Biodiversity, Environmental Challenges and Conservation Measures*; and *Southern Iraq's Marshes: Their Environment and Conservation*. These books were published by Springer in 2021. He also published over 100 papers on fish fauna of Oman, Iraq, Kuwait, and Saudi Arabia. In 2013, he broadened his scientific contact and started to collaborate with over 60 scientists from more than 50 countries around the world in research dealing with different aspects of fish taxonomy and ichthyology.

1 Introduction

At present, there is an expanding interest in aquatic environments, due to leisure and career activities. This can cause nasty confronts between animals and humans, who are not trained to prevent traumas and envenoming caused by the natural defenses of aquatic fauna specifically fishes. The main casualties of these dangers are swimmers, professional and sporting fishermen, surfers, and scuba divers.

In the aquatic environment, the main causes of human injuries are dangerous, poisonous, and venomous animals. Dangerous animals can attack and cause different levels of impairment to humans. Poisonous animals present toxins that cause deleterious effects when ingested or placed in surface contact with the victim. Animals such as puffer fish and toads cause this type of envenoming. Venomous signifies those animals that can introduce toxins through an apparatus such as spines or stingers. These toxins have diverse effects, causing proteolysis, myotoxicity, hemotoxicity (mainly hemolysis), cytotoxicity, and neurotoxicity. The main aquatic animals that cause emergencies in humans include the phyla Porifera (sponges), Cnidaria (anemones, corals, jellyfish, and Portuguese man-of-war), Annelida (marine worms), Mollusca (octopuses and Conus shells), Echinodermata (sea urchins, starfish, and sea cucumbers), Crustacea (crabs and mantis shrimp), and Chordata (fish and reptiles).

Beneath the shimmering surface of the world's oceans, rivers, and lakes lies a vast and diverse ecosystem teeming with life. While many aquatic species coexist harmoniously with humans, there exists a subset that commands caution and respect due to their potential danger and harmful attributes. From venomous spines to potent toxins, certain fish species pose significant risks to human health and safety.

Understanding these dangerous and harmful fish species is crucial for anyone venturing into aquatic environments, whether for recreational purposes, fishing, or scientific exploration.

Throughout history, encounters with dangerous fish have been documented, often resulting in injury or even fatalities. From the notorious box jellyfish to the formidable bull shark, the ocean is home to creatures whose presence demands vigilance and respect. Moreover, freshwater habitats host their own hazardous species, including electric eels and piranhas, capable of inflicting severe harm.

Moreover, human activities such as overfishing, habitat destruction, and climate change can exacerbate interactions between dangerous fish and humans. Understanding the ecological dynamics at play is essential in developing strategies to mitigate these risks and ensure the safety of both humans and ecosystems.

In this book, through the underwater realm of perilous fish species, the author aims to shed light on the often-overlooked hazardous lurking beneath the surface. By gaining insight into their behavior, habitats, and potential dangers, individuals can better navigate aquatic environments

L. A. Jawad, *Dangerous Fishes of New Zealand*, https://doi.org/10.1007/978-3-031-69698-5_1

and minimize the risk associated with encounters with these formidable creatures.

Damages caused by marine animals are mainly in the summertime when the coastal town population rises notably. Bathers constitute more than 90% of the victims, and the incidence of this type of accident is 0.1% or 1 in 1000 patients in emergency rooms. Nearly 50% of the victims are bathers who step on sea urchins and present traumatic or venomous injuries, 25% are bathers injured by cnidarians (jellyfish and Portuguese man-of war), and 25% are fishermen wounded by venomous fish, such as catfish and stingrays.

It is critical to know that the early stages of the injuries will always create an emergency for the victim, because of pain and bleeding associated with wounds. Pain can be very severe, and systemic manifestations can be present, such as settlement of the cardiac, respiratory, and urinary systems. Sporadically, there is risk of death on the victim.

In New Zealand, fish species that cause hazard to humans can be classified into two main groups, the dangerous and the poisonous and venomous groups. Dangerous fishes can be subdivided into several sets such as biting and predator fish species, which include sharks, eels, barracuda, triggerfish, and the ribbonfish; harmful fishes, which include needlefish and surgeonfish; and electric fish such as electric rays. The second group of dangerous fishes is poisonous and venomous fishes. In this large section, several cartilaginous and bony fish species are included.

Poisonous fishes can cause several intoxications to humans such as ciguatera fish poisoning (CFP) is the most customary nonbacterial disease linked to fish consumption (Friedman et al. 2008), influencing between 50,000 and 500,000 people per year (Fleming et al. 1998). Ciguatera is a common ichthyosarcotoxaemia with outstanding clinically vital neurological aspects. Large-sized fishes eat dinoflagellates and usually harbor the neurotoxin in their viscera. Ingestion of herbivorous and carnivorous fishes that have orally accumulated effective levels of ciguatoxins (CTXs) and possible maitotoxin (MTXs) can cause CFP in humans (Bagnis et al. 1979; Sims 1987). Latest investigations and studies have shown global upsurge in the occurrence and intensity of harmful algal events (Glibert et al. 2005; Hallegraef 2010). Despite being considerably underreported, CFP existence worldwide is increasing, with reports of a 60% increase in CFP in the Pacific Islands over the last decade (Skinner et al. 2011). Hallucinogenic fish poisoning or ichthyoallyeinotoxism is a rare form of ichthyosarcotoxism and is indicated by development of central nervous system disturbances, specifically hallucinations and nightmares. Ichthyoallyeinotoxism has been known in many regions around the world (Halstead 1988; Château-Degat 2003). Biotoxication occurs as a result of consuming either the head or flesh of certain fish. Clupeotoxin, a specific type of biotoxication, is caused by clupeoid fish such as sardines and herrings (family Clupeidae), anchovies (family Engraulidae), tarpons (family Elopidae), and bonefishes (family Albulidae). It is widespread in tropical and subtropical areas of the world. Some fatal cases were reported from different parts of the world due to consumption of sardines. Clupeotoxin is a sporadic and unpredictable public health problem in the Atlantic Ocean, Caribbean Sea, and Pacific Ocean. Gempylotoxication happens when fish species of the family Gempylidae are consumed. The incident is also known as keriorrhea (Greek word that means "flow of wax"), which is a gastrointestinal disorder that involves an oily orange rectal discharge that occurs after eating fish with high content of non-saponifiable lipids (wax esters), which are nondigestible by humans (Yohannes et al. 2002; Ling et al. 2009; Caro et al. 2011). The condition has been characteristically linked with the ingestion of escolar (*Lepidocybium flavobrunneum*), rudderfish (*Centrolophus niger*), or oilfish (*Ruvettus pretiosus*) (Givney 2002; Shadbolt et al. 2002; Ling et al. 2009), Scombrotoxin is a toxin usually produced by members of the family Scombridae and can initiate scombroid poisoning. It is also recognized as histamine fish poisoning, with signs and symptoms related to seafood allergies (Hungerford 2010). Mismanaged fish samples will lead to scombroid poisoning because of their intake. Histamine [2-(1H-imidazol-4-yl] ethanamine] and other decomposition products are formed in raw fish that has been mishandled in terms of time and temperature due to the bacterial and enzymatic conversion of free histidine (Rawles et al. 1996). The scombroid fish species have in common a high level of free histidine in their muscle tissues (Suyama and Yoshizawa 1973; Perez-Martin et al. 1988; Ruiz-Capillas and Moral 2004). It has been known that non-scombroid fish species of a wide range of fish groups are also included in initiating scombrotoxism (Taylor 1986; Hwang et al. 1997) as their flesh is rich in histidine (Lukton and Olcott 1958; Taylor 1986). Tetrodontoxin (TTX) is considered among the greatest potent natural marine toxins, which carefully obstructs the act potentials of voltage-gated Na + channels along nerves and skeletal and cardiac muscle membranes, without altering the resting membrane potentials. The name of this toxin came from the Tetraodontidae pufferfish family, from which it was originally isolated and separated. It is the most lethal toxin found in the marine environment (Jal and Khora 2015; Lago et al. 2015). Ichthyotoxin is less acknowledged compared with the other well-known poisoning such as ciguatera and tetrodotoxin. In fish species causing ichthyotoxin, the poison is contained to the gonads, where there is a certain link between gonad activity and toxin production. Freshwater species are mainly the source of this poisoning, but there are some other anadromous brackish water or marine species that are included within this category. Ichthyogallotoxin is a type of intoxication from poisonous fishes belonging to the family Cyprinidae. In New Zealand, two of these species are present, the grass carp *Ctenopharyngodon idella* and the silver carp (Silver bighead) *Hypophthalmichthys molitrix*. The grass and the silver carp are basically species of the aquaculture industry. The habit of eating the gallbladder of the fish is well-known, and many people from different ethnicities have adopted this habit. In rural Asia, people eat gallbladders from various fish species for the treatment of various disorders, including arthritis and decreased visual acuity (Lin and Lin 1999). When linked with toxic effects, it represents a form of ichthyotoxism resulting in acute renal failure. The toxin is believed to be cyprinol sulphate or cyprinol, a C27 bile acid (Hwang et al. 2000). Gallbladders of certain freshwater fish are increasingly recognized as an important source of ichthyosarcotoxism in Asian populations. The venomous fish can inject poison into the body of the victim through an apparatus such as the spine. In this group, different species of stingrays are included, in addition, several species of chimaeras, sharks, and catfish and several other members of the family Scorpaenidae.

Recently, several books were published discussing several issues surrounding dangerous, poisonous, and venomous fishes from different parts of the world, reflecting their relevance in today's world. Among these books are those by Habermehl (2012), Bücherl et al. (2013), Rusoke-Dierich and Rusoke-Dierich (2018), Santhanam (2015, 2017a, b, 2018, 2020, 2024), White (2017), White and Meier (2017), and Haddad (2021).

Poisons and toxins originating from venomous and poisonous organisms have been the center of many investigations over the past 70 years, most of which has been focused at comprehending the biochemical and physiological procedures by which they provoke their affected pathological results (Fox and Serrano 2007). Considerable knowledge has been acquired in terms of how poisons and venoms and their complex toxins give rise to conditions related to envenoming and poisoning, and in some isolated cases, there have been a few such factors promoted for therapeutic use (Zhang 2015). However, it has only been in the past decade that an explosion of interest has occurred in mining these natural, highly evolved libraries of bioactive toxins and poisons for use in pharmacotherapeutics as drugs or drug leads as well as in diagnostic applications (Fox and Serrano 2007).

Some of the toxins, toxin derivatives, or poisons from animal venoms and secretions are in several stages of progress for use as pharmaceuticals or diagnostics in human diseases. These said toxins may be used for human medical cases like hypertension, where snake venoms, mainly Viperidae venoms, have long been identified to be rich supplies of peptides as well as proteins (Robeva et al. 1991; Zhang et al., 1994;

Hayashi and Camargo 2005), heart disease (Calvete et al. 2005; Marcinkiewcz 2005), pain (Garber 2005), and diabetes (Young et al. 1999). These poisons and toxins have also been introduced as diagnostic tools (Marsh 2001). There is a rising implausible development in comprehension toward venoms at the genomic, transcriptomic, and proteomic stages, and consequently it will soon have a rather complete biomolecular valuation of many types of venom. In these cases, information and identification of new structures may provide relevant "leads" for use against appropriate drug targets (Fox and Serrano 2007).

References

Bagnis R, Kuberski T, Laugier S. Clinical observations on 3,009 cases of ciguatera (fish poisoning) in the South Pacific. Am J Trop Med Hyg. 1979;28:1067.

Bücherl W, Buckley EE, Deulofeu V, editors. Venomous animals and their venoms: venomous vertebrates. Elsevier; 2013.

Calvete JJ, Marcinkiewicz C, Monleón D, Esteve V, Celda B, Juárez P, Sanz L. Snake venom disintegrins: evolution of structure and function. Toxicon. 2005;45(8):1063–74.

Caro RJ, Cosculluela AM, Beltran LF, Rihuete HMA. Diarrea oleosa anaranjada. Keriorrhea inducida por pescado. An Pediatr. 2011;74:67–8.

Château-Degat ML. Les toxines marines: problèmes de santé en émergence. VertigO. 2003;4:1–11. https://doi.org/10.4000/vertigo.4698.

Fleming LE, Baden DG, Bean JA, Weisman R, Blythe DG. Seafood toxin diseases: issues in epidemiology and community outreach. In: Reguera B, Blanco J, Fernandez ML, Wyatt T, editors. Harmful Algae. Zunta de Galicia and Intergovernmental Oceanographic Commission of UNESCO; 1998. p. 245–8.

Fox JW, Serrano SM. Approaching the golden age of natural product pharmaceuticals from venom libraries: an overview of toxins and toxin-derivatives currently involved in therapeutic or diagnostic applications. Curr Pharm Des. 2007;13(28):2927–34.

Friedman MA, Fleming LE, Fernandez M, et al. Ciguatera fish poisoning: treatment, prevention and management. Mar Drugs. 2008;6:456–79.

Garber K. Peptide leads new class of chronic pain drugs. Nat Bio Tech. 2005;2005(23):399.

Givney RC. Illness associated with rudderfish/ escolar in South Australia. Commun Dis Intell. 2002;26:440.

Glibert PM, Anderson DM, Gentien P, Graneli E, Sellner KG. The global, complex phenomena of harmful algal blooms. Oceanography. 2005;18(2):136–47.

Habermehl G. Venomous animals and their toxins. Springer Science & Business Media; 2012.

Haddad VH Jr. Medical emergencies caused by aquatic animals: a biological and clinical guide to trauma and envenomation cases. Springer Nature; 2021.

Hallegraeff GM. Ocean climate change, phytoplankton community responses, and harmful algal blooms: a formidable predictive challenge. J Phycol. 2010;46:220–35.

Halstead BW. Poisonous and venomous marine animals. Princeton: Darwin Press Inc; 1988.

Hayashi MA, Camargo AC. The bradykinin-potentiating peptides from venom gland and brain of *Bothrops jararaca* contain highly site-specific inhibitors of the somatic angiotensin-converting enzyme. Toxicon. 2005;45:1163–70.

Hungerford JM. Scombroid poisoning: a review. Toxicon. 2010;56(2):231–43.

Hwang DF, Chang SH, Shiau CY, Chai T. High-performance liquid chromatographic determination of biogenic amines in fish implicated in food poisoning. J Chromatogr B. 1997;693:23–30.

Hwang DF, Yeh YH, Lai YS, Deng JF. Identification of cyprinol and cyprinol sulphate from grass carp bile and their toxic effects in rats. Toxicon. 2000;39:411–4.

Jal S, Khora SS. An overview on the origin and production of tetrodotoxin, a potent neurotoxin. J Appl Microbiol. 2015;119:907–16.

Lago J, Rodriguez LP, Blanco L, Vieites JM, Cabado AG. Tetrodotoxin, an extremely potent marine neurotoxin: distribution, toxicity, origin and therapeutical uses. Mar Drugs. 2015;13:6384–406.

Lin Y-F, Lin S-H. Simultaneous acute renal and hepatic failure after ingesting raw carp gall bladder. Nephrol Dial Transplant. 1999;14(8):2011–2.

Ling KH, Nichols PD, But PP-H. Fish-induced Keriorrhea. In: Taylor S, editor. Advances in food and nutrition research, vol. 57. Burlington: Academic Press; 2009.

Lukton A, Olcott HS. Content of free imidazole compounds in the muscle tissue of aquatic animals. Food Res. 1958;23:611–8.

Marcinkiewcz C. Functional characteristic of snake venom disintegrins: potential therapeutic implication. Curr Pharm Des. 2005;11(7):815–27.

Marsh NA. Diagnostic uses of snake venom. Haemostasis. 2001;2001(31):211–7.

Perez-Martin R, Franco J, Aubourg S, Gallardo J. Changes in free amino acids content in albacore (*Thunnus alalunga*) muscle during thermal processing. Z Lebensm Unters Forsch. 1988;187:432–5.

Rawles DD, Flick GJ, Martin RE. Biogenic amines in fish and shellfish. Adv Food Nutr Res. 1996;39:329–65.

Robeva A, Politi V, Shannon JD, Bjarnason JB, Fox JW. Synthetic and endogenous inhibitors of snake venom metalloproteinases. Biomed Biophys Acta. 1991;50:769–73.

Ruiz-Capillas C, Moral A. Free amino acids, and biogenic amines in red and white muscle of tuna stored in controlled atmospheres. Amino Acids. 2004;26:125–32.

Rusoke-Dierich O, Rusoke-Dierich O. Venomous and dangerous marine animals. Diving Med. 2018:337–55.

Santhanam R. A handbook of harmful marine life. Cambridge: Cambridge International Scientific Publishing; 2015. p. 270.

Santhanam R. Biology and ecology of toxic pufferfish. Apple Academic Press; 2017a.

Santhanam R. Biology and ecology of venomous stingrays. Apple Academic Press; 2017b.

Santhanam R. Biology and ecology of venomous marine scorpionfishes. Academic Press; 2018.

Santhanam R. Biology and ecology of venomous marine cnidarians. Singapore: Springer; 2020.

Santhanam R. Biology and ecology of the venomous catfishes. CRC Press; 2024.

Shadbolt C, Kirk M, Roche P. Diarrhoea associated with consumption of escolar (rudderfish). Commun Dis Intell Q Rep. 2002;26(3):2002.

Sims JK. A theoretical discourse on the pharmacology of toxic marine ingestions. Ann Emerg Med. 1987;16:1006.

Skinner MP, Brewer TD, Johnstone R, Fleming LE, Lewis RJ. Ciguatera fish poisoning in the Pacific islands. PLoS Negl Trop Dis. 2011;5:e1416.

Suyama M, Yoshizawa Y. Free amino acid composition of the skeletal muscle of migratory fish. Bull Jpn Soc Sci Fish. 1973;39:1339–43.

Taylor SL. Histamine food poisoning: toxicology and clinical aspects. Crit Rev Toxicol. 1986;17(2):91–128.

White J. Poisonous and venomous animals-the Physician's view. In: Handbook of clinical toxicology of animal venoms and poisons. CRC Press; 2017. p. 9–26.

White J, Meier J. Handbook of clinical toxicology of animal venoms and poisons. CRC Press; 2017.

Yohannes K, Dalton CB, Halliday L, Unicomb LE, Kirk M. An outbreak of gastrointestinal illness associated with the consumption of escolar fish. Commun Dis Intell. 2002;26:441–5.

Young AA, Gedulin BR, Bhavsar S, Bodkin N, Jodka C, Hansen B, Denaro M. Glucose-lowering and insulin-sensitizing actions of exendin-4: studies in obese diabetic (ob/ob, db/db) mice, diabetic fatty Zucker rats, and diabetic rhesus monkeys (Macaca mulatta). Diabetes. 1999;48(5):1026–34.

Zhang Y. Why do we study animal toxins? Zool Res. 2015;36(4):183.

Zhang D, Botos I, Gomis-Rüth F-X, Doll R, Blood C, Njoroge G, et al. Structural interaction of natural and synthetic inhibitors with the matrix metalloproteinase, Atrolysin C (Ht-d). Proc Natl Acad Sci USA. 1994;91:88447–51.

Part I

Harmful Fishes

2 Biting and Predator Fish Group

2.1 Cartilaginous Fishes

Order: Heterodontiformes
Family: Heterodontidae
Heterodontus portusjacksoni (Meyer 1793) (Fig. 2.1).
Common name: Port Jackson shark.
Etymology: *Heterodontus:* Greek, *heteros* = other + Greek, *odous* = teeth.

Diagnostic Features

- Molariform teeth in rear of mouth greatly expanded and rounded, without medial ridges.
- The origin of the first dorsal fin is located over pectoral bases.
- The tip of the anal fin is situated well short of the lower caudal fin origin.
- Anal fin base is about 3 times distant from anal fin insertion to lower caudal fin origin.
- Body color is gray to light brown or whitish with a prominent black bar across the head and down the cheeks, with a triangular black harness-like set of bars converging on the sides and back from the first dorsal origin, pectoral and pelvic fins bases, and a horizontal black stripe on the sides of the back and tail.

Distribution

This species is a common inhabitant on the Australian continental shelf south of 30°S from Byron Bay (New South Wales) to the Houtman Abrolhos (Western Australia), including Tasmania, but records from York Sound (northern Western Australia) and Moreton Bay (Queensland) are questionable (Huveneers and Simpfendorfer 2015). In New Zealand, one specimen of this species is reported from waters near Wellington (Cox and Francis 1997; Compagno 2001; Last and Stevens 2009). The specimen from New Zealand is a female caught in 1954 (Cox and Francis 1997) and a voucher specimen held at the NMNZ (Paulin et al. 1989).

Habitat, Biology, and Ecology

Heterodontus portusjacksoni is a night feeder found abundantly of coastal reefs throughout its range down to depths of 275 m (Last and Stevens 2009). A detailed description of the habitat this species living in was given by McGaughlin and O'Gower (1971) and many other investigators who studied various facets of the biology of this shark such as Rodda (2000), Tovar-Avila (2006), Powter (2007), Ramos (2007), and Bass (2021).

O'Gower (1995) reported on tagging experiments done on this species and showed that this shark has the ability of vast movement between central New South Wales and Tasmania (O'Gower 1995), with individuals having been recorded up to 850 km from the reproductive areas. Studies on reproduction and breeding of this shark have shown that this species migrates south at the end of the breeding season to spend the summer and autumn in the deeper waters of the Bass Strait (O'Gower and Nash 1978). This shark, during the breeding season, displays a high level of site reliability, with males and females outlaying

L. A. Jawad, *Dangerous Fishes of New Zealand*, https://doi.org/10.1007/978-3-031-69698-5_2

Fig. 2.1 *Heterodontus portusjacksoni* (Meyer 1793). (Courtesy of Kate Hutson, Adelaide, Australis)

about 90 and 85% of their time, respectively, on a single breeding reef and coming back to the same sites year after year (Huveneers and Simpfendorfer 2015).

This shark can reach a size of 165 cm total length (TL) (Huveneers and Simpfendorfer 2015). Males reach maturity at a smaller size than females across the species' range (Powter and Gladstone 2008; Izzo and Rodda 2012). In addition, there is a geographical variation in size at maturity between the two sexes. The size difference between the two sexes is greater in Western Australia specimens, where males mature at 58–65 cm TL and females at 80–90 cm TL. In South Australia, males mature at 55–57.5 and females at 65–75 cm TL. In Victoria, males mature at 67.5–82 and females at 85–99 cm TL, and in New South Wales, males mature at 60–77 and females at 90–91 cm TL (Izzo and Rodda 2012). Moreover, there is a variation in size at birth too. In Western Australia, the size varies between 18 and 21 cm TL, while in Victoria, it is up to 29.5–31.5 cm TL. The number of eggs produced by females range between 10 and 16 eggs each year, with a mean ovarian fecundity of 16 (Powter and Gladstone 2008). Eggs usually laid in crevices and cavities of shallow rocky reefs during the Austral winter to spring, climaxing in August–October, with hatching occurring 10–11 months later (Powter and Gladstone 2008; Rodda and Seymour 2008). Geographical variation in the age of maturity between the two sexes is present in distribution range of this shark. In South Australia, males mature at about 6 and females at about 7–8 years old, while in Victoria and New South Wales, males mature at about 7–12 and females at 12–17 years old (Powter 2007; Tovar-Avila et al. 2009; Izzo and Rodda 2012).

Economic Value

Important species in the aquarium trade on an international level.

Conservation Status

IUCN Red List status: Least Concern. Huveneers and Simpfendorfer (2015) reported on the conservation status of this shark and indicated that even though the bycatch component of this shark is a large bycatch in the several fisheries across its range, utmost individuals are reverted to the water alive, and post-release stress studies have shown that the species is very resistant to capture stress from gillnet, trawl, and longline gear, signifying high post-release survival rates. The assessed decline of this shark in Bass Strait between 1973–1974 and 1998–2001 has now stopped and has been inverted as fishers no longer aim the species. In addition, a rapid semi-quantitative ecological risk valuation method revealed that this shark is at low risk from several fisheries because of its low catch vulnerability. There is presently no indication to propose that this shark confronts any danger of extermination, qualifying a listing of Least Concern.

Human Interaction

This shark is deliberately harmless to people; however, apparently, it can bring a painful bite when provoked as in *H. francisci* (Huber et al. 2005), as it is a genuie durophagy (Compagno 1984a, b, 2001; Summers et al. 2004). Hardiman et al. (2020) in their article on shark-human interactions as reported in major Australian newspapers mentioned one case reporting on the attack of this shark on humans. In October 2011, a man was bitten by a Port Jackson shark at Elwood Beach near Melbourne (Buttler 2011). The bite did not pierce the skin, and the man was able to swim away while the shark was latched on to his calf. The man, a skilled scuba diver who was wearing a wetsuit, did not need to go to the hospital. The 50-year old made his way to shore, still with the shark attached to his leg, before it finally let go after about 20 min. The shark is believed to have been a 1-m-long Port Jackson shark. The victim told the paramedic that he had spent hundreds of hours diving but had never experienced anything like this. The newspaper reporter was informed by the victim that he was swimming along when he felt something latch onto his leg, and when he turned around, he saw the shark fixed to his leg. The victim started to swim for shore and called out for help. The reporter said that by the time the victim made it to shore, the shark was still hanging on, but after a while, he managed to dislodge it, and it swam back out into the bay. The paramedic assessment showed that there wasn't any damage to his wetsuit, but he did have a small bruise to his calf. The victim didn't need any medical attention, and the shark was free to go on his way. Southcott (1970, 1977) suggested that the common Port Jackson shark of Australia and New Zealand, *H. portusjacksoni*, is commonly harmless, but a few stings have been recorded from puncture by a dorsal spine as these are found one in front of each of the dorsal fins. More information about this incidence will be given in Part 2, section 7 of this book.

Order: Lamniformes
Family: Alopiidae
Alopias vulpinus (Bonnaterre 1788) (Fig. 2.2).
Common name: Common thresher.
Etymology: *Alopias:* Greek, *alopex* = fox; *vulpinus:* name comes from the Latin *vulpes*, meaning "fox" (Hart 1973).

Diagnostic Features

- Eyes moderately large at all sizes, orbits not expanded onto dorsal surface of head.
- Dorsal profile of head convex and forehead strongly convex in lateral view.
- Space between dorsal edges of eyes broadly convex.
- Snout relatively short, conical, and pointed.
- No grooves on head above gills; labial furrows present; teeth small, over 29 rows in either jaw.
- Pectoral fins falcate and narrow-tipped.
- Terminal lobe of caudal fin moderately large.
- Body almost uniformly dark brown with abdomen slightly whitish.

Distribution

The shark *A. vulpinus* is present worldwide in tropical to cold-temperate seas (Last and Stevens 2009; Ebert et al. 2013). Presence of this shark in the equatorial and northern tropical Indian Ocean could be misidentified with the Pelagic Thresher (Rigby et al. 2019a). This shark is recorded from the Western Atlantic: New York to Florida (USA), Bahamas, Cuba, Venezuela, and southern Brazil; in the Eastern Atlantic at Portugal, Madeira, Senegal, and Guinea to Sierra Leone and Angola; and also the Mediterranean Sea. It is also recorded

Fig. 2.2 *Alopias vulpinus* (Bonnaterre 1788). (Courtesy of Mark A. Steele, California, USA)

to be present in the Western Indian Ocean at South Africa, Madagascar, Arabian Sea, and the Western Pacific at Southern Japan, Taiwan, New Caledonia, Australia (northwestern coast), and New Zealand. Its presence in the Central Pacific is confirmed in the North and south of Hawaiian Islands and in the Eastern Pacific at Southern California, the Gulf of California, and west of the Galapagos Islands (Compagno 1984a).

Habitat, Biology, and Ecology

This shark is oceanic and coastal living at depths of 650 m (Ebert et al. 2013; Weigmann 2016). It is more often found nearby land and in temperate waters (Ebert et al. 2013). Individuals of this shark reach a maximum size of 573 cm total length (TL) and possibly 635 cm TL. Size at birth is 120–150 cm TL; males mature at 260–420 cm TL, and females mature at 260–465 cm TL (Ebert et al. 2013; Young et al. 2016). The probable annual rate of population increase under sustainable fishing is higher than that of the Pelagic Thresher and Bigeye Thresher and has been estimated at 0.254 (Dulvy et al. 2008). Females reach maturity at age of 13, and maximum age is 38 years based on bomb radiocarbon-validated ages from the Northwest Atlantic (Natanson et al. 2016). There is a possibility of geographical variation in life history parameters, but assuming a cautionary method, the legalized bomb radiocarbon ages are used for both regions to give a generation length of 25.5 years.

Economic Value

This shark has several applications. It is used for its meat, fins, liver oil, and skin (Compagno 2001; Goldman 2005; Jabado et al. 2015). Three species of thresher shark, common thresher, bigeye thresher, and pelagic thresher, collectively accounted for 2–3% in 1991–2001 and 0.5% in 2014, for its fin imported in Hong Kong (Clarke et al. 2006a, b; Fields et al. 2018).

Conservation Status

This shark is caught as both target and bycatch in pelagic and coastal commercial and small-scale longline, purse seine, and gillnet fisheries (Rigby et al. 2019a). The species is frequently used for its meat and fins, except for domestic or international rule bans. Nevertheless, high post-release mortality stays a threat where usage is banned. The species is assessed to be deteriorating in the North Atlantic and Mediterranean Sea and increasing in the Eastern Pacific, following regulations instigated in the mid-1980s. Based on levels of exploitation and the estimated global population reduction of 30–49% over the last three generations (76.5 years), globally, this species of shark is assessed as Vulnerable A2bd (Rigby et al. 2019a).

Human Interaction

Despite their fairly large size, thresher sharks do not create a danger to humans as it is not seen as a possible source of food, but they are animals that in general must be dealt with attention as they are accomplished of imposing deep wounds with their teeth and to break bones with powerful tail (Cacutt 2000). To date, only one attack on humans and four on boats have been recorded (probably due to individuals caught accidentally and chiefly combative), as well as an alleged attack on a New Zealand free diver (Cacutt 2000). The other story on the attack of this shark on human is about an American fisherman decapitated by a blow of the tail of a 5 m thresher shark (Cacutt 2000). Ricci et al. (2016) listed this species of shark among the species of shark showing minority of attack cases globally.

Family: Cetorhinidae

Cetorhinus maximus (Gunnerus 1765) (Figs. 2.3 and 2.4).

Common name: Basking shark.

Etymology: *Cetorhinus:* Greek, *ketos* = a marine monster, whale + Greek, *rhinos* = nose; *maximus:* After *Selache* Cuvier, 1816, a synonym for *Cetorhinus* Blainville, 1816.

Diagnostic Features

The trunk is fusiform and moderately stout. The head is moderately long but much shorter than the trunk; snout moderately long, pointed, and conical, not depressed, flattened, and bladelike; eyes small; mouth large and arcuate, ventral on head, gill openings extremely large, extending onto dorsal and ventral surfaces of head, all anterior to pectoral fin bases; teeth very small, hooklike, not blade-shaped, and with over 200 rows in either jaws; several rows of small anterior teeth in upper jaw, separated from the laterals by a broad gap. The first dorsal fin is large, high, erect, and angular; second dorsal and anal fins moderately large but less than half the size of first dorsal, with broad bases; pectoral fins long and moderately broad, much shorter than head in adults; pelvic fins smaller than first dorsal fin but larger than second; caudal fin lunate, upper lobe moderately long but less than one-third length of rest of shark, lower lobe nearly as long as upper lobe. Precaudal pits are present and caudal peduncle depressed and with strong lateral keels.

Distribution

The basking shark is present mainly in the Atlantic and Pacific Oceans; in the Indian Ocean, it is reported only from southern Australia, Indonesia, and South Africa (Ebert et al. 2013; Fahmi and White 2015). Therefore, it is a cosmopolitan species, frequent in cold to warm temperate waters; rare in equatorial waters (Ebert 2003). Highly migratory species (FAO 1992).

Fig. 2.3 *Cetorhinus maximus* (Gunnerus 1765). (Courtesy of Mario Demelas, Sardinia, Italy)

Fig. 2.4 Head of *Cetorhinus maximus* (Gunnerus 1765). (Courtesy of Samih Belouahem, Ain Barbar Bay (Gulf of Annaba), Algeria)

Habitat, Biology, and Ecology

This species of shark is the second largest shark, reportedly reaching 1220–1520 cm TL (Compagno 1984a), and is thought to live up to 50 years (Parker and Stott 1965). As semi-oceanic or oceanic species, they are highly migratory (Compagno 2001). Individuals of this species prefer living on continental and insular shelves, offshore, and frequently adjacent to land. It has been observed entering bounded bays (Compagno 1984a). This species of shark lives in a coastal-pelagic at 1 m to unknown depths, probably epipelagic (Mundy 2005). It usually lives in single individuals, in pairs, and, in rare occasions, in groups of three or more. A huge school containing up to 100 individuals has been also reported (Last and Stevens 1994).

Individuals of this species can make wide-ranging horizontal and vertical movements along the continental shelf and shelf edge to utilize productive feeding areas (Sims et al. 2003). It becomes near the surface of boreal to warm-temperate areas during the summer months (Compagno 2001). Its food is composed mainly of zooplankton, and it gets its meal by filtering (Sims and Quayle 1998). In winter months of the year, individuals of this species move deeper in water (Last and Stevens 1994; Sims et al. 2003; Mundy 2005). This shark showed a long trans-oceanic migration (e.g., from the British Isles to Newfoundland, Canada (Gore et al. 2008)) and moves between the northern and southern hemisphere in tropical mesopelagic water (Skomal et al. 2009). These migrations have been found to cover distances of over 9000 km.

Individuals of this type of shark may become a part due to their sizes (Narberhaus et al. 2012). It is ovoviviparous (aplacental viviparity), with embryos feeding on other ova produced by the mother (oophagy) after the yolk sac is absorbed (Dulvy and Reynolds 1997). Assessed gestation period is 12–36 months (Compagno 2001; Pauly 1978, 2002; Parker and Stott 1965). TL at birth is estimated between 150 and 200 cm, and mating is thought to take place during early summer; a female giving birth to a young was captured in August in Norwegian waters (Kunzlik 1988). A distinct pairing with embrace was observed in individuals of this shark (Breder and Rosen 1966). Courtship behavior forms linked with elasmobranch reproduction were documented from an aerial observation, which include "parallel swimming," "pectoral biting," "nudging," and "male on top of female." Other mating behaviors particular to this species from the same aerial observation include nose-to-tail following; flank approach; close approach involving rostral contact or proximity with the gill, pectoral fin, vent, and dorsum; and echelon swimming (Harvey-Clark et al. 1999). Among the parasitic animal seen attached to this shark is lamprey, and it has been observed of such an activity in Bay of Fundy, Canada (Bird 2009).

Economic Value

This species is utilized fresh, frozen, dried, or salted (Frimodt 1995) and is also valued for its liver for oil, fins for soup, hide for leather, and carcass for fishmeal (Compagno 2001). It may be a potential source of anticarcinoma drugs (Lee et al. 1984; Langer 1989) and is used in Chinese medicine (Tang 1987). This species is threatened due to bycatch fisheries (HELCOM 2007), but Rigby et al. (2019b) suggested that no utilization or commercial trade of this species is currently known to exist, although prior to 2012, meat and fins were exported from Norway and fins from New Zealand. Also, they reported on no commercial trade records having been noted in the Convention on International Trade in Endangered Species (CITES) trade database since 2012, although not all trade records may be reported to CITES. However, it is likely that if caught, the meat would still be used fresh for either local consumption or exported internationally along with the fins (Rigby et al. 2019b).

Conservation Status

Rigby et al. (2019b) assessed the basking shark (*Cetorhinus maximus*) as Endangered A2bd in the IUCN Red List and gave a set of reasons for considering this rank. They wrote:

> It is a very large (to ~1100 cm total length), highly migratory. The species has low biological productivity with limited fecundity and late age-at-maturity. Although no longer targeted, it is still

caught as bycatch in trawl, trammel nets, and set-net fisheries, and becomes entangled in pot lines. The large fins are extremely valuable in the fin trade. Across regions, there have been some severe historic decline, however there are indications of some stability and possible slow recovery since cessation of target fishing and high levels of protection. The global population may now be beginning to stabilize, with signs of that from the Northeast Atlantic, although elsewhere there is little information upon which is assess stability. However, abundances are still estimated to be well below historic levels and there is ongoing demand for the high-value fins, and a global population reduction of 50–79% is suspected over the past three generations (102 years).

Human Interaction

This species is regarded as ordinarily harmless and inoffensive but potentially dangerous if attacked (particularly when harpooned) (Compagno 2001). Quigley (2007) wrote about this shark: the basking shark is potentially dangerous if provoked. Indeed, during September 1937, the species was involved in two non-fatal attacks on boats in Scottish waters. In addition, contact with its skin should be avoided, as its large dermal denticles have been known to inflict damage on divers and scientists (Florida Museum 2021). Ricci et al. (2016) considered the frequency of attack on humans from this shark as minority compared with the other dangerous shark species.

Family: Lamnidae.

Carcharodon carcharias (Linnaeus, 1758) (Figs. 2.5 and 2.6).

Common name: Great white shark.

Etymology: Carcharodon: Greek, *karcharos* = sharpen + Greek, *odous* = teeth; carcharias:

From the Greek *karcharios*, which refers to man-eater sharks.

Diagnostic Features

The body is heavy, huge, and spindle shaped. Snout is large, blunt, and conical with black eyes. Gill slits are large. First dorsal fin is large, and anal and second dorsal fins are very small. Tail

Fig. 2.5 *Carcharodon carcharias* (Linnaeus 1758). (Courtesy of De Maddalena, A., South Africa)

Fig. 2.6 Front view of *Carcharodon carcharias* (Linnaeus 1758). (Courtesy of De Maddalena, A., South Africa)

has crescent shape and conspicuous keel on caudal peduncle. Its mouth is subterminal with huge triangular teeth. Body is grayish brown in color with white abdomen. First dorsal fin has dark free rear edge because of the presence of black spots.

Distribution

This shark has a cosmopolitan type of distribution and is widely distributed in most oceans. It is found in the western and eastern Atlantic regions. It is also reported from the Seychelles, South Africa, Reunion, and Mauritius (Fricke 1999). In the Western Pacific, it is recorded from Siberia in the north to New Zealand in the south (Smith 1997).

Habitat, Biology, and Ecology

This species of shark is pelagic and prefers temperate continental shelf waters but also reported in estuaries and the open ocean and inhabits depths of 1200 m (Francis et al. 2012). The white shark takes on long-distance trans-oceanic movements, for example, between South Africa and Australasia (Pardini et al. 2001; Bonfil et al. 2005) and California and the Hawaiian Islands (Boustany et al. 2002; Domeier and Nasby-Lucas 2008; Jorgensen et al. 2010). Therefore, its dispersal is not deliberately discontinuous; rather, that interchange between some areas may be limited (Rigby et al. 2019c).

There is no actual estimation of size of this shark but assessed at 600–640 cm total length (TL) (Compagno 2001). Males mature at 310–410 cm TL (Pratt 1996; Compagno 2001; Tanaka 2011), while females mature at 400–500 cm TL (Francis 1996; Tanaka 2011). Reproduction is aplacental viviparous with oophagy and histrophy, with litter sizes of 2–17 and a suspected two-to-three-year reproductive cycle (Francis 1996; Uchida et al. 1996; Mollet and Cailliet 2002; Bruce 2008; Domeier 2012; Sato et al. 2016). There is a geographical variation in the maturity of females of this shark. It has been shown that in the Northwest Atlantic, Northeast Pacific, and Western Indian Oceans, female age at maturity of 30–33 years and maximum age of

30–73 years were reported (Hamady et al. 2014; Andrews and Kerr 2015; Natanson and Skomal 2015; Christiansen et al. 2016). The maximum age suggested for this shark is likely >30 years, with a minimum of 44 years and possibly a maximum of 73 years (Rigby et al. 2019c). They usually go solitary or in pairs but can be found in feeding aggregations of ten or more and do not form schools (Compagno 2001). They feed on bony fishes, sharks, rays, seals, dolphins and porpoises, seabirds, carrion, squid, octopi, crabs (Compagno 1998), and whales (Long 1991).

Economic Value

Both fins and jaws of this shark can fetch high prices, with large fins used as display items (Compagno 1998). Not only large-sized individuals are involved in the fin's trade but also small-sized ones (Shivji et al. 2005). In some cases, collectors keep jaws as an ornament (Compagno 1998). The other usage of this shark is its meat, which can either be used for local consumption or exported internationally. Flesh is usually eaten fresh, dried-salted, and smoked. The skin is utilized as leather, liver for oil, carcass for fishmeal, fins for shark-fin soup, and teeth and jaws for decorations (Compagno 1998).

In some parts of the Arabian-Persian Gulf, the liver is used for waterproofing dhows, and carcasses are retained and used as fertilizer (Jabado et al. 2013). In Iran, there has been considerable recent research activity into the pharmaceutical applications of shark products, particularly for cartilage in anticancer treatment (Rabbani-Chadegani et al. 2008; Rabbani et al. 2007; Razmi et al. 2008; Shahrokhi et al. 2009) and for liver oil in combating fungal infections (Hajimoradi et al. 2009). In addition, gelatin extraction from rays for industrial use has been examined (Jalili 2004). In some states of the Arabian Persian Gulf, vertebrae of this type of shark are used as earrings (Moore et al. 2011).

Conservation Status

The conservation status of the white shark (*Carcharodon carcharias*) was assessed according to the RED List of IUCN by Rigby et al. (2019c) and has been given the rank of Vulnerable A2bd. The justification for assuming such rank is as follows: this species is caught globally, mostly as bycatch in inshore fisheries by a range of gears, is rarely caught in offshore pelagic fisheries, and is targeted in beach protection programs in Australia and South Africa; however, in some instances, these programs release live sharks. In most regions of the geographical distribution of this shark, declines occurred during the 1980s followed by slow recovery since the 1990s when protection was implemented. Globally, based on long-term abundance data and protections originated in the 1990s that have since reduced catches and allowed some recovery, the white shark population is estimated to have reduced by 30–49% over the last three generations (159 years).

Human Interaction

This shark has smart, complex social communications. With the presence of many sensitive cells, this shark is facilitated to discover the electromagnetic field produced by the movement of living animals up to half of billionth of a volt and can perceive the faint electrical pulse (Wikipedia 2015). It is one of a few sharks recognized to lift their head above water to locate prey and known to be a very inquisitive animal. The great white shark has the jaw power of 18,000 newtons (1814 kg) (Wroe et al. 2008). This shark is equipped with a hunting method known as breaching behavior, which is the result of a high-speed approach to the surface with the resulting momentum taking the shark clear out of the water (Martins et al. 2012). It is accountable for the highest number of shark attacks on humans ever conveyed. In 2012, records of 272 of unprovoked attacks on humans are given (SAF 2016). Peter Benchley's bestselling novel, *Jaws*, and the subsequent 1975 film adaptation directed by Steven Spielberg represent the best example of a malicious attack of this shark on humans. In the Mediterranean Sea and during the last two centuries, there were over 30 attacks stated against this shark, most of which were nonfatal. They were included under test bites, where this shark usually test-bites different objects such as buoys and flotsam and might snatch a human or part of his or her body (Benchley 2000).

Mortalities might seem to be low in comparison with the size and spiteful behavior of this species. It has been assumed that such a low proportion of fatalities could be because humans are able to escape after the first bite unlike other mammals. This is clear in the attacks of this shark when divers seemed to be partially consumed. The shark usually waits for the prey to be weakened after the first attack and then endures devouring it. In this aspect, human capability to move out of the scene with the help of others will not complete the attack state. Death usually happens because of great blood loss from the initial bite rather than for any other reason (Tricas and McCosker 1984). The great white shark is stated to attack boats and sometimes sink them. Such attacks are frequently made on kayakers, but it has happened that it attacks boats up to 10 m and bumps the boat from the stern, knocking people overboard. Shark attacks on boats are credited to the electrical field they produce (Tricas and McCosker 1984). Brading (2006) gave a total of 518 cases of shark attacks caused by white shark in the Western; among these, there were 89 fatal cases.

Ricci et al. (2016) classified *C. carcharias* among the five most common species involved in attacks against human and reported 389 attack cases against this shark. They reported that there was a trend toward increased fatality associated with great white shark attacks that did not reach statistical significance (9.0% vs. 7.3%; $p = 0.07$). Hardiman et al. (2020) showed that the number and percentage of shark articles featuring at least one photograph and/or graphic between April 1 and September 30, 2015, in four major Australian newspaper titles (Broadsheet, *Australian*, *Herald*; tabloid, *Mail*, *Telegraph*) is 1. 48 cases for great white shark. They mentioned that after photographs of the near-miss human subject "Fanning," the white shark was the second most-featured subject among all photographs.

Clua and Séret (2010) reported on a case of a fatal, unprovoked shark attack. The case took place on the 30th of September 2007, in the lagoon of Luengoni Bay, Lifou Island (Loyalty Islands, New Caledonia). A young French woman who was snorkeling was sternly bitten on the right thigh and died of hemorrhage. An analysis based on the size and color of the shark, the appearances of the wounds, and the behavior of the shark before and after the bite suggests that the attacker was a great white shark, *Carcharodon carcharias*. Also, and from the same area, Clua and Reid (2013) reported on an unprovoked case of fatal shark attack on a 19-year-old male surfer in New Caledonia. Several severe bites removed the right arm and all flesh from the right thigh, provoking a quick hypovolemic shock that was fatal. The information provided by a witness and the analysis of a partial bite on the right calf allowed us to identify a 2.7 m TL (est. length) white shark as responsible for this attack. Features of the attack are consistent with a young predator motivated by hunger and the development of its predatory skills.

In New Zealand, Reid (2020) reported several white shark attacks on humans. The first case the author reported on is that of a scuba diving near Stewart Island in the Foveaux Straight in 2013. The diver was attacked by a 2.5 m white shark that attempted an ambush from below. The shark caught hold of the diver's catch bag and left arm before the diver fended the shark off from the safety of a boat. The diver's left arm endures serious injuries, severing all tendons in his left hand and permanently limiting arm movement (Wade 2013). Reid (2020) went on talking about the cases of attack recorded against the white shark. In Otago and on February 4, 1964, at approximately 6:15 a.m., a 19-year-old male was swimming off St. Clair beach when a 3–4 m white shark bit both of his legs with the right leg being amputated at the knee. The swimmer was brought upon shore unconscious and called dead by medical personnel. Three years later, on March 9, 1967, another fatal human-shark interaction took place in very close proximity to where 1964 accident happened, where a 21-year-old was bitten. He was active in a surf lifesaving drill at 7:45 p.m. when what is thought to be a white shark took his life. Another swimmer, who was racing victim at the time of the interaction, recalls seeing a shark fin near the body of the casualty, followed by a pool of blood. The second swimmer was also injured and took to the hospital, and rather unusually, the body of the first victim was never found. For details of more shark attack cases recorded against the white shark, see Reid (2020).

A surfer in New Zealand survived a shark attack on Friday October 19, 2018, and had to be airlifted to the hospital after being bitten on his hand, elbow, and mouth. The man was at Baylys Beach, about 100 km south of Auckland, when the attack happened, according to police. He was left in pain and bleeding but was still able to walk and talk (New Zealand Herald 2018).

On January 8, 2021, the Guardian News and Media Limited (2021) reported on a shark attack case, where women die after suspected shark attack near Auckland. She was pulled from the waters at Waihi Beach, a 2-h drive southeast of Auckland. Emergency services were called, but the woman died at the scene. New Zealand police on Friday night confirmed a 19-year-old woman, from Hamilton, had been killed at the beach. A witness told the news website Staff he saw the woman swimming directly in front of the lifeguard flags, quite far out from the beach. "I could hear some screams and then I saw the IRB [inflatable rescue boat] go out to get her," he said. "Obviously her friends were extraordinarily distressed. The lifeguards got her in quickly and tried their hardest to keep her alive with CPR," he said. Another witness said paramedics performed CPR on the woman for more than 20 min as her friends gathered around.

Isurus oxyrinchus (Rafinesque, 1810).
Common name: Shortfin mako.
Etymology: Isurus: Greek, *isos* = equal + Greek, *oura* = tail (Figs. 2.7, 2.8, and 2.9).

Diagnostic Features

This type of shark has a large, spindle-shaped body with long pointed snout, and its eyes are large and black.

Fig. 2.7 *Isurus oxyrinchus* (Rafinesque 1810). (Courtesy of José Antonio Gil Martínez)

Fig. 2.8 Jaws of *Isurus oxyrinchus* (Rafinesque 1810). (Courtesy of Didier Descouens)

Fig. 2.9 Teeth of *Isurus oxyrinchus* (Rafinesque 1810). (Courtesy of Joxerra Aihartza)

Its mouth is U-shaped with long, slender, smooth-edged anterior teeth and posterior teeth smaller and triangular. Head is longer than pectoral fin. Caudal fin is lunate in shape with a strongly developed lower lobe. Second and anal fins are very small. Body is bright dark blue in color, and abdomen is white. Anterior half of pelvic fin is dark (Compagno 1998; Randall 1995; Compagno 1998; Ebert et al. 2013).

Distribution

This shark has a cosmopolitan type of distribution. It is reported from the western Atlantic region from the Gulf of Maine to southern Brazil and Argentina (Menni and Lucifora 2007). It is reported from several areas in the eastern Atlantic from Norway to South Africa including the Mediterranean Sea. It is widely distributed in the Indo-Pacific region from the east of Africa to Hawaii and south to Australia and New Zealand. It is also found in the eastern Pacific (Froese and Pauly 2021). There is no clear record of this species from the Arabian-Persian Gulf. Froese and Pauly (2021) reported this species from the Arabian-Persian Gulf based on Compagno (1984a). Checking this reference showed that the author did not mention Iranian waters but instead gave a generalized distribution of this species from the Red Sea westward to Pakistan, India, and Japan. Therefore, the presence of this species in the Arabian- Persian Gulf cannot be acknowledged. It is having been recorded from both the Sea of Oman and the southern coasts of the Arabian Peninsula (Randall 1995; Manilo and Bogorodsky 2003; Henderson and Reeve 2011).

There are comparatively few records of pregnant females, especially in the eastern North Pacific. However, there are many juveniles and young of the year fish in the southern California Bight indicating that it is a nursery area. In summer, the Southern California Bight is home to a large population of Shortfin Mako sharks (Klimley et al. 2002). Abundance surveys and fishing reports indicate that juvenile mako sharks appear in the Bight in spring when water temperatures rise above 16 °C and may depart from the area in fall when water temperatures decline.

Habitat, Biology, and Ecology

This species of shark is an active, offshore littoral and epipelagic species, inhabiting tropical and warm-temperate seas at a depth reaching 500 m, rarely found where water temperature is <16 °C (Compagno 2002). It is perhaps the fastest shark and is among the utmost active and powerful of fishes. Like other members of the family Lamnidae sharks, this shark is endothermic using a heat-exchanging circulatory system to maintain muscle and visceral temperatures above that of the surrounding seawater, allowing a higher level of activity (Carey et al. 1981; Bernal et al. 2001).

The shortfin mako lives well offshore but enters the inshore littoral just off the surf zone in some areas such as parts of KwaZulu-Natal, South Africa, where the continental shelves are narrow. In the eastern north Pacific, juveniles range into southern California waters and incline to be seen and caught near the surface (Cailliet et al. 1983). They appear to use these offshore continental waters as nursery areas (Taylor and Holts 2001). It was previously thought that they stay near the surface above 20 m depth, in waters between 20 and 21 °C, rarely descending into cold subsurface waters below the thermocline (Holts and Bedford 1993).

Tagging studies in the western north Atlantic show that this shark makes widespread movements of up to 3433 km with 36% of recaptures caught at greater than 420 km from their tagging site (Casey and Kohler 1992). However, only one fish crossed the mid-Atlantic ridge suggesting that trans-Atlantic migrations are not as common as in blue sharks *Prionace glauca* (Cailliet et al. 1983). Sepulveda et al. (2004) demonstrated that seven tagged juveniles remained near the surface at night and went as deep as 200 m, frequently during the day. Moreover, stomach temperatures were measured, showing feeding happened during the daytime, with meals taken during a dive causing stomach temperatures to drop noticeably.

This shark reaches a maximum size of about 4 m (Compagno 2001). Initial age and growth studies in the western north Atlantic suggested that two pairs of growth bands are laid down each year in their vertebral centra, at least in young shortfin makos (Pratt and Casey 1983). However, recent evidence using marginal increment analysis in Mexico (Ribot-Carballal et al. 2005) and bomb radiocarbon (Campana et al. 2002; Ardizzone et al. 2006) indicates that the other theory (one pair of growth bands per year; Cailliet et al. 1983) is valid. Age at maturity has been determined recently in several populations, including New Zealand (7–9 years for males and 19–21 years for females; Bishop et al. (2006)) and the western north Atlantic (8 years for males and 18 years for females (Natanson et al. 2006)). Longevity has been estimated as 29–32 years (Bishop et al. 2006; Natanson et al. 2006). There is a large difference in size at sexual maturity between the sexes. In the northwest Atlantic, males reach maturity at about 195 cm and females at about 265–280 cm (Pratt and Casey 1983; Stevens 1983; Cliff et al. 1990). In New Zealand, males mature at 198–204 cm and females at 301–307 cm (Francis and Duffy 2005). Compagno (2001) reports males mature between 203 and 215 cm, reaching a maximum size of 296 cm, and females mature ranging from 275 to 293 cm, reaching a maximum of almost 4 m.

This shark is ovoviviparous and oophagous, but what little is known of its reproductive cycle indicates the gestation period is 15–18 months, with a 3-year reproductive cycle (Mollet and Cailliet 2002). Litter size is 4–25 pups (possibly up to 30, mostly 10–18), which are about 60–70 cm long at birth (Garrick 1967; Compagno 2001). There are relatively few records of pregnant females. Among 26 shark species, this has an intrinsic rebound potential (a measure of its ability to recover from exploitation) in the mid-range (Smith et al. 1998).

The diet of this species of shark has been reported to consist mainly of teleost fishes (including mackerels, tunas, bonitos and other scombrids, anchovies, herrings, grunts, lancet fishes, cod, ling, whiting and other gadids, salmon, yellowtails and other carangids, sea basses, porgies, swordfish) and cephalopods in studies from the Northwest Atlantic and Australia (Stillwell and Kohler 1982; Stevens 1984), while elasmobranchs were the most common prey category from Natal, South Africa (Cliff et al. 1990).

Large-sized individuals of this species (over 3 m in length) have very broad, more flattened, and triangular teeth, perhaps better suited to cutting large prey than the awl-shaped teeth of smaller individuals (Compagno 1984a).

Economic Value

This is one of the most valuable shark species for its high-quality meat. The meat is utilized fresh, frozen, smoked, and dried-salted for human consumption and eaten broiled and baked (Frimodt 1995), the oil is extracted for vitamins (Compagno 1998), fins are used for shark-fin soup, hides are processed into leather, and jaws and teeth are used for ornaments. It is valued for its fine-quality fins and skin (Compagno 2001). Jaws and teeth are also sold as ornaments and trophies (Frimodt 1995). Kabasakal and de Maddalena (2011) reported a historical record of a larger specimen caught in the Mediterranean Sea off Turkey, about 585 cm (TL estimated from photographs) (Weigmann 2016).

Big-game sports angling for mako sharks is widespread, e.g., in New Zealand, South Africa, and California. This species of shark has become the subject of ecotourism diving in recent years, with most of the mako dive sites being off southern California from the Los Angeles Basin to San Diego but with sites also in South Africa and the Maldives (Compagno 2001).

Conservation Status

Rigby et al. (2019a) assessed the conservation status of this species of shark based on the IUCN Red List rules and granted a "Near Threatened" status. Their justification for giving such a state is described below.

Analysis of longline survey data of mostly juvenile individuals off southern California advises that the Shortfin Mako (*Isurus oxyrinchus*) catch per unit effort (CPUE) may be decreasing marginally, though latest tagging and tracking data reveal that it is also highly migratory, both vertically and horizontally, consequently creating accurate or precise population estimates problematic. There is no sign to propose that the eastern north Pacific subpopulation has been sufficiently depleted to warrant "Vulnerable" status at the present time, and this subpopulation is considered near threatened as a precautionary measure.

Human Interaction

The ISAF statistics (ISAF 2018) showed nine attack cases on human registered against the shortfin mako shark between 1580 and 2017, one of which was fatal, along with 20 boat attacks. This shark is regularly blamed for attacks on humans, and due to its speed, power, and size, it is certainly capable of injuring and killing people. However, this species will not generally attack humans and does not seem to treat them as prey. Most modern attacks involving shortfin mako sharks are deliberated to have been motivated owing to annoyance or the shark being caught on a fishing line. Sharks can be enticed to spear fishermen carrying a stuck fish and may slap them with cavitation bubbles from a swift tail flick. Divers who have encountered shortfin mako note, prior to an attack, they swim in a figure-eight pattern and approach with mouths open. Shortfin makos frequently damage boats and injure fishers after being hooked (Florida Museum 2021).

Brading (2006) gave several fatal and non-fatal cases reported in California, USA, for the short tail Mako, 5 and 24, respectively. The mako shark is AN opportunistic feeder and showed lower incidence of attack but higher average injury rates (Compagno et al. 2005). It is possible that this pelagic species has few opportunities to capture prey and may occasionally see humans as a potential source of prey; however, one would expect a higher fatality rate under these circumstances.

Hardiman et al. (2020) wrote about how sharks and shark-human interactions are reported in major Australian newspapers. They mentioned that the number and percentage of articles mentioning an identified shark taxon between April 1 and September 30, 2015, in four major Australian newspaper titles (Broadsheet, *Australian*, *Herald*; tabloid, *Mail*, *Telegraph*) was five articles on mako shark, with 1.6%, and a total of tabloid of five, with 2.4%. On the other hand, Ricci et al. (2016) in their article on shark attack-related injuries said epidemiology and implications for plastic surgeons that published in *Journal of Plastic, Reconstructive & Aesthetic Surgery* stated that identification of the attacking shark species by victims or observers was difficult, with offending sharks identified by witnesses or forensic evidence in only 1378 cases (27%). They added that the five most common species of shark identified in attacks were the great white shark (*Carcharodon carcharias*), tiger shark (*Galeocerdo cuvier*), bull shark (*Carcharhinus leucas*), nurse shark (*Ginglymostoma cirratum*), and the mako Shark (*Isurus oxyrinchus*). On the other hand, Randall and Levy (1976) reported on a fatal shark attack case by mako shark in the northern Red Sea. They wrote:

> On 7 September 1974 a girl swimmer, aged 20, and a male companion swam about 500 m offshore from North Beach, Hat, Red Sea. At about 5 p.m. the girl was attacked by a shark of about 2 m length. She was bitten approximately 12 times by the shark before being rescued. Most of the wounds were on the distal half of the limbs; amazingly, almost no tissue was removed by the shark. The girl barely survived; ultimately it was necessary to amputate her lower left arm. Two days later a shortfin mako shark *(Isurus oxyrinchus)*, 2.33 m total length, was caught in a weir at Eilat. Mako teeth are long and prong-like without serrations. Evaluation of an imprint of the shark's teeth in plastic to the tooth marks from a bite on the victim's right thigh revealed that a mako of the same size, very possibly the same shark, had bitten her.

Lamna nasus (Bonnaterre, 1788) (Fig. 2.10).

Common name: Porbeagle.

Etymology: *Lamna:* Greek, *lamna*, -es = a shark (Ref. 45335); *nasus:* Latin: *nasus*, nose.

Diagnostic Features

A porbeagle is a robust, spindle-shaped shark with large black eyes; a total preoral length of 5.9 to 9.0% (adults 5.9–7.3%) and space from eye to first gill slit 1.7 to 2.5 times preorbital length; a sharp, conical snout; long gill slits; and small, smooth-edged, narrow teeth with side cusps and first upper lateral teeth with nearly straight cusps (Compagno 1998; Narberhaus et al. 2012). Its cranial rostrum has enlarged but discrete hypercalcified rostral cartilages, not forming a massive knob, with strong keels on the caudal peduncle, short secondary keels on the caudal base, and a crescentic caudal fin; the insertion of the small second dorsal fin is above the insertion of the anal fin (Compagno 1998). Color: gray or bluish-gray to blackish above, white below, with white abdominal color terminating at rear end of pectoral bases; first dorsal fin with an abruptly white or grayish-white free rear tip; ventral surface of head white and abdomen without dusky blotches in adults of typical Northern Hemisphere porbeagles but underside of the head dark and abdomen blotched in some Southern Hemisphere adults (Compagno 2001).

Distribution

This species is coastal and oceanic, with centers of distribution in the North Atlantic, and amphitemperate, in a circumglobal band of temperate water of the southern Atlantic, southern Indian Ocean, southern Pacific, and Antarctic Ocean (Compagno 2001). It was recently caught at the mouth of a brackish estuary in Argentina but does not penetrate fresh water and does not live in equatorial seas as far as is known.

Fig. 2.10 *Lamna nasus* (Bonnaterre, 1788). (Courtesy of De Maddalena, A., South Africa)

In the Western Atlantic, it is reported to be present in Greenland, Canada (Newfoundland Banks, Gulf of St. Lawrence, and Nova Scotia), the USA (Maine, Massachusetts, Rhode Island, rarely in New York, New Jersey, and possibly in South Carolina), and Bermuda, as well as southern Brazil and Uruguay to southern Argentina.

In the Eastern Atlantic, it is recorded in Iceland and the western Barents Sea to the Baltic and North Seas, the English Channel, the Straits of Gibraltar, and the Mediterranean Sea, including Russia, Norway, Sweden, Denmark, Germany, Holland, Scotland, England, Wales, Ireland, the Orkney Islands, France, Portugal, Spain, and Gibraltar; the entire coast of the Mediterranean Sea but not in the Black Sea; Morocco, Madeira, Azores, and possibly the Gulf of Guinea; and off South Africa (Western Cape).

In the Indo-West Pacific, this shark is reportedly found in the south-central Indian Ocean from South Africa (Eastern Cape and possibly KwaZulu-Natal), eastward to between Prince Edward and Crozet Islands, between Kerguelen and St. Paul Islands, and the southern coast of Australia (southern Western Australia and South Australia, Victoria, Tasmania, New South Wales, and southern Queensland) and New Zealand (including Stewart Island), also in the sub-Antarctic waters off South Georgia, Marion, Prince, and Kerguelen Islands.

In the Eastern South Pacific, it distributed in the southern Chile south to Cape Horn.

Habitat, Biology, and Ecology

This species of shark lives inshore to offshore fishing banks and infrequently to open ocean areas. It is migratory, moves normally along continental shelves (inshore to surface during simmer and offshore in deeper water during winter) (Compagno et al. 2005). It is most abundant on continental offshore fishing banks but also found far from land in ocean basins and sporadically near inshore (Compagno 1984a). The greatest depth recounted reaches to 1360 m. Its age reached at least 26 years in the North Atlantic and probably 65 years in the Southern Hemisphere (Mecklenburg et al. 2018).

This is one of the most cold-tolerant sharks; in the northwest Atlantic, it is mostly found in 5–10 °C (Campana and Joyce 2004), with extremes of −1 °C and 23 °C (Mecklenburg et al. 2018). It is known to temporarily tolerate salinities as low as 10 to follow its prey (Fricke et al. 2007) and is found singly and in schools and feeding aggregations (Compagno 1984a).

Individuals of this species feed on small- and medium-sized pelagic schooling species and demersal fishes (cod, white hake, red hake, haddock, and cusk (Scott and Scott 1988; Compagno 2001)). This shark is a proverbially voracious feeder on small-to-moderate-sized pelagic schooling fishes, including mackerel (*Scomber*, Scombridae), pilchards and herring (Clupeidae), hake (Merluccidae), ice fishes (Channichthyidae), and John dories (Zeidae). Chondrichthyan prey includes dogfish (*Squalus acanthias*, Squalidae) and tope sharks *Galeorhinus galeus* (Triakidae). Cephalopod prey includes squid and cuttlefish (Compagno 1998). It will forage hooked fishes containing cod from longlines.

They are ovoviviparous species (Compagno 2001; Dulvy and Reynolds 1997) and a uterine cannibal (oophagous), with litters of one to five young, but most litters are of four young. The fetuses grow enormously by feeding on fertilized eggs and develop grotesquely expanded abdomens and branchial regions. Small porbeagle embryos possess fang-like functional teeth to tear open egg capsules and release the contained ova; the fangs are shed at 34–38 cm FL (Francis and Stevens 2000). Females grow larger than males (Gauld 1989). The gestation period has been assessed at about 8 or 9 months for North Atlantic and South Pacific populations. The length of the whole reproductive phase is not known. Preliminary studies in this region suggest that males mature at about 175 cm FL (Age 7) and females at around 212 cm FL (Age 14) (Compagno 1999). Pupping and nursery regions may be in continental waters, but little is known about it. In the western North Atlantic, mating is thought to happen off southern Newfoundland. Porbeagles breed on both sides of the North Atlantic, off the Atlantic coast of Europe and the

British Isles, where females have embryos during most of the year, except July through September, and off North America from Massachusetts to Maine, where females can be found with young at all times of the year. Youngs are deceptively born in the spring off Europe, in either winter-spring or late summer off North America, and perhaps from April to September (peak June–July [winter]) in the Southern Hemisphere. Mating in European waters happens in late summer, and breeding there perhaps takes place every year. A prolonged mating time appears to happen for Southern Hemisphere populations around Australia and New Zealand (Compagno 2001).

The porbeagle is designated as active and strong-swimming in chasing prey but when hooked is moderately sluggish and inactive in contrast to the shortfin mako (*Isurus oxyrinchus*) and does not have remarkable jumps like that species.

Economic Value

This species is considered by some as a sport fish (Schultz 2000). Its flesh is of good quality and texture and is said to taste like swordfish (Schultz 2000) and is utilized fresh, dried, salted, or frozen for human consumption and for oil and fishmeal and fins for shark-fin soup (Compagno 1984a. It may be pan-fried and broiled (Frimodt 1995).

This species has been heavily fished commercially and exploited as a food for human in the temperate North Atlantic and the Mediterranean but is also captured as bycatch in the Southern Hemisphere, where it is the second most common shark as bycatch of the New Zealand longline fishery.

In the past, porbeagles were deliberately a pain to commercial fishermen since they break light gear set for bony fishes (such as cod nets) and bit fish off hooks, but perhaps not so much at present with greatly exhausted porbeagle stocks and reduced stocks of some of its prey species (including cod).

A substantial bycatch fishery for porbeagle by Japanese longliners and maybe the pelagic fishing fleets of other countries has been found in the southern Indian Ocean and possibly to another place in the Southern Hemisphere. Little is known of its catch, and it may be not utilized fully, except for fins.

It has considered as complementary bycatch (fins utilized) of the Japanese longline fishery for southern bluefin tuna off Tasmania. It is used fresh and dried-salted for human consumption and for oil and fishmeal for fertilizer and fins for shark-fin soup.

This shark species has been frequently required by sports fishing in the UK, Ireland, and the USA and is a strong fighter (particularly on light tackle from a small boat). It is listed as a record game fish by the International Game Fish Association.

Conservation Status

This species of shark has been granted a Vulnerable A2bd status in the Red List Category & Criteria based on the following causes as stated by (Rigby et al. 2019a):

> The Porbeagle (*Lamna nasus*) is a large (to 357 cm total length) pelagic shark with disjunct distribution. There are two discrete subpopulations, the North Atlantic and the Southern Hemisphere. The species has little biological productivity with small litters, late age-at maturity, and a briefer life span of 26 years in the North Atlantic related to 60 years in the Southern Hemisphere. Global fisheries activities are usually targeting this species and frequently caught by the way in coastal and pelagic commercial and small-scale longline, purse seine, and gillnet. Used for human consumption due to the high-value meat as well as its fins. Sharp drops have happened in the North Atlantic, with failures also obvious, though not as steep, in the Southern Hemisphere.

Human Interaction

This species is regarded as potentially dangerous to people because of its size and activity but has never or very seldomly attacked people or boats (Compagno 1984a).

Quigley (2007) reported that porbeagles have accounted for a total of five non-fatal attacks on humans. At least two, and possibly three, non-fatal attacks have been accredited to this species in UK waters: the first in 1876, between Hastings and Fairlight, Sussex; the second on July 27, 1969 (no locality); and the third on June 1, 1971, off Beesands, South Devon. Porbeagles have also been filmed making fast passes at divers on oil platforms in the North Sea without attacking. This is perhaps agonistic (defensive) or exploratory behavior.

The International Shark Attack File (ISAF) has only three probeagle attacks on record (Roman 2010).

Family: Triakidae.
Galeorhinus galeus (Linnaeus, 1758) (Fig. 2.11).
Common name: Tope shark.
Etymology: *Galeorhinus:* Greek, *galeos* = a shark + Greek, *rhinos* = nose (Romero 2002).

Diagnostic Features

Body slender, with moderately long snout, stunted anterior nasal flaps, no nasoral grooves, large horizontally oval eyes with internal nictitating lower eyelids, subocular ridges superseded, mouth arched and long, reasonably long upper labial furrows shorter than the lower symphysis, bladelike compressed teeth with oblique cusps and distal cusplets in both jaws, second dorsal fin much smaller than first and about as large as anal fin, and an extremely long terminal caudal lobe about half the dorsal caudal margin.

Distribution

This species of this type of shark is widely distributed in cold to warm temperate waters of most oceans. It inhabits the Northeast, Eastern Central, Southwest, and Southeast Atlantic, the Mediterranean Sea, the Eastern Indian, and the Southwest, Southeast, Western Central, Eastern Central, and Northeast Pacific. It is absent from the Northwest and Western Central Atlantic, the Northwest Pacific, and the Western Indian Sea (Compagno et al. 2005: Ebert et al. 2013).

Fig. 2.11 *Galeorhinus galeus* (Linnaeus, 1758). (Courtesy of Clinton Duffy, New Zealand)

Habitat, Biology, and Ecology
The tope shark is benthopelagic in temperate waters on continental and insular shelves and upper to mid-slopes from shallow inshore to well offshore to depths of 826 m; however, it is regularly found to depths of 200 m (Walker et al. 2006; Weigmann 2016; Thorburn et al. 2019). Some large individuals travel long oceanic distances offshore well away from the continental shelves and slopes, but they do not cross ocean basins (Walker 1999; Walker et al. 2008; Colloca et al. 2019). The species has pupping and nursery areas in shallow, protected bays and estuaries where the young can remain for up to 2 years (Stevens and West 1997; Walker 1999; Walker et al. 2006; Bovcon et al. 2018). The maximum size differs regionally with the largest maximum size of 200 cm total length (TL) in the Mediterranean Sea (Capapé and Mellinger 1988) and the smallest maximum size of 155 cm TL in the Southwest Atlantic (Peres and Vooren 1991). There is also regional difference in size at maturity: males mature at 107–170 cm TL, and females mature at 118–185 cm TL (Walker et al. 2006; Ebert et al. 2013). Reproduction is aplacental viviparous with litter sizes 6–52 (average 20–35), a reproductive cycle that apparently varies regionally from annual to triennial, although studies with intensive sampling indicate triennial cycles and a gestation period of ~12 months and size at birth that varies regionally from 26 to 40 cm TL (Peres and Vooren 1991; Walker 1999, 2005; Ebert et al. 2013). Female age at maturity varies from 10 to 15 years (average 12.5 years), and maximum age is estimated as 40 years (tag returns suggest a possible maximum age of 60 years); generation length is therefore 26.3 years (Ferriera and Vooren 1991; Freer 1992; Francis and Mulligan 1998; Walker 1999).

Economic Value
This shark species is targeted for its meat and fins and historically for the squalene oil (Walker et al. 2006; Dent and Clarke 2015; Fields et al. 2018) and fins for soup (Compagno 1984a). It is also utilized as fishmeal (Compagno and Niem 1998). It is marketed fresh, dried-salted, and frozen (Frimodt 1995) and adapts well in captivity if carefully captured and handled (Michael 1993).

Conservation Status
Walker et al. (2020) assessed the IUCN Red List status of this shark species and assigned Critically Endangered A2bd criterion based on the following facts: the genetic and tagging data support up to six separate subpopulations of this species, and while it creates widespread movements within each of the subpopulations, there is no indication of mixing between them. It is of low biological productivity with a late age at maturity and triennial reproductive cycle. It is caught globally as target and bycatch in industrial and small-scale demersal and pelagic gillnet and longline fisheries and to a lesser degree in trawl and hook-and-line fisheries. This species is often taken for its meat and fins but is thrown away or released in some areas, in line with regional management measures. Sharp subpopulation and stock reductions of >80% over the past three generation lengths (79 years) have happened in the Southwest Atlantic, southern Africa, and Australia. In the Northeast Atlantic, the subpopulation is assessed to have experienced a decrease of 76% over the past three generation lengths (79 years). New Zealand stock is appraised to have suffered a reduction of 30–49% over the past three generation lengths (79 years). In the Northeast Pacific, a dramatic decline in the subpopulation occurred in the early 1940s, with no recovery until 1997–2004 when localized management led to a restricted increase in abundance. The steady steep subpopulation drops across most of the analyzed subpopulations and stocks together with the lack of movement between the subpopulations are cause for serious concern.

Human Interaction
There is only one non-fatal and unprovoked incidence of attack on human as reported by the international Shark Attack File (ISAF). This means that this shark is harmless but has the potential to inflict an attack on humans. Therefore, it is important especially for divers to take extra precaution in case any confrontation happens with this shark.

Family: Carcharhinidae
Carcharhinus brachyurus (Günther 1870) (Fig. 2.12).
Common name: Copper shark.
Etymology: *Carcharhinus:* Greek, *karcharos* = sharpen + Greek, *rhinos* = nose (Romero 2002).

Diagnostic Features

The following description is after Compagno (1984a). This shark is large in size and fairly slender, with a reasonably long narrowly rounded snout; eyes circular and moderately large; anterior nasal flaps low and poorly developed; upper labial furrows short and inconspicuous; hyomandibular line of pores just behind mouth corners not conspicuously enlarged; gill slits moderately long, usually no interdorsal ridge; long pectoral fins; first dorsal fin large and falcate, with pointed or narrowly rounded apex and posterior margin curving ventrally or posteroventrally from fin apex; origin of first dorsal fin over or slightly anterior to pectoral rear tips; inner margin of first dorsal moderately long, a third of dorsal base or less; second dorsal fin small and fairly low, a small second dorsal with a short rear tip; and no conspicuous markings on fins. Color bronzy to olive gray above and white below; most fins with inconspicuous darker edges and dusky to black tips but fin markings not conspicuous; a reasonably conspicuous white band on flank.

Distribution

The copper shark has a circumglobal but patchy distribution (Ebert et al. 2013).

Habitat, Biology, and Ecology

This species of shark is found in warm temperate and subtropical waters on the continental shelf from close inshore to depths of 145 m (Ebert et al. 2013), as well as in brackish or freshwater in the lower reaches of large rivers and estuarine bays, appearing singly and in loose schools of hundreds of individuals and having inshore nursery grounds (Cliff and Dudley 1992; Ebert et al. 2013). It reaches a maximum size of 325 cm total length (TL). Males mature at 200–255 cm TL, and females mature at 215–270 cm TL, with regional variation in these parameters; size at maturity is smaller in Argentina than in South Africa and Australia (Cliff and Dudley 1992; Last and Stevens 2009). Reproduction is viviparous with yolk-sac placenta and litter sizes of 7–26, a likely biennial reproductive cycle, and size at birth of 59–70 cm (Ebert et al. 2013). Female age at maturity is 16 years, and maxi-

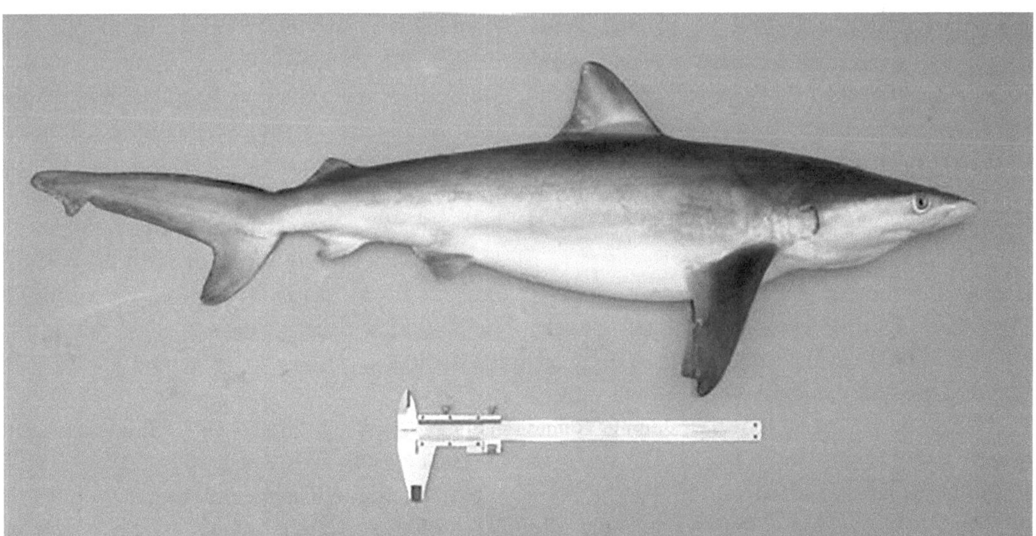

Fig. 2.12 *Carcharhinus brachyurus* (Günther, 1870). (Courtesy of Clinton Duffy, New Zealand)

mum age is 31 years, resulting in a generation length of 23.5 years (Ebert et al. 2013).

Economic Value

This shark is used for human consumption in several places around the world. In New Zealand, biochemical identification methods have established the landing of this species for fin and meat markets mislabelled as commercially managed species such as the rig shark (Smith and Benson 2001). Despite restricted knowledge of stock status and the influences of current harvest levels, their high-value meat endures to be exported to meet the needs of the Australian market—primarily in the fish and chip trade (Da Silva and Bürgener 2007).

Conservation Status

This species of shark faces several kinds of threat. Among the reasons for such threats is its life history, which makes it vulnerable to overfishing by directed fisheries and as bycatch in fisheries targeted at more productive species (Walker 1998; Castro et al. 1999; Vidthayanon 2002).

Even though it is widespread, regional populations appear to be discrete and movement of individuals between them considered scarce or absent. It also does not appear to be naturally abundant anywhere. Therefore, Duffy and Gordon (2003) assessed this species as Vulnerable in East Asia due to intensive fisheries and the obvious prevalent failure of fisheries for large coastal sharks. Fisheries activities in the region will endure influencing the population of this species of shark by taking pregnant females and juveniles. In addition, coastal nursery areas in this region are also at jeopardy from development and pollution.

Human Interaction

There are 15 incidences of human attacks reported on this species. Among these, there are 14 non-fatal and unprovoked cases and only 1 fatal case as reported by the International Shark Attack File (ISAF), involving a kitesurfer in Australia (Clua et al. 2014).

Khan (2016) suggested that even a slight bite of this shark would result in a non-fatal attack up to a case of fatality due to the presence of marine bacteria that infect the wound and thus may lead to a complicated case. Moreover, there is restricted information on record about the antibiotic resistance of bacteria associated with marine top predators. Khan (2016) in her thesis examined oral swabs taken from different shark species in South Africa and isolated two species of bacteria belonging to the families Micrococcaceae (*Micrococcus luteus* and *Arthrobacter creatinolyticus*) and one species belonging to the family Pseudomonadaceae (*Pseudomonas fulva*). The species of the genera *Micrococcus* and *Pseudomonas* are shown to be resistant to the following: 50% to cefuroxime, 38% to ampicillin, 18% to nalidixic acid, 14% to tetracycline, 11% to erythromycin, 10% to ceftriaxone, and 2% to ciprofloxacin (Khan 2016). Therefore, shark attack is mild and non-fatal, but the bacterial infection resulted from such attack can be severe and could lead to death.

Shark and human encounters happen because of fishing activities such as by beach seine. The best example of this event is that of the False Bay on the south coast of South Africa (Lamberth 2006). Because of these fishing activities, a growing public awareness that large catches of this collective group of "keystone" predators were causing ecological discrepancies in False Bay (Lamberth et al. 1994). Among the public concerns is the attraction of large predators such as white shark and other large carcharhinid species like *C. brachyurus* into nearshore. Such attraction is due to beach seine operations and/or blood bringing them into contact with bathers, surfers, and other water users (Lamberth 2006). In his study, Lamberth (2006) showed that the percentage of *C. brachyurus* is 15.1% in the False Bay beach-seine fishery during the period of 1974–1987.

Carcharhinus galapagensis (Snodgrass and Heller 1905) (Fig. 2.13).

Common name: Galapagos shark.

Etymology: Carcharhinus: Greek, *karcharos* = sharpen + Greek, *rhinos* = nose.

Fig. 2.13 *Carcharhinus galapagensis* (Snodgrass and Heller 1905). (Courtesy of Pedro Niny Duarte, Portugal)

Diagnostic Features

This shark has large body with long, broadly rounded snout. Anterior nasal flaps are low with large eyes and large erect teeth. Presence of an interdorsal ridge is noted. Pectoral fin is large and wide. Body is dark gray in color with white abdomen. There are no visible marking on fins, but their tips are dusky (Compagno and Niem 1998a, b; Ebert et al. 2013).

Distribution

This shark has a circumtropical type of distribution with patchy distribution. It is reported from the eastern Atlantic, western Indian Ocean, and western and eastern Pacific (Lavenberg et al. 1994; Ebert et al. 2013). There is no record of this species from the Arabian-Persian Gulf, but it has been reported from the Sea of Oman (Henderson and Reeve 2011).

Habitat, Biology, and Ecology

This shark is a marine species that lives in association with reefs and found at depths ranging from the surface down to 286 m, but it is usually seen at 30–180 m. It favors areas with strong currents (Myers 1999; Mundy 2005; Ebert et al. 2013).

Individuals of this shark are found in small aggregations but does not form organized schools (Ebert et al. 2013). It reaches a maximum size of 3500 mm in total length, with males reaching maturity in 2050–2500 mm in total length and age of 6.5–9 years. Females reach maturity at 2150–2500 mm in total length with age of 6–8 years (Bass et al. 1973; De Crosta et al. 1984; Last and Stevens 1994; Wetherbee et al. 1997). The number of young at birth ranges from 4 to 16, with 600–810 mm in total length (Bennett et al. 2003). Mating happens during winter to spring, and females breed every other year (Smith et al. 1998).

Economic Value

This shark, like other shark species, is used for human consumption, specifically its liver oil, skin, and jaws as a tourist commodity.

Conservation Status

Kyne et al. (2019) assessed the conservation status of this shark and granted a Near Threat status in the IUCN Red List based on the following reasons.

The Galapagos shark (*Carcharhinus galapagensis*) is a large shark (with at least 300 cm total length) with a circumglobal but patchy distribution. This shark has limited biological productivity and, consequently, limited capability to withstand exploitation. It presents as bycatch in commercial, small-scale longline, purse seine, and gillnets fisheries operated in pelagic oceanic waters, around islands, and seamounts. The depletion of the Galapagos shark been reported at several fisheries grounds around the world. Population monitoring is recommended where the species is fished, and enforcement of regulations in marine reserves where illegal fishing is remarked as an issue will be important to secure local populations (e.g., Ecuador's Galápagos Marine Resources Reserve). The aggressive nature and its presence in shallow areas increase the pressure to eradicate this species from local populations (Bennett et al. 2003).

Human Interaction

This shark is aggressive and has shown to display a hunch position once divers come close to its territory and regraded potentially dangerous to humans (Ebert et al. 2013). Unprotected divers should not attempt to enter areas where individuals of this shark are abundant. It might start a feeding frenzy if weapon is used against them (Compagno 1984a). This shark has two attack records, one is non-fatal unprovoked and the other fatal unprovoked (ISAF 2009), but other incidences were also reported against this shark.

Individuals of this shark species are known to reach close to swimmers, showing interest in swim fins or hands, and are drawn in large numbers by fishing activities. Edwards and Lubbock (1982) wrote about the experience of Fitzroy in 1839 with this species of shark. Fitzroy observed off St. Paul's Rocks "as soon as a fish was caught, a rush of voracious sharks was made at him, notwithstanding blows of oars and boat hooks, and the ravenous monsters could not be deterred from seizing and taking away more than half the fish that were hooked." Limbaugh (1963) reported from Clipperton Island "at first, the small sharks circled at a distance, but gradually they approached and became more aggressive ... various popular methods for repelling sharks proved unsuccessful." The situation eventually escalated to the point at which the divers had to retreat from the water. Excited Galapagos sharks are not easily discouraged; driving one away physically only results in the shark circling back while attracting others to follow, whereas using weapons against them could trigger a feeding frenzy (Compagno 1984a). As of 2008, the Galapagos shark has been confirmed to attack two people: one fatal attack in the Virgin Islands, at Magens Bay on the north shore of St. Thomas, and a second non-fatal attack off Bermuda (Compagno 1984a).

Carcharhinus longimanus (Poey 1861).
Common name: Oceanic whitetip shark.
Etymology: Carcharhinus: Greek, *karcharos* = sharpen + Greek, *rhinos* = nose.

Diagnostic Features

This shark has a very big body, with short and blunt snout. Eyes are small. Teeth are in the upper jaw, triangular with strong serration; lower jaw have narrow cusp, slightly serrated and erect. The interdorsal ridge is low. Pectoral fins are large with rounded end, first dorsal fin rounded. Caudal keel is present, but not clear. The body is dark gray to black in color with bronze shades. Ventral side is whitish. Tips of first dorsal, pectoral, and lower lobe of caudal fin are white or with white spots. Black spots are present at tip of anal and second dorsal fin and ventral lobe of caudal fin, with dark saddle marks anterior to dorsal fin (Compagno and Niem 1998a, b; Randall 1995; Compagno 1998; Ebert et al. 2013).

Distribution

This species of shark has a cosmopolitan type of distribution. It has been reported to be present in the western and eastern Atlantic regions. It is also found in the Indo-Pacific and the eastern Pacific (Claro 1994; Smith 1997; FAO Fisheries Department 1994). There is no record of this shark from the Arabian-Persian Gulf and the sea of Oman (Henderson and Reeve 2011; Moore et al. 2012), but it is reported from the coasts of the south Arabian Peninsula at the coasts of Yemen and Oman (Manilo and Bogorodsky 2003).

Habitat, Biology, and Ecology

This species of shark inhabits the pelagic region of the world oceans. It is oceanodromous and found living from surfaces down to 230 m but regularly at 152 m (Reide 2004; Mundy 2005; Florida Museum of Natural History 2021). It is one of the most widespread sharks, ranging across entire oceans in tropical and subtropical waters (Young et al. 2017). It prefers water and has a habit of restricting its presence to the central equatorial belt, perhaps going farther to include 45°N and 43°S. This shark has a habit of going up to the surface and putting its heads outside the water to locate its prey (Shark Facts 2021).

It reaches a maximum size of 350 cm total length (TL), possibly 395 cm TL; males mature at 168–198 cm TL and females at 175–224 cm TL (Ebert et al. 2013; Weigmann 2016; D'Alberto et al. 2017). Reproduction is placental viviparous with litter count of 1–15 pups. Litter size increases with female size; gestation period is 10–12 months with most likely a biennial reproductive cycle and size at birth of 57–77 cm TL (Seki et al. 1998; Bonfil 2008; Last and Stevens 2009). There is regional variation in age estimates: female age at maturity is 4.5–8.8, 6.5, and 15.8 years, and maximum age is 11, 17, and 24.9 years in northwest Pacific, southwest Atlantic, and western central Pacific, respectively (Seki et al. 1998; Lessa et al. 1999; Liu and Tsai 2011; Joung et al. 2016; D'Alberto et al. 2017).

Economic Value

This shark has commercial value for its meat, fins, and vitamins in its liver oil. Its skin is used to make leather (Shark Facts 2021). A percentage of 1.8% of oceanic white shark fins has been imported in 1991–2001 and 0.6% in 2014 in Hong Kong (Clarke et al. 2006a, b; Fields et al. 2018). In some areas, the meat of juvenile oceanic whitetip shark is sold fresh for human consumption at local markets.

Conservation Status

Rigby et al. (2019a) assessed the conservation status of this shark as Critically Endangered A2bd for the following reasons: this species of shark is large (to 350 cm total length), wide-ranging, and oceanic, living in tropical and temperate seas worldwide, with a preference for surface waters, although it has also been reported to be at depths of 1082 m. It has litters of 1–15 pups and likely has a biennial reproductive cycle. The species is fished globally as target and bycatch in pelagic large- and small-scale longline, purse seine, and gillnet fisheries and frequently kept for its meat and fins, unless rules forbid holding. The species has a high catchability due to its preference for surface waters and its nosey nature. Steep population decrease has happened in all oceans. This shark was once one of the most abundant pelagic shark species in tropical seas worldwide but is now rare in some regions. The global population is assessed to have undergone a reduction of >98%, with the highest probability of >80% reduction over three generation lengths (61.2 years).

Human Interaction

This species of shark is regarded as the most dangerous of all sharks (Cousteau and Cousteau 1970). The whitetip shark is accountable for many fatal attacks on humans together with the great white shark and other species of sharks that live in shallow waters as it is reported to attack survivors of shipwrecks or floored aircraft (Bass et al. 1973). It has a record of 15 attack incidences; among these, 12 were non-unprovoked and 3 were fatal unprovoked cases (ISAF 2021). This same species is responsible for the attack on the initial survivors of the USS Indianapolis on July 30, 1945 (Martin 2006). Only a total of 192 people survived, with many deaths attributed to this shark (Bass et al. 1973). The whitetip shark is also responsible for the famous Red Sea shark attack in 2010 where a diver's back and thigh were snatched by sharks while the diver was handfeeding them (Wikipedia 2016). It is always said that if you are in deep water and you come face to face with the whitetip shark, watch its white-tipped elongated dorsal fin, but if you are in such depth, your chances of escaping unharmed from this shark are very slim. This shark becomes very aggressive during feeding frenzy, bites, and snaps; wounds from its continuous bites will be fatal (Shark Facts 2021).

Carcharhinus obscurus (Lesueur 1818) (Figs. 2.14, 2.15, and 2.16).

Common name: Dusky shark.

Etymology: Carcharhinus: Greek, *karcharos* = sharpen + Greek, *rhinos* = nose.

Fig. 2.14 *Carcharhinus obscurus* (Lesueur 1818). (Courtesy of Tassapon Krajangdara)

Fig. 2.15 Head of *Carcharhinus obscurus* (Lesueur 1818). (Courtesy of Hugo Bornatowski)

Fig. 2.16 Close-up of one tooth of Gilbertese shark tooth weapon (FMNH99071) showing how the teeth of C. obscurus were attached using braded cord

Diagnostic Features

This shark has a large body with a broad and rounded snout. Teeth in upper jaw are triangular with serrated edges. The dorsal fin is large and wide. Pectoral fin curved and moderate size. The body has an interdorsal ridge. The body is gray to blue in color above and white below. Tips of pectoral and pelvic fins are dark, as well as the lower lobe of caudal fin (Compagno 1998; Compagno and Niem 1998a, b; Ebert et al. 2013).

Distribution

This shark has an almost worldwide distribution. It is found in the western and eastern Atlantic regions. It also recorded from several localities in the Indo-Pacific region (Bass et al. 1986a, b; FAO Fisheries Department 1994; Menni and Lucifora 2007). This species of shark has not been recorded from either the Arabian-Persian Gulf or the Sea of Oman (Compagno 1984a;

Randall 1995; Henderson and Reeve 2011; Moore et al. 2012; Ebert et al. 2013), but it has been reported in the southern coasts of the Arabian Peninsula in the coasts of Yemen (Manilo and Bogorodsky 2003).

Habitat, Biology, and Ecology

This shark species inhabits coastal and pelagic areas throughout its range, where it is found in the surf zone to well offshore and from the surface down to depths of 500 m (Weigmann 2016). It prefers water with a temperature range of 19–28 C and avoids estuaries (Ebert 2003; Hoffmayer et al. 2009). It has a maximum size of 420 cm total length (TL); males mature at 265–280 cm TL; females mature at 257–310 cm TL (Simpfendorfer et al. 2002; McAuley et al. 2007; Weigmann 2016). Reproduction is placental viviparous with litter sizes of 2–18, with a long gestation of 22 months, a two- or three-year reproductive cycle, and size at birth of 70–100 cm TL (Branstetter and Burgess 1996; Romine 2004; Dudley et al. 2005). Females usually move to shallow areas to give birth where such areas provide shelter for the young ones (Smale and Goosen 1999; Ebert 2003). In the Northwest Atlantic, female age at maturity is 17.6 years, with a maximum age up to 42 years based on bomb radiocarbon-validated ages; generation length is therefore 29.8 years (Natanson et al. 2016). Female age at maturity of 27–32 years was validated in Western Australia and the eastern Indian ocean, as well as maximum age of 40–53 years (Simpfendorfer et al. 2002). The estimate of intrinsic rates of population increase are low in the Indo-Pacific subpopulation, estimated from Western Australian stocks at about 0.025 per year in the absence of fishing (Simpfendorfer et al. 2002).

Economic Value

This shark is usually used for its valuable fins and meat. Its meat is utilized for human consumption as fresh, dried-salted, frozen, and smoked. The skin is used for leather manufacturing and the liver to extract oil for vitamins (Compagno 1984a).

Conservation Status

The dusky shark faces several threats with it being fished globally. It is caught targeted and bycatch in commercial, small-scale pelagic longline, purse seine, and gillnets fisheries. Most of the catch is obtained as bycatch of commercial pelagic fleets in offshore and high seawaters (McCandless et al. 2014). It also captured in coastal longlines, gillnets, trammel nets, and sometimes trawls, particularly in areas with narrow continental shelves (Martínez-Ortiz et al. 2015). The species is often kept for its meat and fins (Clarke et al. 2006a, b; Fields et al. 2018), unless rules forbid retention. Underreporting of catches in pelagic and domestic fisheries is likely (Dent and Clarke 2015). At-vessel mortality is estimated at 34% in the US pelagic longline fishery, 81% in the northwest Atlantic bottom longline fishery, and 1.3% on Western Australia demersal longlines (McCandless et al. 2014; Braccini and Waltrick 2019). Post-release mortality is estimated up to 67% in the northwest Atlantic bottom longline fishery, though markedly reduced when kept for less than 3 h on the line (McCandless et al. 2014).

The conservation status of this species of shark was assessed by Rigby et al. (2019a) as Endangered A2bd with the following reasons: The dusky shark (*Carcharhinus obscurus*) is a large (to 420 cm total length) coastal and pelagic shark with a patchy distribution in tropical and warm temperate seas from the surface down to depths of 500 m. The species has low biological productivity with late age at maturity and a long reproductive cycle. Sharp declines have happened in the northwest Atlantic and eastern Indian ocean and to a lesser degree in the western Indian ocean; trends in the Eastern Atlantic off West Africa are suspected to be declining. The global population reduction is assessed to have undergone a reduction of 71.8%, with the highest probability of >80% reduction over three generation lengths (89.4–114 years). Areas with available abundance trend data are relatively small compared to the global distribution, and abundance trends are unknown in parts of its distribution. The overall estimated population reduction is abrupt, and management in many parts of its range is lacking, including areas of intense exploitation, likely leading to sharp inferred population reduction. However, management action in the northwest Atlantic and eastern Indian ocean resulted in the stabilization and possible slow recovery in these regions.

Human Interaction

Its large size makes it potentially dangerous to humans. Six attacks were reported for this shark in 2009. These attacks were on human and on boats, and three of these six attacks were fatal (ISAF 2021). Shark nets used to protect beaches in South Africa and Australia entangle adult and larger juvenile dusky sharks in some numbers.

Galeocerdo cuvier (Péron and Lesueur 1822) (Figs. 2.17 and 2.18).
Common name: Tiger shark.
Etymology: Galeocerdo: Greek, *galeos* = a shark + Latin, *cerdus* = the hard hairs of pigs.

Diagnostic Features

The head is big. The anterior part of the body is narrower toward the posterior end. The snout is round, big, and blunt. The upper lateral furrow is very long and lateral keel on each side low. The spiracles are small and situated behind eyes. The gill slits are small. The mouth is big with teeth similar in both jaws, strongly serrated and convex anteriorly. The body is gray dorsally with gray bars and spots, white ventrally (Compagno 1998; Randall 1995; Ebert et al. 2013).

Distribution

This species of shark has a circumtropical type of distribution and found in temperate and tropical waters. It is reported from the western Atlantic regions. It is also recorded from the Indo-Pacific and eastern Pacific regions (FAO Fisheries Department 1994). This shark is reported from several localities in the Arabian-Persian Gulf (Compagno 1984a; Moore et al. 2012). It is also recorded from the Sea of Oman (Henderson and Reeve 2011) and from the southern coasts of the Arabian Peninsula at the coasts of Yemen and Oman (Manilo and Bogorodsky 2003; Henderson and Reeve 2011).

Habitat, Biology, and Ecology

The tiger shark prefers to live on the shelf, reef, and slope habitats, is sometimes associated with coral reefs, and infrequently makes longer-distance excursions into the pelagic zone. This species of shark has the habit of swimming slowly. This type of swimming and its cryptic coloration make this shark unnoticeable by preys (Heithaus 2001). The species is mostly found at 100 m depth but also dives to depths greater than 1000 m (1136 m maximum recorded) and has been noted to move hundreds, and in some cases thousands, of kilometers (Holmes et al. 2014; Werry et al. 2014; Afonso and Hazin 2015; Lea et al. 2015).

Fig. 2.17 *Galeocerdo cuvier* (Péron and Lesueur 1822). (Courtesy of Muazzam Khan, Pakistan)

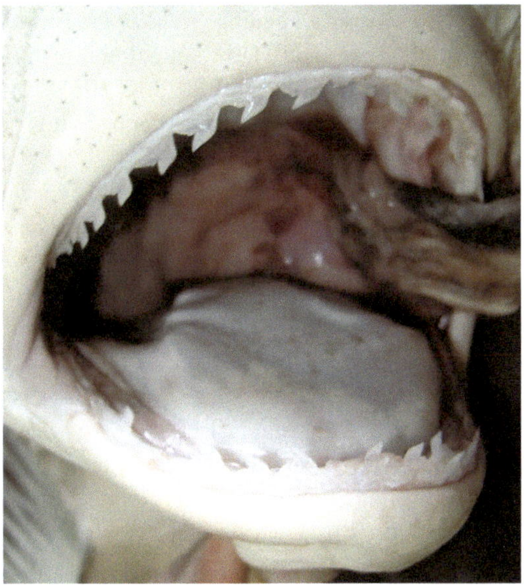

Fig. 2.18 Head of *Galeocerdo cuvier* (Péron and Lesueur 1822). (Courtesy of Hamid Badar Osmany, Pakistan)

The species has a higher intrinsic rate of population increase than many other carcharhinid sharks (Dudley and Simpfendorfer 2006). Maximum-recorded size is 740 cm total length (TL), but individuals larger than 500 cm TL are rarely seen (Ebert et al. 2013). There is regional variability in sizes and ages at maturity with females maturing at 274–345 cm TL and males at 250–305 cm TL (Compagno 1984a; Randall 1992; Simpfendorfer 1992; Wintner and Dudley 2000; Whitney and Crow 2007; Holmes et al. 2015). The maximum age is estimated to range between 27 and 37 years (Natanson et al. 1999; Kneebone et al. 2008; Holmes et al. 2015), with age at maturity of 4–13 years (Branstetter et al. 1987; Smith et al. 1998; Natanson et al. 1999; Afonso et al. 2012; Holmes et al. 2015). This species of shark is the only species of the family Carcharhinidae that is lecithotrophic viviparous. Litter sizes are large, with a maximum of 82 embryos recorded and an average litter size of 26–33 pups (Tester 1969; Bass et al. 1975; Simpfendorfer 1992). The size at birth is 51–90 cm TL (Randall 1992, Simpfendorfer 1992). Gestation is approximately 15–16 months with a triennial reproductive cycle in Hawaii and Australia (Whitney and Crow 2007; Holmes et al. 2015).

Economic Value

Its flesh and fins are used for human consumption, while the liver is used for its high vitamin A content. The thick skin is used as a leather commodity.

Conservation Status

Ferreira and Simpfendorfer (2019) assessed the conservation status of this shark and allocated the status of Near Threatened (close to meeting Vulnerable A2bd + 3d) based on the following reasons: The tiger shark (*Galeocerdo cuvier*) is a large highly mobile shark (740 cm total length) with a global range throughout the world's tropical and warm temperate oceans. This species has relatively fast growth rates and large litters (on average 26–33 pups), but the likely triennial reproductive cycle reduces its ability to recover from fishing pressure. The tiger shark is usually caught by commercial, recreational, and artisanal fisheries as a target species or bycatch. Nevertheless, fishery activities from around the world have reported about this shark. The species is one of the major constituents in catches from shark control programs that target large shark species in Australia, South Africa, and Reunion Island. Information on stock assessment, population structure, and overall population trends is lacking, for much of the species' distribution and available trends in catch rates are based on analyses of long-term shark control programs and on observer surveys from the western North Atlantic. In Australia, the Queensland Shark Control Program showed declines in catch rates since the 1980s, and in New South Wales, evidence of a decline is also seen in the last 20 years. Furthermore, a noteworthy reduction of the proportion of mature sharks in catches over 60 years proposes changes in population structure and raises worries over possible effects from this program on the tiger shark population off New South Wales. In Arabian seas, the tiger shark has been assessed as Vulnerable based on a suspected decline of at least 30% over the past three generations based on levels of exploitation, with the continued demand from the shark fin market suspected to result in further population reduction in the future. A further population reduction is suspected over the future three generation lengths (2018–2086) based on current levels of exploitation that are likely to continue.

Human Interaction

It is a dangerous species of shark and must be avoided and watched. Reported attacks on humans ranked as second to those of the white shark. There are a total of 131 cases of attacks against humans, of which 97 incidences were of nonfatal and unprovoked and 34 were of fatal and unprovoked nature (ISAF 2021).

Its large body size and voraciousness make it one of the most dangerous sea creatures. All recorded attacks by this shark involved divers, usually those spearfishing or attracting sharks for photography, swimmers, and boats (Randall 1995; Marine Species Identification Portal 2016).

The tiger shark is said to be accountable for a large share of fatal shark-bite incidents and regarded as one of the most dangerous shark species (ISAF Statistics on Attacking Species of Shark 2012). It frequently visits shallow reefs, harbors, and canals, creating the potential for encounter with humans. This shark also dwells in river mouths and other runoff water (Marinebio 2021). It is categorized second on the list of a number of recorded bites on humans, behind only the great white shark (ISAF Statistics on Attacking Species of Shark 2012). Typically, three to four shark bites are reported per year in Hawaii, but they are infrequently fatal; one notable survivor of such an attack is surfing champion Bethany Hamilton, who lost her left arm at age 13 to a tiger shark in 2003. This shark's bite rate is very low, considering that thousands of people swim, surf, and dive in Hawaiian waters every day (ISAF Statistics on Attacking Species of Shark 2012). Human interactions with tiger sharks in Hawaiian waters have been shown to increase between September and November, when tiger shark females are believed to migrate to the islands to give birth (Marinebio 2021).

Between 1959 and 2000, 4668 tiger sharks discarded in an aim to guard the tourism industry. Despite damaging the shark population, these efforts have shown to be unsuccessful in decreasing the number of interactions between humans and tiger sharks. Feeding sharks in Hawaii (except for traditional Hawaiian cultural or religious practices) is illegal (WPRFMC 2013), and interaction with them, such as cage diving, is discouraged.

West (2011) reported that attack incidences against humans by the tiger shark showed a decrease from 14 incidents for the previous two decades to 10 incidents, including 3 fatalities, 2 injuries, and 5 uninjured incidents. On the other hand, Chapman and McPhee (2016) showed shark bite incidence by species between 1982 and 2013 in Volusia County, Florida, USA, where the identification of the species definitively recorded is 46 (5.1%).

Clua et al. (2014) reported on a case of a non-provoked fatal shark attack on a 15-year-old male kitesurfer in New Caledonia. The victim lost his board and was pulled by the sail along the water surface in a reef passage when a shark attacked. The shark inflicted at least two bites on the left leg, comprising a severe one around the knee, resulting in a quick hypovolemic shock that was fatal. Analysis of one of these bites indicated that a 2.8 m TL (est. length) tiger shark was accountable for this attack. Features of the attack are the same with those of a predator's response to a surface feeding stimulus.

In terms of classification, the case reported by Clua et al. (2014) should be considered as an unprovoked attack, demarcated as incidents where an attack on a live human by a shark occurs in its natural habitat without human provocation of the shark (such as people spearfishing, feeding sharks, etc.). The attack described by Clua et al. (2014) resulted in one major injury that resulted in rapid blood loss and induced hypotension. Combined with the fatal result, these features directly classify it as level 5 in the Shark-Induced Trauma (SIT) Scale (a 1 to 5 scoring system based on the severity of the attack) introduced by Lentz et al. (2010). The analysis of wounds indicated a 2.8 m TL (est. length) tiger shark as responsible for this attack. It is possible to consider it as juvenile, as the average length at first maturity in male and female tiger sharks in the Central Pacific is greater than 3 m (Whitney and Crow 2007). The estimates of total length of the shark based on IDD and bite circumference vary by less than 5%, which lends confidence to the estimate of TL. Our finding is also agreeable with the presence of tiger sharks in quite high densities in the coastal waters of New Caledonia (Clua et al. 2014).

The activity of kitesurfing has been recorded relatively rarely in reports of shark incidents in the short period since this sport commenced. Reported shark incidents with kitesurfers are 2 in 2012 in Florida (USA), 1 in Papua New Guinea (among 41 incidents) in 2011, and 2 incidents each in Western Australia and Florida (among 38 incidents) in 2010. Apart from the current additional case from New Caledonia and the Florida case in 2010, all these incidents were non-fatal.

Apart from the lower number of practitioners of this sport, which could explain why kitesurfing accounts for significantly less shark attacks than other types of surfing sports, some features of this sport may be a factor in explaining the relative frequency of shark-related incidents. Kitesurfing includes the use of a board moving along the water surface, but the sail contributes to much higher speeds than in surfboard riding, in addition to multiple jumps (sometimes of several tens of meters). Those features might be significant limiting factors for a shark to attack a kitesurfer, compared to a surfer. However, when a kitesurfer does lose his board and is pulled by his sail along the surface, such as in the case reported by Clua et al. (2014), with relatively high speed and intermittent touching down on the surface, it could represent a strong feeding stimulus for a shark. However, the skimming motion of a kiteboard could be a possible factor under certain conditions. In the case described by Clua et al. (2014), a kitesurfer could resemble a seabird moving over the surface and could constitute a feeding stimulus for a tiger shark, triggering some form of active response (Barlow 1977). This hypothesis is agreeable with the feeding behavior of tiger sharks in New Caledonia. Data presented by Rancurel and Intes (1982, New Caledonia), Simpfendorfer (1992, Queensland, Australia), Lowe (1996, Hawaii), and Simpfendorfer et al. (2002, Western Australia) indicate that there are some differences in tiger shark diet between widely separated areas of the Indo-Pacific Ocean. Rancurel and Intes (1982) showed that sea snakes and sea birds were the most commonly occurring prey items in New Caledonia, while teleost fishes were most frequent in Hawaii and Australia, with teleost fish, sea turtles, and dugongs being the most frequent prey in Western Australia. The fact that tiger sharks in New Caledonia usually feed on seabirds might have been a facilitating factor for the attack of this kitesurfer in a reef passage, particularly where large birds, such as Boobies (Clua et al. 2014), are quite common.

Tirard et al. (2015) reported an accident of shark attack on human in New Caledonia. They identified the species of the shark by examining the patterns of the bite of the shark left on humans. Accordingly, they identified the shark as a tiger shark, not a great white shark, based on the large and homodont jaws. The case happened in Luengoni (Lifou Island, New Caledonia) on September 30, 2007. The wound was a single large bite with a maximum length of about 40 cm. Diagnostic features of the wound were its large size and its dorsoventral symmetry, indicating homodont jaws. These two features combined clearly designate the tiger shark as the only possible attacker in this case.

Hardiman et al. (2020) in their paper on how sharks and shark-human interactions are reported in major Australian newspapers gave the number and percentage of articles mentioning an identified shark taxon between April 1 and September 30, 2015, in four major Australian newspaper titles (Broadsheet, *Australian*, *Herald*; tabloid, *Mail*, *Telegraph*). The tiger shark represents 33 in tabloids, with 14.9%; 13 in total broadsheets, with 12.7%; and 46 among all articles, with 14.9%.

Prionace glauca (Linnaeus 1758) (Figs. 2.19, 2.20, 2.21, and 2.22).

Common name: Blue shark.

Etymology: Prionace: Derived from Greek, *prio*, *prion*, saw; *glauca*: name comes from the Latin *glauca*, meaning blue.

Fig. 2.19 *Prionace glauca* (Linnaeus 1758). (Courtesy of Muazzam Khan, Pakistan)

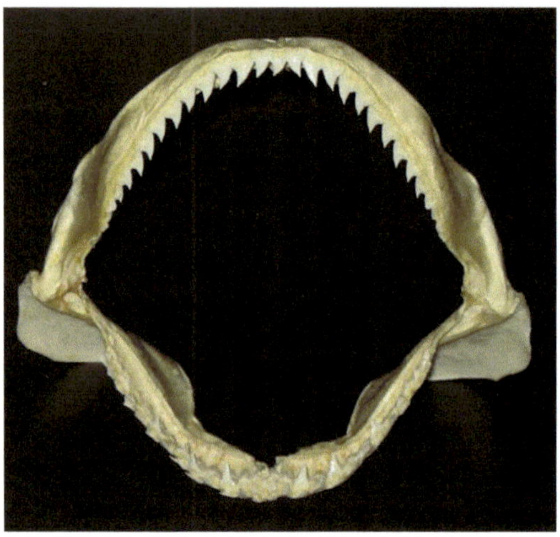

Fig. 2.20 Jaws of *Prionace glauca* (Linnaeus 1758). (Courtesy of Muazzam Khan, Pakistan)

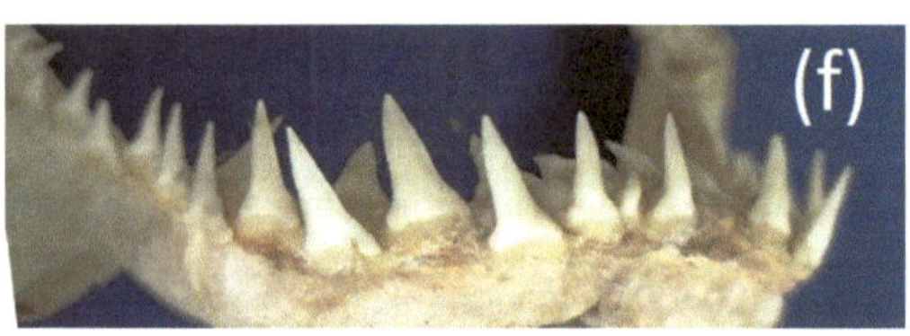

Fig. 2.21 Teeth of *Prionace glauca* (Linnaeus 1758). (Courtesy of Muazzam Khan, Pakistan)

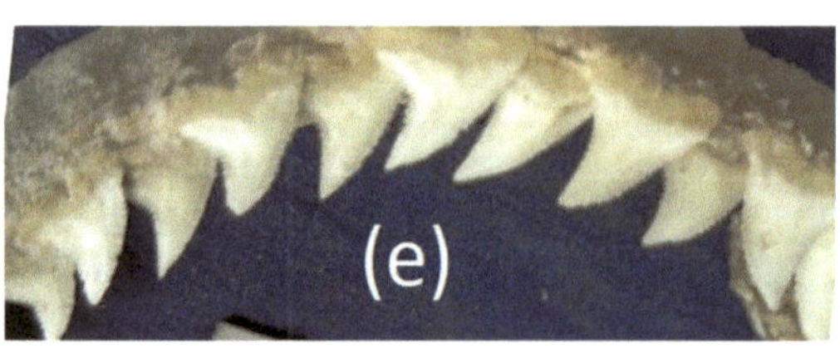

Fig. 2.22 Teeth of *Prionace glauca* (Linnaeus 1758). (Courtesy of Muazzam Khan, Pakistan)

Diagnostic Features

The body is slim, with conical long snout and large eyes. Spiracles and interdorsal ridge are absent. Teeth in upper jaw are serrated with narrow cusp and broad base, with those in the lower jaw less serrated and less oblique with narrow cusp. Keel on both sides of the caudal peduncle is low. Pectoral fin is long, narrow, and slender. The second dorsal fin is slightly smaller than the anal fin. Body is dark blue at dorsal side and bright blue ventrally with dark tips of pectoral and anal fins (Compagno 1998; Randall 1995; Compagno and Niem 1998a, b; Ebert et al. 2013).

Distribution

This shark has a circumtropical type of distribution in temperate and tropical waters. It is found in the western and eastern Atlantic regions. It is also recorded from the Indo-Pacific and eastern Pacific regions too (Froese and Pauly 2021). There are no records of this species from the Arabian-Persian Gulf (Compagno 1984a; Moore et al. 2012; Ebert et al. 2013). The record of this species from Iranian waters given by Froese and Pauly (2016) is based on (Compagno 1984a). Checking this reference revealed that Compagno (1984a)does not record this species from Iranian waters. It is not reported from the Sea of Oman either (Henderson and Reeve 2011), but Manilo and Bogorodsky (2003) have reported this shark to be from southern coasts of the Arabian Peninsula at the coasts of Yemen and Oman.

Habitat, Biology, and Ecology

This shark is a marine species that inhabits pelagic-oceanic habitats (Reide 2004). It prefers surfaces to 1000 m but usually lives at depths 80–220 m (McMillan et al. 2011; Florida Museum of Natural History 2021). It is usually found in the shores and can see divers in temperate seas, but in tropical areas, it lives in deeper water (Compagno 1984a). The blue shark is highly migratory with movement patterns associated with reproduction and prey distribution (Nakano and Stevens 2008). It may travel considerable distances (one specimen tagged in New Zealand was recaptured 1200 km off the coast of Chile) (Cox and Francis 1997). The maximum size attained by this shark is somewhat over 3800 mm in total length. Males and females reach maturity in 1820 and 2210–3230 mm in total length, respectively. This species has an age of maturity of 5–6 years, and they live for more than 20 years. This shark is viviparous, and females give birth to about 35 pups with a gestation period of 9–12 months. The maximum number of pups recorded is 135. At birth, pups are 350–500 mm in total length (Pratt 1979; Stevens 1984). Females can stand the bites of males during the hard courtship as they have skin three times thicker than that of males (Pratt and Carrier 2001).

The blue shark has the highest known population growth rate for pelagic sharks, that is, 0.287–0.314 in the North Atlantic, 0.299 in the South Atlantic, and 0.264–0.331 in the Indian Ocean (Dulvy et al. 2008; Cortés et al. 2015; Murua et al. 2018).

Economic Value

It is marketed fresh, dried, salted, and frozen for human consumption, hides for leather, and fins for soup (Frimodt 1995). It comprises most shark fins imported in Hong Kong: 17.3% in 1991–2001 and 49% in 2014 (Clarke et al. 2006a, b; Fields et al. 2018). There is a report of high concentration of heavy metals (mercury and lead) in the edible flesh (Lopez et al. 2010). This shark is usually used for recreation. It is not possible to keep the meat of this shark for a long time as it ammoniates very quickly, and shark specimens have their fins cut and sold in the shark fin market for shark fin soup. The meat of the blue shark is of low value, but its fins are valuable, and there is a great demand for them (Jawad 2017).

Conservation Status

There are several threat issues facing this species of shark. It is caught globally as target and bycatch species in commercial and small-scale pelagic longline, purse seine, and gillnets fisheries. Most of the catch is obtained as bycatch of industrial pelagic fleets in offshore and high seas waters (Camhi et al. 2008). It also captured in coastal longlines, gillnets, and sometimes trawls, mainly in regions with narrow continental shelves (Camhi et al. 2008). Underreporting of catches in the pelagic and domestic fisheries is likely (Dent and Clarke 2015). The species is highly valued by big-game recreational fishers, and although many practice catch and release, recreational fishing could be a threat due to post-release mortality (Camhi et al. 2008). At-vessel hooking mortality from commercial Portuguese longlines in the Atlantic is relatively low (13.3%), but smaller individuals have higher probabilities of mortality than larger adults (Coelho et al. 2013). The post-release mortality of only a few pelagic shark species was studied and showed 10–35% for the blue shark (Campana et al. 2009, 2016). The species is included in beach protection programs that target large sharks (Dudley and Simpfendorfer 2006; Simpfendorfer et al. 2010; Reid et al. 2011).

Rigby et al. (2019a) assessed the conservation status of this shark and granted the criterion Near Threatened ver 3.1 according to the IUCN Red List based on the following reasons. They stated that the blue shark (*Prionace glauca*) is a large (to 380 cm TL) oceanic pelagic shark wide ranging throughout tropical and temperate waters from the surface down to 1000 m deep. The species has the highest known population growth rates among pelagic sharks; it matures young and has large litters. It is caught globally as a target and bycatch species in commercial and small-scale pelagic longline, purse seine, and gillnets fisheries. However, most of the take is bycatch of industrial pelagic fleets in offshore and high seawaters. Generally, this species of shark is caught for its fins, although the demand for meat is likely increasing. Sharpest population decreases have happened in the North and South Atlantic, with lesser drops in the Indian Ocean and increases in the Pacific.

Human Interaction

The blue shark is an aggressive species with records of 13 attacks on humans reported in the International Shark Attack File (ISAF Statistics on Attacking Species of Shark 2012), with nine nonfatal and unprovoked incidences and four fatal and unprovoked accidence. There are also four incidences of attacks on boats. In addition, this shark is usually attacking sailors on floating ship wreckage. During its circulation, the shark plans for an exploratory bite as a feeding test (Jawad, 2017).

Barreiros et al. (2014) reported on an attack incidence by the blue shark on angler operating on the South coast of Terceira Island, Azores, Portugal, Northeast Atlantic. This incidence happened on July 24, 2013, near a shallow water rocky reef approximately 50 m from the coastline; a blue shark that bit his right flipper surprised a 48-year-old spear angler. The victim was familiar with the Azorean ichthyofauna, particularly the Chondrichthyans, and easily identified the shark species based on its slender body, dark-blue color, and elongated snout and fins. Late afternoon, the spear angler entered the water and swam away from the adjacent fishing port toward the tip of the adjacent rocky reef. There, he saw a school of Atlantic Bonito, *Sarda sarda*, and straightway leaped to place himself in a position suitable to target one of the fish swimming in front of him. As he descended to a depth of approximately 10 m, he felt his right foot being pushed and turned his head to see an approximately 2.1 m total length blue shark biting his flipper. The teeth marks left in the upper flipper surface indicate, at least, two almost simultaneous incomplete bites by the shark lateral jaws, possibly trying to grab and then remove a piece, as is usually observed in the bite kinematics of large Carcharhinidae sharks (Moss 1977). At the same time, he reported that at least five more similar-size blue sharks were close by, and he was able to chase the first specimen away simply by touching it with his left leg. The sharks disappeared, and the victim continued to fish in that same area with no further sightings.

Chapman and McPhee (2016) reported on the global shark attack hotspots. In their study, they gave data on the shark bite incidence by species between 1982 and 2013 in Volusia County, Florida, USA. The percentage of attack by the blue shark was 0.11% in the USA.

Family: Sphyrnidae.

Sphyrna zygaena (Linnaeus 1758) (Figs. 2.23, 2.24, and 2.25).

Common name: Smooth hammerhead.

Etymology: Sphyrna: Greek, *sphyra* = hammer.

Fig. 2.23 *Sphyrna zygaena* (Linnaeus 1758). (Courtesy of Hamid Badar Osmany, Pakistan)

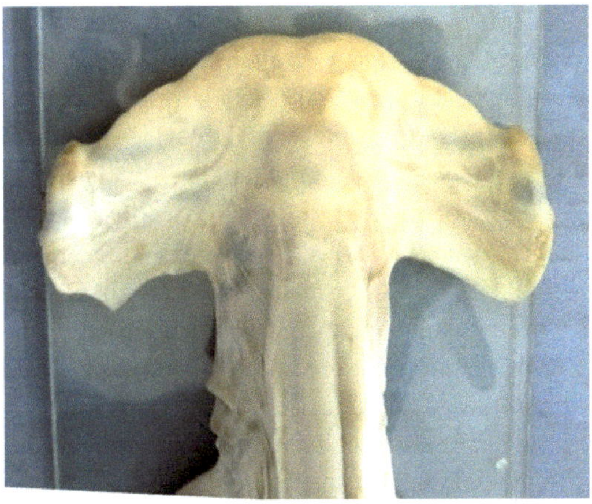

Fig. 2.24 Dorsal side of the head of *Sphyrna zygaena* (Linnaeus 1758). (Courtesy of Hamid Badar Osmany, Pakistan)

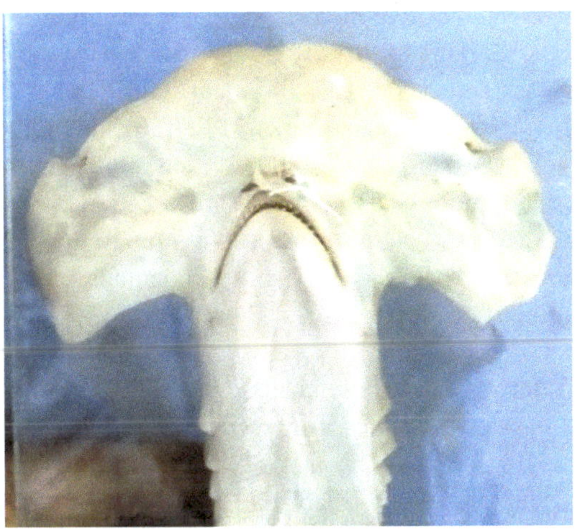

Fig. 2.25 Ventral side of the head of *Sphyrna zygaena* (Linnaeus 1758). (Courtesy of Hamid Badar Osmany, Pakistan)

Diagnostic Features

The body is large, with the absence of a notch at the center of its curved head. The first dorsal fin is high, and second dorsal is low. The head is wide and very broadly arched with lateral indentation but short longitudinally. The prenarial grooves are well developed and situated anterior to nostrils. The serrated anterior teeth are long, while the posterior ones are molariform. The upper part of the body is olive-gray to gray-brown in color with white abdomen (Compagno 1984a, 1998; Compagno 1998; Ebert et al. 2013).

Distribution

This species of shark is distributed in the area from the western Atlantic from Canada to Brazil and Argentina, from the eastern Atlantic at the coasts of Côte d'Ivoire including the Mediterranean Sea, and in the Indo-Pacific from South Africa to Sri Lanka to southern Australia and New Zealand (Compagno 1998; Ebert et al. 2013). Compagno (1984) and Ebert et al. (2013) have reported this species to be present in the Arabian-Persian Gulf, but not Moore et al. (2012). It is present in the Sea of Oman (Henderson and Reeve 2011) and in the southern coasts of the Arabian Peninsula at the coasts of Oman (Henderson and Reeve 2011).

Habitat, Biology, and Ecology

This species of shark is chiefly marine, but it enters brackish waters. It inhabits a depth range of 0–200 m but is frequently found at a depth of 0–20 m (Compagno 1998; Compagno 1998; Florida Museum of Natural History 2005; Ebert et al. 2013). It moves northward to the pole during summer to stay in cooler water and returns to the equator in wintertime (Ebert 2003). The smooth hammerhead reaches a maximum size of 3700–4000 mm in total length (Compagno 1984a). Males and females mature at 2500–2600 and 2650 mm in total length, respectively, in the eastern coasts of Australia. Females and males reach maturity at 2700 and 2500 mm in total length, respectively, but this depends on the location (Bester 2008). It is a viviparous species and uses the yolk sac to get nutrition from the mother's body. Once the yolk is finished, the yolk sac changes into a placental connection and delivers food to the young. The gestation period is about 11 months, and then females give birth up to 50 pups (Ebert 2003). At birth, the young measure up to 610 mm in total length, and this shark reaches 20 years of age (Bester 2008). This species reported to feed on fish specimens captured by sport fishers in California (Compagno 1984a.

Economic Value

This species is usually obtained for its meat, which is used fresh, dried, salted, and smoked. Fins are used for shark fin soap. The liver is used for its valuable oil content. The skin is used for leather manufacturing and the remaining meat as fishmeal (Bester 2008).

Conservation Status

Rigby et al. (2019a) suggested that this species of shark faces different levels of threat. Such issue ranges from the capture of this hammerhead shark globally as targeted and bycatch species using different fishing gears. Fishing operations in areas with narrow continental shelves and in some areas, such as Peru, usually capture pregnant females and juveniles (White et al. 2006; Camhi et al. 2008; Gonzalez-Pestana et al. 2014; Martínez-Ortiz et al. 2015). The species is often used for its fins (Clarke et al. 2006a, b; Dent and Clarke 2015; Fields et al. 2017). Underreporting of catches in pelagic and domestic fisheries is possible (Dent and Clarke 2015). Released individuals after capture showed high percentage of injuries, reaching up to 100% for the closely related scalloped hammerhead in purse seines (Eddy et al. 2016). The smooth hammerhead is included in the beach protection programs that target large sharks with high mortality in the beach mesh nets in New South Wales of ~95% prior to 2010, where approximately all smooth hammerhead catch is juveniles (Dudley and Simpfendorfer 2006; Simpfendorfer et al. 2010; Reid et al. 2011).

Rigby et al. (2019a) assessed the conservation status of this shark as Vulnerable A2bd ver 3.1 based on the following set of reasons: the size of this shark is large and inhabit coastal areas. It is distributed widely in temperate and tropical seas to depths of at least 200 m. Individuals of this species of shark showed low population growth (0.225 per year). This shark is usually caught by different fishing gears globally. Such activities are enhancing the decline of its populations across the geographical distribution of this species. The greatest decline in the population of this species has seen in several regions of the Indian Ocean but a lesser decline observed in the South Pacific area.

Human Interaction

The smooth hammerhead is potentially dangerous to humans. As of 2008, the International Shark Attack File lists 34 attacks attributable to large hammerhead sharks, 17 of them unprovoked (1 fatal) (ISAF 2009). Nevertheless, owing to the smooth hammerhead's presence in temperate regions where people are less likely to enter the water, it was probably accountable for a minority of these attacks (Bester 2008).

Family: Chlamydoselachidae

Chlamydoselachus anguineus (Garman 1884) (Figs. 2.26, 2.27, and 2.28).

Common name: Frilled shark.

Fig. 2.26 *Chlamydoselachus anguineus* (Garman 1884). (Courtesy of Matsuura, Keiichi, Japan)

Fig. 2.27 Head of *Chlamydoselachus anguineus* (Garman 1884). (Courtesy of Matsuura, Keiichi, Japan)

Fig. 2.28 Jaws of *Chlamydoselachus anguineus* (Garman 1884). (Courtesy of Nesnad, Tokyo, Japan)

Etymology: *Chlamydoselachus:* Greek, *chlamys*, -idos = cloak, cape + Greek, *selachos* = shark (Romero 2002).

Diagnostic Features

The body is slender and eel-like. There is obvious keel on the abdomen, with presence of six pairs of gill slits on the side of the head and lower ends of the first gill slits connected to each other across the throat. The snout is very short and truncated, with a very long terminal mouth on the head. Teeth are alike in upper and lower jaws, with three strong cusps and a pair of intermediate cusplets. The anal fin is larger than the dorsal fin, with no subterminal notch.

Distribution

The distribution of this species of shark is wide, but it is patchy. It is reported from the Indian Ocean on the coasts of South Africa. It also recorded from the Western Pacific in the Japanese and New Zealand waters. In the Eastern Pacific, it is found in the waters of southern California, USA, to northern Chile, while in the Eastern Atlantic, it is reported from northern Norway to northern Namibia, possibly the Eastern Cape of Good Hope, South Africa (Uyeno et al. 1983).

Habitat, Biology, and Ecology

A marine and bathydemersal shark species, it inhabits a total depth range of 1570 m (Springer 1990), usually 120–1280 m (Compagno 1984a). It is an uncommon primitive shark (Cox and Francis 1997) found on outer continental and insular shelves and upper slopes (Compagno 1984a). Individuals of this species of shark feed on other sharks, squid, and bony fish (Cox and Francis 1997). It is ovoviviparous (Cox and Francis 1997) and likely matrotrophic, with a probable gestation period of 1–2 years. Litter size ranges from 2 to 10 (Tanaka et al. 1990).; young are born 39 to 60 cm in total length (TL), and 2 to 15 pups are produced per litter, which are nourished by large uterine eggs (11–12 cm diameter). Size at maturity is estimated to be within 92–163 cm TL for males and 130–135 cm TL for females; maximum size, known only for females, is approximately 196 cm TL (Gudger and Smith 1933; Tanaka et al. 1990; Last and Stevens 2009; Ebert et al. 2013).

Economic Value

Incidental in bottom trawl catches and utilized as fishmeal and as food fish (Compagno 1984a).

Conservation Status

Smart et al. (2016) assessed the conservation status of this shark and allocated the rank of Least Concern ver 3.1 for the following reasons. The distribution of this shark is wide but patchy. It has low flexibility to utilization. It is rarely present in the commercial catch, as it inhabits deeper waters. In case it is caught via bycatch, then it usually is retained for its meat or fishmeal or even discarded. Expansion of the deepwater fisheries might affect the population of this species. Several countries like Australia and New Zealand, where this species of shark reside in their waters, have put a management plan to restrict catching this species.

Human Interaction

A potentially dangerous shark, with sharp teeth enough to inflict lacerations on the hands of the unwary scientist examining its mouth (Compagno 1984a).

Family: Hexanchidae

Notorynchus cepedianus (Péron 1807) (Fig. 2.29).

Common name: Broadnose sevengill shark.

Fig. 2.29 *Notorynchus cepedianus* (Péron 1807). (Courtesy of Clinton Duffy, New Zealand)

Etymology: *Notorynchus:* Greek, *noton* = back + Greek, *rhyngchos* = snout (Romero 2002), *cepedianus: maculatus* meaning "spotted" (Hart 1973).

Diagnostic Features

This species of shark is the only extant member of the genus *Notorynchus*, in the family Hexanchidae. It is identifiable for its seven gill slits, with the exception of the members of the order Hexanchiformes and the sixgill saw shark. The body is large and thick, with a broad head and blunt snout. The teeth shape in the upper and the lower jaws is different. Those of the top jaw has jagged, cusped teeth, while the bottom jaw has comb-shaped teeth. The single dorsal fin is set far back nearer the caudal fin and located behind the pelvic fins. The upper lobe of the caudal fin is much longer than the lower and is somewhat notched near the tip. The dorsal surface of the body is silver-gray to brown in color in order to blend with the dark water and substrate when viewed from above. On the contrary, the ventral surface is very pale, blending with the sunlit water when viewed from below. Small black and white spots cover the body and fins (Compagno 1984a).

Distribution

This species of shark has a circumglobal distribution: it is reported from the tropical to temperate waters, except North Atlantic and Mediterranean Sea (Ebert et al. 2013). In addition, it is recorded from the Southwest Atlantic in southern Brazil to northern Argentina. Southeast Atlantic records of this species include Namibia to East London, South Africa (Compagno 1998), while the Western Pacific presence includes southern Japan to New Zealand. In the Eastern Pacific, it is reported from British Columbia, Canada, to Chile (Ruiz-Campos et al. 2010; Ebert and Mostarda 2016). They are also recorded in tropical waters off the Peruvian and Ecuadorian coastlines (Baerez 1996; Compagno 2009) that experience seasonal cooling by the upwelling Humboldt Current (Martínez-Ortiz et al. 2015). Recently, it is reported to be found in the Galápagos Marine Reserve (Buglass et al. 2020).

Habitat, Biology, and Ecology

This shark is a marine species found also in brackish water and prefers demersal areas at 570 m (Cox and Francis 1997), usually at 80 m (Compagno 1998). Young individuals of this species are found close to inshore and in bays and estuaries (Last and Stevens 1994), while adults inhabit deeper waters offshore and deep channels in bays (Compagno 1984a).

This shark feeds on a wide variety of food, comprising other sharks, rays, chimaeras, bony fish, hagfish, dolphin and porpoise meat, seals, shark egg cases, sea snails, and mammalian carrion, including rats and humans (Compagno 1998). Studies in 2003 showed that its food menu consisted of 30% mammals with a frequency of occurrence of 35% (Crespi-Abril et al. 2003). This shark also hunts shallow waters (Castro et al. 1999). Several individuals of this shark may join a hunting adventure to share a larger prey, using tactics such as stealth to succeed. It takes this shark several hours to slowly digesting the food, and in case the meal was too large, this shark would refrain from eating for days (Castro et al. 1999). The females are ovoviviparous (Dulvy and Reynolds 1997) and can give birth to 82–95 young in a litter (Smith et al. 1998). This shark lives for about 30 years (Finucci et al. 2020) although the Washington Department of Fish and Wildlife in Washington State lists a maximum of 49 years (WDFW 2021).

In New Zealand, it is also one of the most common inshore sharks (NIWA 2012). While it is mainly a nocturnal forager, it may unscrupulously feed on prey casually found during the day; nevertheless, studies in 2010 showed that this shark has similar amount of activity in both day and night. It is also found that as Norfolk Bay does not have adequate shelter cover, this species may use group formation to avoid predation (Barnett et al. 2010).

Economic Value

This shark is often caught by anglers from the shore (Bass et al. 1986a, b). The meat is utilized for human consumption, its skin for leather, and its liver as a source of oil (Compagno 1984a, b). Its meat and fins are in demand in countries such as the USA, Brazil, Spain, Germany, the Netherlands, and Israel and packaged for frozen food (Vannuccini 1999). This shark is also a source of vitamin A and utilized by South African sport anglers for winter tournaments; however, this shark is not easy to land despite being readily hooked (Van der Elst 1993).

Conservation Status

The broadnose sevengill is allocated the rank of Vulnerable A2bd in the IUCN Red List throughout its range (Finucci et al. 2020a). It faces a continuous impact originated from several kinds of fisheries activities in addition to catch as a bycatch (Finucci et al. 2020). In Argentina, this shark is included in a fishing completion and caught by line and rode (Cedrola et al. 2009). Pollution is another impact factor affecting the duration of this shark. In addition, this shark is caught for its liver oil. In June 2018, the New Zealand Department of Conservation classified this shark as "Not Threatened" with the qualifiers "Data Poor" and "Secure Overseas" under the New Zealand Threat Classification System (Duffy et al. 2018).

No numerous conservation plans are in action, but it has been reported from one marine reserve in South Africa, and it is present in La Jolla Cove, La Jolla, San Diego, California, with the latter having an obvious population upsurge in 2013 (KPBS 2013). The studies published in 2009 suggested that Bahía Anegada be made a conservation area given the high number of sharks there (Cedrola et al. 2009). In Washington State, recreational fishing of broadnose sevengill shark is closed on all state waters (WDFW 2021). In Victoria, Australia, the Department of Environment and Primary Industries sets a one-bag limit and must be whole or in carcass form (DVG 2015).

Human Interaction

The International Shark Attack File considers this shark potentially dangerous due to its proximity to humans and for its aggressive behavior when provoked. It is also noted as aggressive toward divers and spearfishermen in both public aquariums and the wild. In 2013, in Fiordland, New Zealand, a sevengill bit a diver's regulator and then his head (FLMNH 2013; Turner 2013). Human remains were also supposedly found in one specimen's stomach. In 2020, a 13-year-old girl was bitten while surfing at Oreti Beach in New Zealand. The girl continued to surf for an hour before realizing her leg was bleeding (Damian 2020).

This species is responsible for several attacks on humans in public aquaria—there have been no verified attacks in open water (Fishes of Australia 2021). According to the International Shark Attack File, the sevengill shark has been responsible for five documented unprovoked and nonfatal attacks on humans (ISAF 2009).

In 2017, a French tourist survives after being attacked by a shark while body boarding at Curio Bay on the Southland coast, New Zealand. Shark expert Clinton Duffy thinks the shark was likely to be a Sevengill, a shark that can get as big as 3 m long but which is not known overseas as an aggressive shark. In New Zealand though, they bite prey like humans realistically regularly. Mr. Duffy can't say exactly where swimmers and surfers might encounter the Sevengill. In his statement to *Morning Report* trying to determine where a shark attack might happen, he says, "A bit like trying to predict where lighting is going to strike" (RNZ 2017).

Family: Somniosidae

Scymnodon ringens (Barbosa du Bocage and de Brito Capello 1864) (Fig. 2.30).

Common name: Knifetooth dogfish.

Etymology: *Scymnodon:* Greek, *skymnos*, -ou = little lion + Greek, *odous* = teeth (Romero 2002).

Diagnostic Features

The body is black. The head is rather thick and high, with a broad and short snout. Small spines were found on the dorsal fin. The gill slits are rather long, longest over half-eye length. Anal fin is absent. Teeth are small lanceolate lacking cusplets in those of the upper jaw and huge high, knife-cusped cutting teeth in those in the lower jaw. The mouth very wide and broadly arched. Pectoral fins are narrow and leaf-shaped, apices

Fig. 2.30 *Scymnodon ringens* (Barbosa du Bocage and de Brito Capello 1864). (Courtesy of Alberto, López, Spain)

of pectoral fins nearly reaching the base of the first dorsal spine. The caudal fin has a weak subterminal notch and no lower lobe. Lateral side of trunk with denticles lacking cross-ridges on crowns. Caudal peduncle is short, distance from second dorsal base to upper caudal origin about half-second dorsal base.

Distribution

This shark is reported from the Eastern Atlantic region along the slope from Scotland to Spain, Portugal, and Senegal. It is also recorded from the southwest Pacific in New Zealand waters (Cox and Francis 1997), where it is known from only a handful of records (Roberts et al. 2015). Abundance indices for more commonly recorded New Zealand deepwater sharks show no trend over time between 1992 and 2014 (Francis et al. 2016).

Habitat, Biology, and Ecology

A rare species (Cox and Francis 1997) prefers the continental slopes as a living area (Compagno 1984a). It is found in the mesopelagic region, while it is caught so often near the bottom (Springer 1990). This shark uses its razor-like teeth in the lower jaw in attacking its prey, especially the large ones (Compagno 1984a). Females are ovoviviparous (Breder and Rosen 1966). The maximum size reached by this shark is 110 cm in total length (TL) (Ebert et al. 2013). Individuals of this species of shark feed on teleosts, crustaceans, and cephalopods of the meso- and bathypelagic, bathydemersal regions (Ramos 2018).

Economic Value

The meat of this shark is taken for human consumption and eaten dried or salted and used for fishmeal (Compagno 1984a).

Conservation Status

This species of shark faces a wide range of threats; among these are trawl and longline fishery activities in the northeast Atlantic (Heessen 2003; Walker et al. 2005). It is also captured in demersal fisheries (e.g., Moura et al. 2018). Discard mortality is indefinite but could be high and could be like that of its congener, Owston's Dogfish (*Centroscymnus owstoni*), which reaches to 80% (Ellis et al. 2017).

An absence of a particular species information can limit the capability to recognize which species recorded from the region. Recently, longline fisheries have targeted deepwater sharks in countries such as Mauritania and Ghana (FAO 2016). In Mauritania, deepwater sharks have been reported as bycatch from Black Hake (*Merluccius senegalensis* and *M. polli*) fisheries (Fernández et al. 2005).

In the southwest Pacific, deepwater sharks are regularly reported as bycatch from offshore deepwater fisheries in New Zealand (Finucci et al. 2019). Between 2008 and 2013, the knifetooth dogfish was not recorded in any commercial records (Francis et al. 2016); the species may be recorded under a generic "deepwater shark" reporting code; however, reporting of deepwater shark catch to the species level has improved considerably with the introduction of identification guides.

Finucci et al. (2020) assessed this shark species and is allocated Vulnerable A2bd ver 3.1 rank in the IUCN Red List for the following reasons: This species of shark is small, inhabiting deep water in the northeast and central Atlantic (Scotland to Senegal) and the southwest Pacific (New Zealand) Oceans. It is caught by way of trawl and longline fisheries. A report of significant decrease in the population of this shark in some areas is present, where intense fisheries activity is performed. The knifetooth dogfish is estimated to have undergone a population reduction of 30–49% over the past three generations (111 years) based on abundance data and actual levels of exploitation (Finucci et al. 2020a).

Human Interaction

A potentially dangerous shark, with low number of attack incidences on man (Ramos 2018).

Family: Dalatiidae.

Isistius brasiliensis (Quoy and Gaimard 1824) (Figs. 2.31, 2.32, and 2.33).

Common name: Cookiecutter shark.

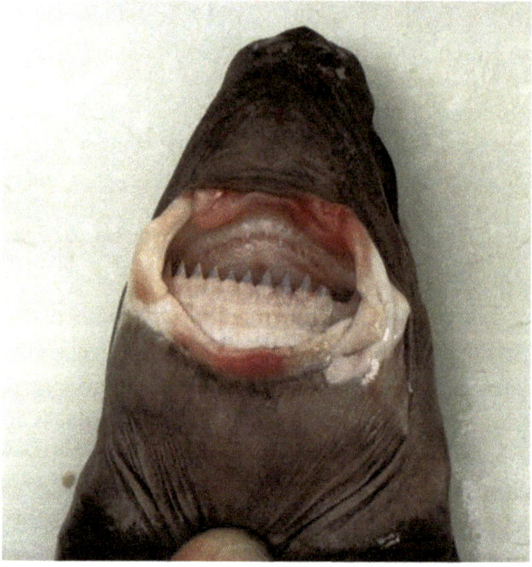

Fig. 2.32 Ventral view of the head of *Isistius brasiliensis* (Quoy and Gaimard 1824). (Courtesy of Mark Harris)

Fig. 2.31 *Isistius brasiliensis* (Quoy and Gaimard 1824). (Courtesy of Mark Harris)

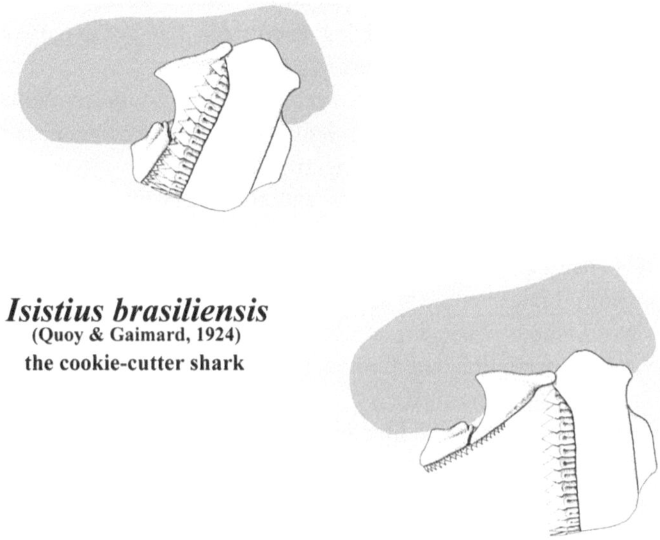

Fig. 2.33 The skull of *Isistius brasiliensis* (Quoy and Gaimard 1824). (Courtesy of Jaime A. Headden)

Etymology: *Isistius:* Greek, *isos* = equal + Greek, *istios* = sail (Romero 2002).

Diagnostic Features

The name "cookiecutter shark" denotes to its feeding habit of extracting round plugs, as if cut out with a cookiecutter, out of larger animals. Cuts made by cookiecutter sharks have been found on a wide variety of marine mammals and fishes, as well as on submarines, undersea cables, and even human bodies. It also completely consumes smaller prey such as squid (Wikipedia 2022).

This species of shark has a characteristic small cigar-shaped body with two small close-set spineless dorsal fins located posterior on its back. There is no anal fin. There is the presence of huge and triangular-cusped teeth that lack blades. The snout is short and bulbous, with unique suctorial lips (Compagno 1984a). This species of shark is distinctive for its conspicuous dark collar marks around its throat; large, nearly symmetrical caudal fin with a long ventral lobe over 2/3 length of dorsal caudal margin; and moderately large lower teeth in 25–32 rows. Eyes set anterior of the head but sufficiently far back to lack an extensive anterior binocular field; pectoral fins subquadrate; pelvic fins larger than dorsal fins (Compagno 1984a). Tooth count: 30–37/25–31. Vertebral count: 81–89. Spiral valve count: 8–10 (Ebert 2003). The body is dark brown dorsally but pale ventrally, except for a blackish band across the throat. Tips of caudal lobe are blackish (Bass et al. 1986a, b).

Distribution

Individuals of this species of shark are distributed in the western Atlantic regions. They have been reported from the Bahamas and southern Brazil. They are also present in the Eastern Atlantic region at Cape Verde, Guinea, to Sierra Leone, southern Angola, and South Africa, including Ascension Island. In addition, they were reported from the Indo-Pacific region at the Mauritius to New Guinea, Lord Howe Island, and New Zealand (Cox and Francis 1997), north to Japan, and east to the Hawaiian Islands. In the eastern Pacific region, they were recorded from Easter Island (Pequeño 1989) and the Galapagos.

Habitat, Biology, and Ecology

This shark is an oceanic species (Compagno 1984a). It lives in the epi- to bathypelagic at 1–3500 m (Mundy 2005). Individuals of this species make diurnal vertical migrations from below 1000 m in the day to or near the surface at night (Last and Stevens 1994; Mundy 2005). They usually travel long vertical distances more than

2000–3000 m on a diel cycle (Compagno 1984a). Their food comprises of free-living deepwater prey such as large squid, gonostomatids, and crustaceans but is also a facultative ectoparasite on larger pelagic animals such as wahoo, tuna, billfishes, and cetaceans (Compagno 1984a). Females are ovoviviparous (Breder and Rosen 1966), with 6–12 young per litter (Ebert 2003). Different pairing with embrace (Breder and Rosen 1966). Males mature at about 36 cm, with a maximum length of 42 cm; females mature at 39 cm, with a maximum length of 56 cm. Size at birth is unknown (Ebert 2003).

This shark has suctorial lips and a strongly altered pharynx that allow it to attach to the sides of its prey (Compagno 1984a). It then drives its sawlike lower dentition into the skin and flesh of its victim, twists about to cut out a conical plug of flesh, and then pulls free with the plug cradled by its scoop-like lower jaw and held by the hook-like upper teeth (Compagno 1984a). Teeth are shed as a complete unit; the lower teeth are swallowed, maybe to keep adequate calcium levels in its body (Compagno 1984a). Interconnection at the bases of individual tooth allows a whole row of teeth to move if one tooth is touched. This shark is reported to radiate light for as long as 3 h after its death.

Cookiecutter sharks have adaptations for hovering in the water column and likely rely on slyness and trick to capture prey that is more active. Its dark collar seems to mimic the silhouette of a small fish, while the rest of its body blends into the downwelling light via its ventral photophores. This species has been known to travel in schools (Wikipedia 2022).

The intrinsic green luminescence of the cookiecutter shark is the strongest known of any shark and has been reported to persist for 3 h after it has been taken out of the water (Mallefet et al. 2021). The ventrally positioned photophores serve to disrupt its silhouette from below by matching the downwelling light, a strategy known as counter--illumination that is common among bioluminescent organisms of the mesopelagic zone. The individual photophores are set around the denticles and are small enough that they cannot be discerned by the naked eye, suggesting they have evolved to fool animals with high visual acuity and/or at close distances (Widder 1998).

Economic Value

No economic value is reported for this shark.

Conservation Status

Kyne (2018) evaluated the conservation value of this species of shark and categorized it as Least Concern in the Red List of the IUCN for the following reasons: This species of shark is likely to be cosmopolitan in oceanic waters of temperate and tropical regions, although distribution records are patchy. This small epipelagic-to-bathypelagic shark (to ~56 cm total length) undertakes vertical migrations on a diel cycle (from depth probably below 1000 m; possibly as deep as 3500 m) during the day to near the surface at night. This shark is an occasional bycatch of oceanic trawl and longline fisheries, including mid-water trawls. Its small size and habitat/behavior probably limits its capture overall and given that the species is extremely widespread. In June 2018, the New Zealand Department of Conservation classified the cookiecutter shark as "Not Threatened" with the qualifier "Secure Overseas" under the New Zealand Threat Classification System (Duffy et al. 2018).

Human Interaction

This species of shark has been involved in a few attacks; in one case, a school of 30-cm (12 in)-long fish with blunt snouts attacked an underwater photographer on an open-ocean dive. Similar reports have come from shipwreck survivors, of suffering small, clean, deep bites during nighttime (Ebert 2003). In March 2009, a victim was bitten by a cookiecutter shark while swimming across the Alenuihaha channel (Perry 2009; Honebrink et al. 2011). Another victim was bitten while he was swimming by a cookiecutter shark on March 31, 2019, while crossing the Kaiwi channel, suffering a large laceration to his stomach (Tracking shark 2019a). A second cookiecutter attack occurred in the same spot 3 weeks later. Another victim was attempting to swim the channel on April 6, 2019, as part of the Oceans Seven challenge when he was bitten on the left shoulder

(Tracking shark 2019b). A third person attempting to complete the swim was bitten in nearly the same area of the channel. On July 29, 2019, a swimmer was bitten once on the stomach and then a second time on the left thigh (Tracking shark 2019c). Two of the three swimmers were using electrical shark deterrents, which did not deter the sharks. In 2017, a 7-year-old boy was bitten in the leg while wading in Alma Bay in North Queensland with his family. The shark caused a 7.3 cm wound that was nearly down to the bone. There are several records of bodies recovered from the water with postmortem cookiecutter shark bites (Martin 2010; Ribéreau-Gayon et al. 2017).

Cookiecutter sharks use specialized pharyngeal muscles, upper jaw, hyoid and branchial arches, and fleshy lips to suck onto their prey, inserting hook-like upper teeth and consistently massive lower teeth, then twisting in a circular motion to remove a plug of flesh (Motta and Wilga 2001; Lowry et al. 2009). The resultant circular and concave wounds have been observed on bodies of a variety of large pelagic species, including such diverse biota as cetaceans, pinnipeds, scombroid fishes (tunas, snake mackerels, billfishes), sharks and rays, and sea turtles (Jones 1971; Papastamatiou et al. 2000). It is possible that nearly any large epipelagic or mesopelagic species is fair game for cookiecutter sharks. There have been dispersed reports of these shark interactions with people, but all involved incidents have complicated bites on deceased humans. These are recognized as "scavenges" by the International Shark Attack File (ISAF), which investigates and analyses shark-human interactions, as opposed to allocation to the "unprovoked" or "provoked" shark attack categories, which include bites on live human victims (Honebrink et al. 2011).

Honebrink et al. (2011) recognized this species of shark involved in attacks against humans. They proposed that the appearance of wounds allowed immediate recognition of the cookiecutter shark, owing to the shape of the bites (*Isistius* sp.). The conditions adjoining the shark attack incident, specifically habitat, depth, and time of day, fit the behavior pattern of *Isistius*, which migrates vertically to surface waters at night to feed (Last and Stevens 1994; Papastamatiou et al. 2000). Moreover, distances between mandibular tooth apices, as evidenced by the puncture wounds on the victim, were more consistent with *I. brasiliensis* than *I. plutodus* (Compagno 1984a). Small round or oval scooped-out wounds on fishes and cetaceans have been reported in the literature as early as 1886, and Jones (1971) summarized reports through 1969. The cause of these wounds had been attributed to bacteria, invertebrate parasites, or lampreys (e.g., Pike 1951). Jones (1971) offered convincing sign that crater and crescentic wounds, and related scars, found on large pelagic fishes and cetaceans were the result of bites inflicted by *I. brasiliensis*.

To estimate the size of the shark, Papastamatiou et al. (2000) used a mouth width to total length regression to estimate sizes of cookiecutter sharks that imposed bites on several commercial fish species. In the attack incidence described by Honebrink et al. (2011), the initial estimate of the leg wound size was 10 cm in diameter. Measurements taken 7 days after the injury provided dimensions of 5 cm by 6.5 cm, but by that time, the size of the wound was somewhat reduced because healing was taking place. Jones (1971) indicated that the largest dimensions reported for oval crater wounds caused by *Isistius* were 4 or 5 cm by 7 cm. It is not known on what type of animal(s) these wounds were observed, nor the degree of healing that already taken place. The maximum length attained by *I. brasiliensis* is reported to be 50 cm (Compagno 1984a). The size of the victim's leg wound is analogous with the largest stated bite dimensions for this species, indicating an individual that was certainly an adult and likely near the maximum size usually encountered (about 40 cm).

2.1.1 Connection of Sharks with Humans

2.1.1.1 Historic Perspective

Humans know about sharks thousands of years ago. It is known to the ancient Mesopotamians, and they gave it different names according to the language they used in different areas and in the different time. Among these names is what Verderame (2017) mentioned, **kušu2/kušû**, which is used in our present time by people in modern Iraq. They usually used a general term (Iraqi Arabic dialect for shark Kausaj = كوسج) for any species of shark. Such similarity with the term used by ancient Mesopotamians (5000 BC) may support a fascinating hypothesis of a local survival of the ancient term in modern times.

The first ever documented shark species is the hammerhead shark. Some cylinder seals from the period of Ubaid culture (Southern Mesopotamia, 6500–3800 BC) were found showing this shark. Goff (1963) described several seals depicted with animal's images, among them is a seal that showed two antlers, a fish, and two vultures (Fig. 2.34). The image of the fish looks like a hammerhead shark judged from the shape of the body, head, location of the fins, and the tail. The shape of the head of the fish depicted in the seal looks extending horizontally like that of the hammerhead shark. The second feature is the shape and position of the pectoral fins, which are shown flat nearly triangular and extending horizontally. The third anatomical character is the shape of the caudal fin. All sharks including the hammerhead sharks have caudal fin called heterocercal fin, where the dorsal lobe is longer than the ventral.

Fig. 2.34 Plate from Ubaid culture (Southern Mesopotamia) 6500–3800 BC showing different animal, with fishes. (Source Goff 1963)

The other hammerhead shark species, whose teeth clearly depicted on the cylinder seal, is a species of the genus *Pristis*. Boehmer (2002), in his work on sawfish and sharks in antiquity, showed images of fossilized saw of a sawfish and seal impression of sawfish from Susa (the southern area located between Iraq and Iran) in the mid of the fourth millennium (Uruk Period 4000–3100 BC). Bossneck (1987, cited in Boehmer 2002) refers to the two well-known stamp seals (Fig. 2.35) from the Middle and Early Late Uruk Period from Susa and recognized them as two sawfishes. The location where these two fishes were observed is Susa, very close to Karun River in south of Iran, which joins Shatt al-Arab River in Iraq further south. The latter joins the Arabian Gulf at City of Fao. It seems that the two sawfish have swam about 200 km up the rivers Karun and Kerkha (Assyrian uknû) to reach Susa, where they were observed (Boehmer 2002). This information gives an indication of the possibility for large vertebrates such as the sawfish to ascend freshwater system, and the impression on the seals and the specimens of fossilized saw of the sawfish is an indication of early range extension for this species of shark.

The image of the upper fossilized saw shown in Fig. 2.35b belongs to *P. pristis* species, while the lower one in the same image could belong to *P. zijsron*. However, the lower saw has many teeth, which are not opposite to teeth on the other side. This case is an asymmetry case in the number of teeth on both sides of the saw. Fish images that appeared on the seal impression shown in Fig. 2.35c, d are clearly sawfishes, judging from the presence of saw at their part of their head.

These fossilized artefacts, with an image of large-sized fishes such as sawfish, represent the first record in the history of this shark in the freshwater system. Such information is important for fish taxonomists when they report on a new record of shark species from a freshwater system around the world.

The other shark species that ancient Mesopotamians have encountered in the freshwater environment is probably the bull shark, *Carcharhinus leucas* (Müller and Henle 1839). Goff (1963) considered Gawra in the northern

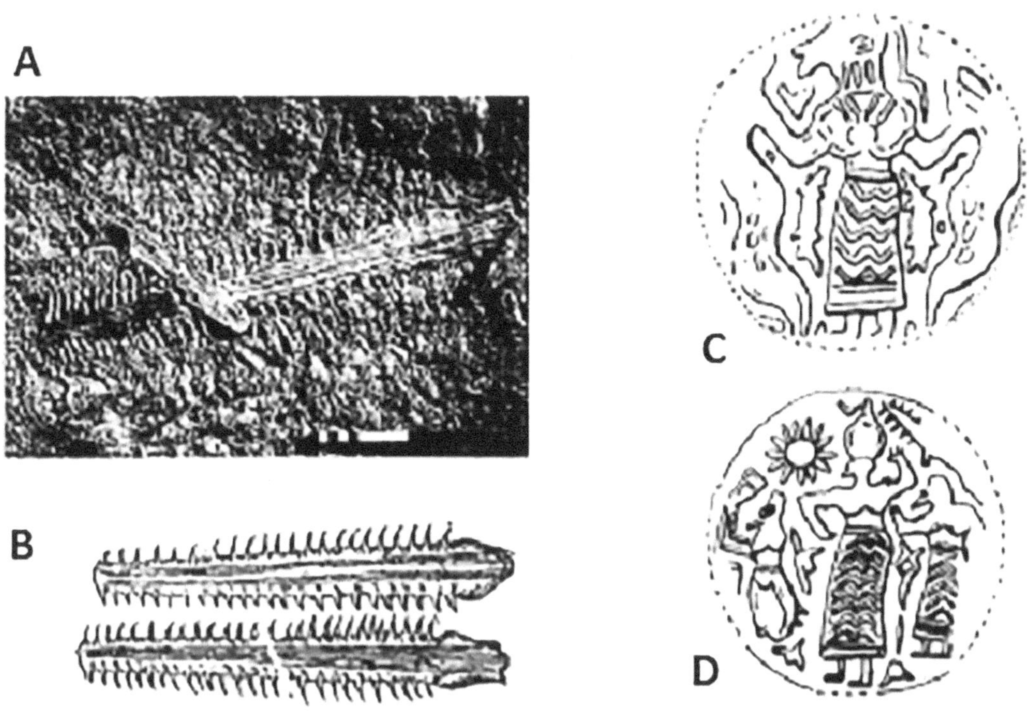

Fig. 2.35 Fossilized saw of a sawfish and seal impression of sawfish from Susa in the mid of the fourth millennium (Uruk Period 4000–3100 BC). (Source: Boehmer 2002 in Bossneck 1987)

Mesopotamia approximately present at the time of the Uruk Period in the South (4000–3100 BC). The Gawra name was used to designate artefacts obtained from Tepe Gawra area as many objects were retrieved from this site. A cylinder seal impression picture given by Goff (1963) showed an ibex, two kneeling animals, and a pair of fish (Fig. 2.36). The image of one of the two fishes can be an image of a shark based on the shape of the body, shape of the tail, and the head. In the image of the fish at the bottom, it is clear the upper jaw is longer than the lower. This characteristic is among distinguished characteristics of sharks. The possible species of shark that this image can relate to is the bull shark *Carcharhinus leucas* (Müller and Henle 1839).

Green and Black (1992) showed an image of a cylinder seal with Samaš, the sun god, in his anthropomorphized sacred boat (Fig. 2.37). This cylinder seal of the Akkadian Period is from Ešnunna (modern Tell Asmar, Diyala Governorate, northeast of Baghdad) (2350–2170 BC). Boat with a human head. The

Fig. 2.36 A seal impression picture from Uruk period 4000–3100 BC. (Source Goff 1963)

boats that are shown carrying people or gods by river or canal are on occasion reduced with a prominent prow terminating in a human head, occasionally also with human torso and arms,

Fig. 2.37 A cylinder seal with Samaš, the sun god, in his anthropomorphized sacred boat from the Akkadian Period from Ešnunna (modern Tell Asmar). (Source: Green and Black 1992)

with which the man-boat might row himself. In the image of another cylindrical seal (Fig. 2.38) given by Green and Black (1992), the water god Ea and his two-faced minister god Usmû were shown. In this cylinder seal, Enki is represented as a seated god with a long beard, wearing a cap with many horns and a long, pleated robe. Streams of 60-water flow from his arms to the ground, sometimes with little fish swimming along the flow.

Examining the fish images in Figs. 2.37 and 2.38 and judging from the shape of their body, head, pectoral fins, and eye, it is possible to say that these images belong to a species of shark. The tail, which is a characteristic of a shark species, is heterocercal (the dorsal lobe is larger than the ventral lobe), but the artist in this image has reversed the two lobes. Scenes in both cylindrical seals shown in Figs. 2.37 and 2.38 are in a freshwater habitat, and the presence of a shark in such an environment might raise the question of how such a marine creature can be illustrated to be present in the freshwater niches. The presence of species of sharks in the freshwater environment is not an unusual event. In modern history, records of the sporadic presence of marine species in freshwater systems of Iraq go back to the time of Günther (1874), who recorded the presence of the shark *Carcharhinus gangeticus* (*C. leucas*) in the Tigris River near Baghdad. This record is a misidentification of a bull shark *Carcharhinus leucas,* as this species can tolerate pure freshwater and is often found hundreds of miles up rivers (A. Moore, personal communication). On the other hand, Kennedy's (1937) list was significant in recording the presence of shark species *Carcharias lamia* Risso (could be misidentified for bull shark *Carcharhinus leucas*) in the Tigris River just north of Baghdad.

Images of shark that appeared in the cylinder seal given by Green and Black (1992) (Figs. 2.37 and 2.38) can prove two issues. First, the presence of shark in the freshwater water is an event that happened nearly 6000 years ago. Therefore, all documentations written in modern history about the presence of cartilaginous fishes in the freshwater environment and their ability to adjust their physiological characters to withstand low level of salinity (Thorson et al. 1983) can be rendered not valued. This is based on images of shark depicted by ancient Mesopotamians on cylindrical seal nearly 6000 years ago. Second,

Fig. 2.38 Depiction of cylinder seal showing Boat with human head. The water god Ea and his two-faced minister god, Usmû, are shown. (Source: Green and Black 1992)

the aforementioned cylindrical seals invalidate any modern historical records that claim to report the first presence of shark species in the freshwater environment of Iraq (Hussain et al. 2012).

Another image depicting a shark is a bronze model from Babylon dating to around 592 BC, which is inscribed with an omen concerning an anomaly (Fig. 2.39). The fish has two fins on its right side but only one fin on the left. The inscription reads: If a fish lacks a left fin, a foreign army will be destroyed. There are not much ichthyological details in this image that can tell the species of the shark. The general body and head shapes look like those of a shark. The ichthyological information given in Fig. 2.39 can be considered the second evidence for the presence of shark in the Euphrates-Tigris River system. The first was noted in images (Figs. 2.37 and 2.38) given by Green and Black (1992) as an impression of a cylinder seal.

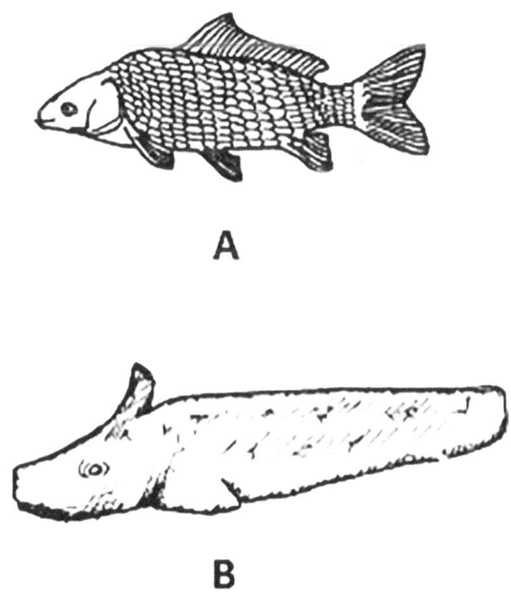

Fig. 2.39 (**a**) hand drawing of *Cyprinus carpio*; (**b**) a bronze model of a shark from the Babylon dating to c. 592 BC. (Source: Gane 2012)

The same bronze shark (Fig. 2.39) was also given by Sahrhage (1999) (Fig. 2.40), but with more detail and with some cuneiform writing, meaning "If no fish does not have its left fin, the hostile country will perish." It is clear from the figure given by Sahrhage (1999) that the bronze shark figurine was missing its pelvic fins. This is clear from the image, where both sides are shown. This evidence suggests that the ancient Mesopotamians made careful observations and noted that the shark figurine is missing its pelvic fins. Therefore, it is possible to say that recording and archiving cases and incidences of fish abnormalities started more than 1000 years ago and not recently as scientists think.

The image given by Sahrhage (1999) in his book is for another species of shark (Fig. 2.41) and dated 593 BC. Ichthyological features present in the image suggest strongly the Arabian carpetshark (*Chiloscyllium arabicum*; Gubanov 1980). The shapes of the head and body, the number of fins, and their location confirm identifying this species. The shark shown in Fig. 2.41 have one pelvic fin missing. It is another incidence of fish abnormality, and it is another evidence to support the suggestion of the discovery of fish anomalies long time ago, older than any date they usually documented.

Fig. 2.40 A bronze model of a shark from the Babylon dating to c. 592 BC. (Source: Sahrhage 1999)

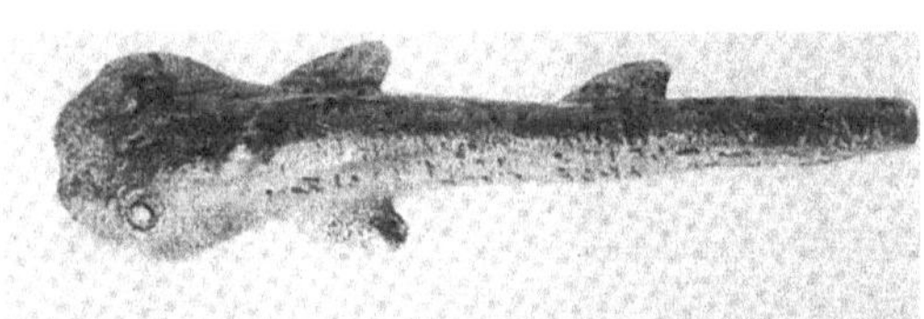

Fig. 2.41 Image of a shark dating 593 BC. (Source: Sahrhage 1999)

Ancient Mesopotamians were also aware of the danger of sharks and were probably the first who gave this species the title of "man-eater." This happened in 2500–2300 BC. The following passage is from the Presargonic riddles of Lagaš (Lagash /ˈleɪgæʃ/; Sumerian, *Lagaš*), also known as Shirpurla. Lagaš was an ancient city-state located northwest of the junction of the Euphrates and Tigris rivers and east of Uruk, about 22 km (14 mi) east of the modern town of Al-Shatrah, Iraq. Lagash (modern Al-Hiba) was one of the oldest cities of the Ancient Near East) (Biggs 1973). It may well refer to the shark: 'Its canal is "(the canal) going to Nina", its deity is Nanše, the mighty lady, its fish is "the man-eater" (ux -lu2 gu7 -[g]u7), its snake is [...]' (2H-T25 ll. i 1–3). It is possible to identify the river monster conjured in the short Akkadian incantation from Susa published by Cavigneaux (2000) with the shark. On the contrary, of the well-documented archive of the ancient Mesopotamians, Ncff and Hueter (2013) suggested that Carl Linnaeus (1758) was behind the genesis for the label of "man-eater" for the shark. Linnaeus's historic volumes redefined the scientific and social world, and white sharks were singled out for their motivation as a man-eater. It is clear that the term "man-eater" was proposed by Ancient Mesopotamians in 2500 BC, depicted in their artistic work and cannot be deserted.

2.1.1.2 Labelling Human-Shark Interactions

Neff and Hueter (2013) suggested a new system of four categories for scientists, the media, policy makers, and the public to use in cataloguing human-shark incidents. These categories are:

1. Shark sightings, where no physical shark-human interaction can take place.
2. A shark meeting is when physical contact occurs between a shark and a person or an inanimate object holding that person but no

injury takes place. For example, shark bites on surfboards, kayaks, and boats would be classified under this label. In some incidences, this could involve narrow escapes; a shark physically "bumping" a swimmer without biting would be known a shark encounter, not as a shark attack. A minor abrasion on the person's skin might occur because of contact with the rough skin of the shark.
3. Shark bites are incidents resulting in minor to moderate injuries. Small or large sharks might be intricate, but typically, a single, nonfatal bite occurs. If more than one bite occurs, injuries might be serious.
4. Fatal shark bites refer to serious injuries that take place in human-shark encounter, causing a significant loss of blood and/or body tissue and fatalities.

Burgess (1990) suggested that shark attacks could be classified into three categories:

1. Sneak: in this type, the attack is unanticipated and frequently occurs to a diver or swimmer in deep water.
2. Bump and bite: the victim is surrounded by the shark and gets a succession of bumps in which the shark tests how dangerous the prey is and at the same time produces injuries that might lead to incapability of the victim to escape (Baldridge 1988).
3. Hit and run: in this type of shark attack, the victim is seized and released before being able to respond. This type of attack occurs in shallow water and usually comes from juvenile sharks.

Attacks by large-sized sharks are within the first two categories where the shark recurrently attacks the victim or the likelihood of having another attack by the same shark is still present with high severity of injuries. Alternatively, the hit-and-run attack includes a first bite pursued by a direct retreat of the shark from the scene.

Sharks are attracted to shallow water areas where schools of mullet and herring are present, and individuals of these schools are characterized to have movement comparable to that produced by the moving hands and legs of swimmers and splashing water. Such activity puzzles the shark and triggers the attack. For surfers, the movement of hands and legs in an area away from the shore attracts large-sized sharks. In addition, hanging targets like surfers' legs hanging in the water while they sit on the board and wait for waves to come entice sharks (Cliff 1991). If the surfing area is near a jetty where fishing activities are going on, then more sharks gather in that area to have their portion of the captured fish. Surfers are more susceptible to shark attacks than waders are as the former have their legs and hands moving, whereas the latter have their legs standing in shallow water (Hazin et al. 2008).

The relationship between the lunar phase and shark attacks is usually linked to the effect of the moon on the tide, where during the new moon phase, a high tide occurs, which further attracts sharks to the coastal area. This relationship is noted in other marine organisms such as the different teleost fishes (Naylor 1999), invertebrates (Gliwicz 1986), and mammals (Wright 2005). With the new moon, light intensity is low, and not much light penetrates the water, making organisms follow the light to the surface and shore areas. Sharks do this; they follow the prey to their new locations and remain there until the next moon phase when the moonlight becomes stronger and penetrates deeper into the water. Such changes in moon intensity causes marine organisms to move deeper, and sharks follow them there and leave shallow and shore areas (Lowry et al. 2007).

2.1.1.3 Polluted Shark Products as a Kind of Human-Shark Interaction

Polluted shark products might be considered as a kind of human-shark interaction. Therefore, in this section, a short review of this problem will be discussed.

In the last few decades, worldwide shark fisheries have extended (Myers et al. 2007). The upsurge chiefly owes to the need for fins in Asia (Walker 1998; Clarke et al. 2006a, b; Worm et al. 2013), though shark flesh (muscle tissue) also offers an imperative supply of protein in many regions of the world (Simpfendorfer et al. 2011). The need for other shark products such as squalene and cartilage for use in a range of applications including cosmetics and pharmaceuticals also contributes to the increase (Walker 1998; Momigliano and Harcourt 2014).

Sharks are considered one of the apex predators; therefore, they bioaccumulate and biomagnify certain metals and metalloids in their tissues (Ratkowsky et al. 1975; Gray 2002; Endo et al. 2008; Gilbert et al. 2015), which in turn cause harm to human health (FSANZ 2002). Vulnerability to bioaccumulation and biomagnification of some metals and metalloids is thought to result in the longevity, size, slow growth, and low fecundity of many shark species (Walker 1988; Branco et al. 2007; Pethybridge et al. 2010). Such fact is linked with the uptake of metals and metalloids in diets of large predatory fish and/or marine mammals such as larger sharks (Cortes 1999; Endo et al. 2008; Hussey et al. 2012). The liver of sharks is rich in lipid and is intended for buoyancy (Last and Stevens 2009); it is also mainly responsible for their rapid intake and accumulation of lipophilic metal species such as methylmercury (MeHg) (Domi et al. 2005); mercury (Hg) and other metals such as cadmium (Cd), copper (Cu), iron (Fe), and zinc (Zn); and metalloids including arsenic (As) and selenium (Se), which occur naturally in the environment in several forms and are considered vital to many biological processes (Islam et al. 2004; Mamtani et al. 2011; Brewer et al. 2012). Although these numerous metals and metalloids are necessary in trace concentrations, they also contribute to possible harmful impacts if taken or present in excess forms and amounts (Haynes and Johnson 2000; Mamtani et al. 2011; van Dam et al. 2011).

The sternness of toxic outcomes relies on the extent of contact, pathway of uptake, and magnitude of exposure (Adams and McMichael 1999). Consumption of shark muscle tissue and shark products is believed to be a major dietary source of metals and metalloids for humans (Adams and McMichael 1999; Pethybridge et al. 2010; FSANZ 2002). Therefore, metal and metalloid concentrations in shark species of commercial value (e.g., blue shark, *Prionace glauca*, and school shark, *Galeorhinus galeus*) have been widely studied, particularly for Hg and MeHg concentrations (e.g., Walker 1988; Branco et al. 2004). Later, a maximum mercury concentration of 0.5–1.0 mg kg-1 wet weight (ww) in fish tissue has been set for human consumption in most countries. Maximum limits also exis for some of the other metals and metalloids analyzed in studies (FSANZ 2002; US EPA 2014).

Despite the historical and ongoing commercial significance of sharks, there is still relatively limited information on metals and metalloids in many species that are commonly fished globally, including commercial, artisanal, and recreational catches (Stevens and Brown 1974). Windom et al. (1973) emphasized the need to identify Hg and other element concentrations in non-commercial species of finfish to comprehend if high levels in predatory commercial species are of an environmental or physiological origin.

Hg concentrations in the liver are higher than muscle concentrations in tuna, swordfish (Branco et al. 2007), and odontocetes (Endo et al. 2008). However, available data on Hg levels in the liver and muscle of sharks are limited. The few studies completed thus far indicate a unique distribution pattern of Hg in those organs (Branco et al. 2007). Positive correlations between Hg concentration in the muscle and body length (weight) have been reported for tuna and swordfish (Kojadinovic et al. 2006) and sharks (Walker 1976; Lyle 1984; Turoczy et al. 2000; Branco et al. 2004, 2007). Researchers hypothesize that the phenomenon is due to the increasing Hg burden in aging predators. Interestingly, the liver has historically escaped scrutiny since it is not regarded as an edible organ. Therefore, only a limited information is available regarding the relationship between Hg concentration in the liver and body length of the predator. Cadmium (Cd) accumulation may be highest in mollusks, particularly cephalopods, due not only to the inherent hierarchy of the marine food web but also to species-specific physiologic mechanisms. The squid is a significant source of Cd for predatory fish and marine mammals. Concentrations of essential metals like zinc (Zn) and copper (Cu) in the liver and muscle of marine mammals and fish tend to decrease during the growth period. Essential metal levels in livers of marine mammals, as well as hepatic Hg and Cd concentrations, tend to increase after the growth period (Endo et al. 2008). However, no studies focus on fluctuating concentrations of these metals in the shark liver nor have there been reports on the correlation of this change with body length.

2.1.1.4 New Zealand's History of Human-Shark Interactions

In Reid's thesis (2019), "Sharks and New Zealand News Media," a comprehensive revision of the history of the human-shark interaction together with the history of New Zealand's policy and management regarding sharks is given. A concise discussion of those issues is given in this section. Interested readers should read on the thesis (2019) to learn more about this subject.

Human-shark interaction in New Zealand is considered substantial as the country is in the top five among worldwide shark attacks after the USA, Australia, South Africa, and Brazil (Midway et al. 2019).

The information provided by Reid (2019) is based on that given in *Te Ara: The New Zealand Encyclopaedia*. In such data, there are a limited number of available and correct records of human-shark interactions, with no official reporting in New Zealand (Reid 2019). Eventually, this indicates there are no present accessible records that can be considered categorically true. In *Te Ara*, there are 51 human-shark interactions since 2014 when archiving these incidences. Later and after 2014, *Te Ara* stopped archiving and recording human-shark incidences and was replaced by other sources such as New Zealand news media outlets, which are now considered the only source reliable for records of human-shark attack incidences.

Based on *Te Ara*, Reid (2019) reported that among 51 human-shark interactions within New Zealand, 12 are fatal, of which the white shark is responsible for. As to the regions in New Zealand where the incidences happened, Otago, Southland, and Auckland showed the highest number, with 4 of the 12 fatal interactions occurring in Otago (Reid 2019).

2.1.1.5 Following the Shark Attack

Not much information was released about the nature of the wound caused during shark attacks because people were focused more on the health of the victim. In this section, the nature and management of wounds caused by sharks are highlighted.

Once the human body is in a marine environment, it is exposed to several bacterial species that cause several diseases, for example, *Vibrio* species, which are chief species available in seawater as well as on teeth of most sharks and are clinically important to humans, owing to their pathogenic nature (Finkelstein and Oren 2011; Buck et al. 1984). Nevertheless, the real profile of oral flora varies from species to species (Unger et al. 2014). Other bacterial species that have clinical significance to humans are *Staphylococcus, Streptococcus, Aeromonas, Clostridium, Erysipelothrix, Mycobacterium, Pseudomonas, Plesiomonas, and Shewanella* (Finkelstein and Oren 2011). In addition to bacteria, distant bodies that contain parts of shark teeth can also be introduced into the wound and may be considered a source of multiple infections. Wound tissue will lose energy, allowing infection to become a stern threat. In addition, neurovascular bundles in bites might lead to substantial long-term deficits (Cooper et al. 2018).

Extra precautions should be performed through a series of imaging in order to make sure whether any pieces of shark teeth are left in the wound, fractures, or periosteal damage. Imaging can be through x-ray, ultrasound, or CT scan to view the inside of the wound (Cooper et al. 2018).

The first step in the management of shark attacks is to remove the victim from any area that can possibly cause further harm to the victim or to treating providers. In most cases, the victim is removed from the water and laid down on the shore to start the initial medical care, which includes assessing and stabilizing the patient's airway, breathing, and circulation (Cooper et al.

2018). Bleeding management is an urgency in this case, with hands and legs being the most possible to be affected. Bandages may be used to stop the bleeding (Unger et al. 2014). In such cases, the victim is usually given intravenous treatment for fluids and/or blood transfusions. To reduce hypothermia, rewarming measures should start as soon as possible. At the clinic, isolated wounds need to be cleaned exhaustively to stop infection. As with any wound produced by an animal bite, foreign objects can remain in the tissue and need to be recognized for removal (Cooper et al. 2018). An exhaustive wound cleaning is important to remove microbes and stop infection. The damaged tissue at the region of the wound should undergo debridement cleaning (Unger et al. 2014; Tomberg et al. 2018). Wounds may be closed, but this decision is usually made after a thorough check for the complete absence of an infection (Lentz et al. 2010).

The role of medication at this stage of management is important. Prophylactic antibiotics are prescribed to the victim even if the wound is small. The first choice of antibiotics is that against *Vibrio* species such as fluoroquinolones, third-generation cephalosporins, and doxycycline, which are highly active against *V. vulnificus*. Furthermore, treatment of infections caused by *Staphylococcus* and *Streptococcus* is done with another set of antibiotics such as cephalexin and doxycycline or cephalexin and ciprofloxacin (Lentz et al. 2010).

Most shark attack injuries are minor and require simple sutures, but in the case of victims with more complicated wounds that lead to wider death rates and incidences of limb loss, more complicated surgical procedures are needed. In most fatal accidents, deaths occur because of heavy bleeding from the wound (Woolgar et al. 2001).

Some complications might occur for the wound originating from a shark attack, such as neurovascular damages and acute fractures. On the other hand, insufficient examination or wound care can cause foreign bodies to remain in the wound and continuing defect of the affected tissue, leading to tissue loss and illness (Cooper et al. 2018).

2.1.1.6 Education toward Decreasing Shark Attacks

Based on proposals set by Global Shark Attack File, Cooper et al. (2021) mentioned these recommendations in his paper. In this section, a brief account of these suggestions is given.

1. Check the area where you intend to do your activity for presence or absence of sharks.
2. Be conscious of what happens around you.
3. Do not postpone leaving the water once you start feeling unwell.
4. In any case, do not harass, stay close to, or touch sharks, no matter of their size.
5. In case a shark is present in the intended swimming or surfing area, do not go in the water. Follow the instructions given by the lifeguards, and leave the water quietly.
6. Always swim, surf, or dive somewhere else.
7. Do not swim far away from the shore or go beyond the marked line if there is any.
8. Avoid swimming at night.
9. Choose clear water areas for swimming.
10. Avoid swimming near river mouths.
11. Do not swim near sandbars.
12. Remember that jetties are attractive areas for sharks; therefore, avoid swimming near such areas.
13. Remember, areas where birds usually dive are an attractive area for sharks; therefore, avoid them.
14. Do not chase any school of fish, as this event is appealing to sharks.
15. Leave water straight away once you notice baitfish are leaping at or above water surface.
16. Remember to attach your catch while you are spearfishing or collecting shellfish to anything other than your body and keep changing position regularly.
17. Remember to not swim or do any water activities near fishing areas.
18. If you notice the presence of porpoises and dolphins, then leave the water immediately as this can indicate presence of hunting sharks.
19. Immediately, leave the water once you see a group of dolphins heading to shore.

20. Always wear a low-contrast swimsuit.
21. Do not splash excess water or move quickly as such events may attract sharks.
22. Take no dogs or horses with you while swimming.

2.1.1.7 Role of Forensic Science in Shark Attack on Humans

Even though shark populations are going toward a decline globally, verified cases of unprovoked attacks by sharks on humans have been growing in latest years, rousing public worry and making fundamental rules such as blind culling. The yearly normal number of fatal attacks increased from 4.3 persons per year (2001–2010) up to an average of 8.0 persons per year between 2011 and 2015 (ISAF 2021). Such an unanticipated drift may elucidate by an important upsurge of the number of sea users that surges the likelihood of meeting between these marine predators and humans as shown in Australia (West 2011).

Forensic accounts of shark bite injury happen for submarine cables, marine turtles, pinnipeds, cetaceans, other sharks, surfboards, and humans (Byard et al. 2000; Caldicott et al. 2001; Woolgar et al. 2001; Heithaus et al. 2002; Zahuranec 2003). On soft objects, such as the skin or blubber, shark bites are characterized by a sequence of ragged-edged, coarsely parallel lacerations that may overlap to form a crescent-shaped perimeter around a mass of tissue that may be completely removed (Woolgar et al. 2001; Auerbach and Burgess 2007) (for images, see Lowry et al. 2009). On hard or rigid objects, such as the bone or metal, shark bites tend to look like a series of roughly parallel scratches or a obviously defined curved array of dimples or dents (Byard et al. 2000; Woolgar et al. 2001) (For images see Lowry et al. 2009). There are some exceptions to these generalities in the morphology of shark bites. Cookiecutter shark (*Isistius brasiliensis*) bites consist of a distinct circular wound from which a plug of tissue is typically removed (Gallo-Reynoso and Figueroa-Carranza 1992; Gasparini and Sazima 1996), while nurse shark (*Ginglymostoma cirratum*) bites appear as a concave wound produced by suction feeding (Lowry et al. 2009).

It is possible to estimate shark size and species from investigation of bite wounds and/or damage. Reliable estimates of size and species are crucial points to understanding feeding and hunting behavior (Stillwell and Kohler 1982), dietary composition (Cortes 1999), and ecological interactions (Heithaus et al. 2002) in addition to assessing potential hazard to humans and human constructs (Burgess 2008). Up until now, no consistent forensic measures for making such estimates are found. Guesses are characteristically grounded on the incidental knowledge of an agent with feeding behavior, geographic distribution, and bite damage from species and size ranges of sharks, or in some cases, they are made by matching the wound pattern to dried jaws in museum collections (Ritter and Levine 2004). Nevertheless, many qualitative and quantitative accounts of shark bite damage occur; only one comprehensive effort to differentiate bites among species and sizes of bony fishes and sharks has been published and was largely confined to bites on ropes (Berteaux et al. 1987). George et al. (1994) commented on the incidence of predation of killer whales (*Orcinus orca*) on bowhead whales (*Balaena mysticetus*). They are resolute that the range of the distance between consecutive tooth tips (interdental distance or IDD) in the skulls of museum specimens matched the range of the distance between slashes on the skin of *B. mysticetus* and was distinct from the IDD range of beluga whales (*Delphinapterus leucas*), another known *B. mysticetus* predator. If species-specific ranges of IDD occur in sharks, recognizing the species accountable for damaging an object or organism should be conceivable by creating a few simple forensic measurements. In addition to IDD ranges possibly being species-specific, considerable differences in tooth morphology are present between the upper and lower jaws of many shark species (Compagno 1984a) such that tooth spacing may differ between jaws within species. Consequently, injuries caused by each jaw are usually distinguishable, allowing for an assessment of the shark's position relative to its prey during a bite (Ritter and Levine 2005). Tooth size can differ directly with total length in some shark species (Compagno 1984a; Shimada

2002, 2004; Shimada and Seigel 2005). Such relationship also relates to IDD and is reliable across a broad size range within species; estimates of total length is usually made after species identification. Otherwise, in the presence of an overlap between IDD ranges for two or more shark species and together with the knowledge of total length ranges, it is possible to eliminate species from consideration.

Shark Jaws as Potential Lethal Weapons

Most sharks are predators that feed on other animals they capture, aided by modified jaws holding several lines and rows of teeth. Contrary to mammalian teeth, shark teeth contain fluorapatite, Ca5 (PO_4) F, which is harder than hydroxyapatite (Frazzetta 1988). Teeth may have diverse functions and therefore different structures and hardness. For example, mako shark (*I. oxyrinchus*) teeth are curved to the interior and are used to puncture prey flesh, while tiger shark (*G. cuvier*) teeth have serrated margins and are chiefly used for cutting prey in a sawing motion. Serrations differ from one species to another in unevenness and in distribution along tooth edges. Serrated teeth can make greater use of the available biting forces, and they have a greater cutting effect than smooth-edged teeth (i.e., mako shark, *I. oxyrinchus*). The latter depend upon friction, which, because the coefficient friction is always less than 1.0 (often very much less), can make use of only a fraction of the total bite force (Enax et al. 2012). Nevertheless, smooth tooth blades can pierce prey with less resistance and are less prone to binding (becoming immobilized) in the prey tissue (Frazzetta 1988). In Carcharinidae (including the bull shark, *C. leucas*), heterodonty is diagnosed by triangular and serrated teeth on the upper jaw, directing for cutting, while teeth from the lower jaw are slender and smoother, acting as puncturing/holding tools before the shark starts moving its head laterally to cut the tissues.

Whatever their shape and however sharp, these jaws and teeth serve a possible danger to humans, also considering that specific species, such as the tiger shark, may produce biting forces of up to 3300 kg/cm^2 (Byard et al. 2006; Whitenack and Motta 2010; Wroe et al. 2008).

Postmortem Reconstruction of Body Length Based on Wound Analysis

Characteristics of jaws and teeth and examination of the wound resulting from a shark attack all help in the identification of the shark species. Wound examinations usually concentrate on the number of bites, margin structure, tooth imprints, wound depth, and tissue loss to assess species identity (Davies and Campbell 1961; Byard et al. 2000). Nonetheless, most of the time, these data are inadequate for consistent identification and need be accompanied with ecological data about the presence and behavior of the species (pelagic vs. coastal affinities, seasonality, etc.) and/or, if available, data delivered by potential witnesses of the attack.

Generally, the size of the animal was judged by matching the length of the individual wound margins with jaws from known-sized animals (Ritter and Levine 2005). In addition, regressions that can offer the size of an animal based on the bite width (BW) are available.

Studies performed on cetaceans (George et al. 1994) and based on anatomical differences of jaws and teeth within shark species, Lowry et al. (2009) suggested a forensic investigation technique of shark bite wound patterns. This technique is established on the determination of a standard relationship between the measure of the average interdental distance (IDD) and the bite circumference (BC) of the jaw with the individual's total length (TL). The IDD and BC are allometric with the global size and are accurate predictors of the TL of a species individual. One of the many elasmobranch characteristics is a nonstop replacement of the teeth throughout life, allowing the jaw growth. The number of teeth will continue to be constant, but each new tooth is somewhat larger than the one it replaces (Moss 1981).

Consequently, in the present time and to guess length of a shark from an autopsy or file pictures, it might be rapid and rational to measure either interdental distances or bite radii from upper or lower jaws, whatsoever measurements are existing, and apply them to established log-log regressions provided by Lowry et al. (2009). If the IDD and BC can provide critical information, they might be insufficient for a consistent result. It is then needed to include other measurements of the wounds.

2.2 Osteichthyes (Teleostean Fishes)

2.2.1 Species of Moray Eel

The species of moray eel are included in the family Muraenidae. In New Zealand, there are 13 species of moray eel belonging to three genera, *Anarchias* (two species), *Enchelycore* (two species), and Gymnothorax (ten species) (Roberts et al. 2015; Trnski et al. 2015).

All species of moray eel are potentially dangerous to humans, able to inflict a severe wound, and probably fatal. There are several bite incidences against humans reported for certain moray eel species from around the world. Some of these cases will be given at the end of this section as an example of the vicious behavior of moray eel. In the following species account, a description of the 12 moray eel species found in New Zealand will be given for enhanced awareness on these types of species for those who use the marine environment for different activities, such as fishermen and aquarium keepers.

Family: Muraenidae
Anarchias cf. *seychellensis* (Smith 1962).
Common name: Seychelles moray.
Etymology: *Anarchias:* Greek, *anarchia*, -as = without a chief.

Diagnostic Features
The body is small and moderately long. The fourth supraorbital pore is closely behind the posterior nostril. The teeth are biserial; teeth on the inner row are well spaced, long, and depressible. There are three median intermaxillary teeth and 5–9 short teeth in a uniserial row on vomer. Body color is highly changeable; it is usually medium reddish brown with a complex irregular stellate spot pattern; some spots are pale brown to dark brown and without pattern. The head has pale spots, about equal to the eye and located over the snout, nape, and behind the rictus. They are also found along the dorsal midline of the trunk by an irregular series of pale spots. The lower jaw and abdomen are sometimes pale or without pattern. The head pores are white in color (McCosker and Stewart 2006).

Distribution
Individuals of this species are distributed in the Indo-Pacific region from East Africa to Easter Island, north to Marcus Island, and south to Lord Howe and Rapa. It is also reported from the Mariana and Marshall Islands in Micronesia (Allen and Erdmann 2012).

Habitat, Biology, and Ecology
This species of eels lives in rocky and coral reefs, crevices of rubble, and coralline algae. It also prefers shallow depths, rarely exceeding 35 m (Myers 1999; Mundy 2005). It is a sluggish eel (Smith and Smith 1963) but may bite when provoked (Randall et al. 1990).

Conservation
Individuals of this species do not face major threats. However, existing increase of habitat devastation owing to coral reef destruction may have a severe impact on this species (Smith et al. 2019a, b, c, d, e, f, g, h). This species is common throughout its range, which overlaps with the ranges of several marine-protected areas. Therefore, Smith et al. (2019a, b, c, d, e, f, g, h) evaluated this species as Least Concern in the IUCN Red List.

Anarchias supremus (McCosker and Stewart 2006).
Common name: None.
Etymology: *Anarchias:* Greek, *anarchia*, -as = without a chief, lack of authority (Romero 2002); *supremus:* name from a Latin word meaning "the uppermost," refers to having the most vertebrae of any known *Anarchias*.

Diagnostic Features
The body is small and reasonably long. The fourth supraorbital pore is contiguous to, above,

and behind the posterior nostril. The teeth are biserial in jaws, but they are uniserial on vomer. The anus is well before the mid-body. Eyes are small and slightly closer to the rictus than to the snout tip. The anterior nostril is tube in shape, reaching but not exceeding the snout tip and the posterior nostril above mid-eye, with a barely raised rim. Gill opening is small and located just below midside. Fins are confined to the tail tip, and the dorsal fin begins slightly before the anal fin. The body reddish brown in color, superimposed with a faint irregular stellate spot pattern posteriorly. Head pores are in small white spots, pale spots along lower half of flanks. The posterior half of the body has diffuse white bands, and the tail tip is white.

Distribution
This species is known only from its original locality, that is, 154 m off Macauley Island, Kermadec Ridge (Pacific Ocean).

Habitat, Biology, and Ecology
This species is newly discovered and is currently only known from Kermadec Ridge, New Zealand (McCosker and Stewart 2006). The only in-depth information known for this species is that of the collection of the holotype, at 154 m. Information about its habitat, biology, and ecology is lacking.

Conservation Status
Population information is also unknown for this species. In addition, threats to this species are unknown. Since it is newly discovered, there are no conservation measures in place for the protection of this species. Therefore, this species is listed as Data Deficient (Smith et al. 2019b).

Human Interaction
As a species of the moray eel group, it has the potential for sever bites to humans.

Enchelycore ramosa (Griffin 1926) (Fig. 2.42).
Common name: Mosaic moray.
Etymology: *Enchelycore:* Greek, *enchelys* = eel + Greek, *kore* = pupil (Romero 2002).

Fig. 2.42 *Enchelycore ramosa* (Griffin 1926). (Curtesy of David Smith, NMNH, USA)

Diagnostic Features
The body is elongated and laterally flattened and tapered toward the head. The jaws are long, curved, and equipped with very large transparent teeth. Pectoral and pelvic fins are absent. Gill opening is small. The skin is thick and leathery, with no scale. The dorsal fin starts just posterior to the head. The body is pale cream in color, with dark mosaic pattern of brownish green that extends over the head and into the mouth. There are 14 teeth in each of the upper and lower jaws.

Distribution
This species is distributed in the Pacific Ocean in Southeast Australia and New Zealand eastward to Rapa and Easter Island. In New Zealand, it is found in Bay of Plenty northward. It is also reported from the Kermadec Islands (Paulin and Roberts 1992).

Habitat, Biology, and Ecology
Individuals of this species inhabit coral and rocky reefs (Lieske and Myers 1994; Hoese et al. 2006). They are more visible to and seen frequently than other moray eel species. They are active during the day and feed on crabs, shrimps, and small fishes (Lieske and Myers 1994).

Conservation Status

Smith et al. (2019c) evaluated the conservation status of this species of moray eel and granted a Least Concerned status in the Red List of the IUCN. They based their decision on the following justifications: There are very few records of this species, and it is considered uncommon. This species of eel is utilized for local subsistence but is not presently considered a major threat. Therefore, it is listed as Least Concern.

Human Interaction

As a species of the moray eel group, it has the potential for sever bites to humans.

Gymnothorax berndti (Snyder 1904) (Fig. 2.43).
Common name: Y-patterned moray.
Etymology: *Gymnothorax:* Greek, *gymnos* = naked + Greek, thorax, *-akos* = breast (Romero 2002).

Diagnostic Features

The body is large and stout, with a tapering tail. The head is long and elevated behind the eye. Eyes are small, located above mid-jaw. The anterior nostril is located in a long tube. The dorsal fin originates in midway between gill opening and the snout tip. Teeth are large, uniserial, and well-spaced. Outer maxillary teeth are slightly larger than those of the inner row. Median fangs are depressible. The body is pale gray in color, with fine dark-brown lines and reticulations. The body has 29–35 irregular narrow dark-brownish bars. The dorsal fin has a wider dark bar and a white margin posteriorly. The anal fin has a broad irregular dark brown basal stripe and a conspicuous white margin. The inside of the mouth has dark speckles.

Distribution

Individuals of this species are widely distributed in the Indo-Pacific region. It is reported from Hawaii, the Marianas, Taiwan, the Maldives, Mauritius, Japan, Reunion, the Tonga Ridge, and the Kermadec Islands (McCosker and Stewart 2006).

Habitat, Biology, and Ecology

This marine species lives in association with reefs at a depth range of 30–303 m (Myers 1999). It reaches up to 1000 mm in TL (Masuda et al. 1984).

Conservation Status

This species has not been evaluated for the Red List of the IUCN.

Human Interaction

As a species of the moray eel group, it has the potential for sever bites to humans.

Gymnothorax eurostus (Abbott 1860) (Fig. 2.44).
Common name: Abbott's moray eel.
Etymology: *Gymnothorax:* Greek, *gymnos* = naked + Greek, thorax, *-akos* = breast (Romero 2002).

Fig. 2.43 *Gymnothorax berndti* (Snyder 1904). (Courtesy of Thaman)

Fig. 2.44 *Gymnothorax eurostus* (Abbott 1860). (Courtesy of Scott Johnson)

Diagnostic Features

The maxillary teeth are biserial, with inner and outer series about equal in length. Intermaxillary teeth are arranged in a peripheral series, a median series, and an intermediate series on each side between the median and peripheral series (Smith et al. 2008). Jaws are slightly curved. The body has dark purplish-brown coloration with small yellow spots that are sparse on the tail. The density of these spots increases with proximity to the head (Castle and McCosker 1986). There are several rows of black-brownish dots on the side of the body. The iris is orange in color, with whitish anterior nostril root, jaw pores, and tail tip margin (Chen et al. 1994).

Distribution

Individuals of this species are found in the Indo-Pacific region, in the Seychelles (Smith and Smith 1963), but Randall and van Egmond 1994 (Randall and van Egmond 1994) believe otherwise. They are also reported from the Eastern Pacific at Costa Rica and Easter Island (Lieske and Myers 1994) and Chile (Pequeño 1989).

Habitat, Biology, and Ecology

There is not much information available about the biology and ecology of this species. It inhabits crevices found in benthic environments (Mundy 2005) and feeds on crustaceans and small fishes (Bacchet et al. 2006).

Conservation Status

Smith et al. (2019a) evaluated the conservation status of this species of moray eel and listed it as Least Concern in the Red List of the IUCN. This species has an anti-tropical distribution found in South Africa, Easter Island, Madagascar, Cocos Islands, Ryukyu Islands, Ogasawara Islands, Minami-Tori-shima, Hawaii, Marquesas Islands, Lord Howe Island, Kermadec Islands, New Caledonia, Austral Islands, Western Australia, and Japan. The distribution of G. *eurostus* is common and moderately abundant. This species is used in aquarium trade; however, it is not usually seen in the trade, which is not considered a major threat to this species. It is believed to not have any major threats. Its distribution also overlaps numerous marine-protected areas in this range.

Human Interaction

As a species of the moray eel group, it has the potential for sever bites to humans.

Gymnothorax *nubilus* (Richardson 1848) (Fig. 2.45).

Common name: Gray moray.

Etymology: *Gymnothorax:* Greek, *gymnos* = naked + Greek, thorax, *-akos* = breast (Romero 2002).

Diagnostic Features

This moray eel species has a moderately long body. The anus is located before the midbody. Head and snout are long. Eye is closer in position to the rictus than to the snout tip. The anterior nostril is in a shape of a long tube, while the posterior nostril has a pore shape located above and behind anterior margin of the eye. Gill openings are at midside. The teeth are large and smooth, uniserial in adults. Mouth has three long median fangs. Vomerine teeth are short, uniserial, or slightly staggered. Body color is pale tan, with large, diffuse pale-brown spots in irregular rows along the base of the dorsal fin and along the midside, sometimes forming irregular bars on tail. The head is pale tan, and the lower jaw, thorax, and abdomen are pale. There is a faint dark spot denoting for the anterior-lateral line papillae (extending to gill opening and almost to anus in some speci-

Fig. 2.45 *Gymnothorax nubilus* (Richardson 1848). (Courtesy of Skipworth, I)

mens). The anterior nostril is brown, and the posterior nostril has a brown rim. Jaw pores have faint brown rings (not visible on pale specimens). The dorsal fin is decorated or dusky basally, with pale margin for its entirety or only posteriorly. The anal fin is dusky or dark basally with an obvious pale margin (Bohlke and McCosker 2001).

Distribution

Individuals of this moray eel species are distributed in the Southwest Pacific. It is reported from the Lord Howe and Norfolk Islands, New Zealand, and the Kermadec Islands (Trnski et al. 2015).

Habitat, Biology, and Ecology

This species of moray eel lives in temperate and subtropical marine habitats on rocky and other structure reefs from 0 to 25 m depth (Hoese et al. 2006). It can reach up to 680 mm (Bohlke and McCosker 2001).

Conservation Status

Smith et al. (2019d) assessed the conservation status of this moray eel species and listed it as Least Concern in the Red List of the IUCN for the following reasons: it is common throughout its range and has no threats known to affect it nor is it utilized for trade. While there are no species-specific conservation actions taken for the protection of this species, the range of this species overlaps with several marine-protected areas in the region.

Human Interaction

As a species of the moray eel group, it has the potential for sever bites to humans.

Gymnothorax obesus (Whitley 1932) (Fig. 2.46).
Common name: Griffin's moray.
Etymology: *Gymnothorax:* Greek, *gymnos* = naked + Greek, thorax, *-akos* = breast (Romero 2002).

Fig. 2.46 *Gymnothorax obesus* (Whitley 1932). (Courtesy of Ian Shaw, Australia)

Diagnostic Features

This moray eel is characterized to having a huge and elongated body. The anus is located just before the midbody. Head is moderate in length. The eye is small and located above gape and closer to rictus than to snout tip. The anterior nostril is in a moderate tube form, while the posterior nostril is in a form of a short tube and located well before the eye margin. The dorsal fin originates above the gill opening and behind the second branchial pore. The gill opening is located at midside.

Teeth in general are smooth. Maxillary teeth are partially biserial. Intermaxillary teeth are 5–6, plus 3–5 outer tiny teeth, and 3 median teeth. The inner row composed of 3–5 thin long maxillary teeth, while the outer row contains 16–17 small, short teeth. Vomerine teeth are up to 14. They are very small, pointed, and hidden in muscle folds. There are two to four strong large inner anterior dentary teeth, enclosed by 4–6 small outer teeth, continuing into a row of 16–19 teeth of varying sizes.

Body coloration is with brown background and small bright-white irregular spots and semicircles. The snout has a uniform pale-brown color, without spots. The lower jaw is pale but mottled.

Distribution

Individuals of this moray eel are distributed in the Southwest Pacific. It is reported in Australia, Norfolk Island, and New Zealand (Francis 1993) including the Kermadec Islands. (Gomon et al. 1994). There is doubt about its presence in the Juan Fernández Islands (Dyer and Westneat 2010).

Habitat, Biology, and Ecology

This moray eel inhibits benthic habitats (Hoese et al. 2006) found inshore (Paxton et al. 1989) in temperate marine waters. It lives at depths up to 120 m (Gomon et al. 2008).

Conservation Status

Smith et al. (2019d) evaluated the conservation status of this moray eel and listed it as Least Concern in the Red List of the IUCN based on the following justifications: There are no population data available for this species. Nevertheless, there are no major threats known to this eel species, nor is it utilized for human consumption. There are no conservation measures in operation for *G. obesus*, although it may be present in marine-protected areas, particularly one of UNESCO's World Heritage Sites, the Great Barrier Reef.

Human Interaction

As a species of the moray eel group, it has the potential for sever bites to humans.

Gymnothorax pictus (Ahl 1789) (Fig. 2.47).
Common name: Paintspotted moray.
Etymology: *Gymnothorax:* Greek, *gymnos* = naked + Greek, thorax, *-akos* = breast (Romero 2002).

Diagnostic Features

The body of this moray eel is moderately long. The anus is near the midlength. The dorsal fin origin is above the gill opening. Jaws are moderately long and of equal length. These are of moderate size and are located in the middle of the upper jaw. The anterior nostril is tubular in shape and short, while the posterior nostril is located above the anterior margin of the eye. Teeth are smooth, moderate in length, and pointed. Intermaxillary teeth are arranged in a single peripheral series, with 6–7 on each side. They are conical in shape, increasing in size posteriorly. Maxillary teeth are uniserial, about 9–13. The lower jaw has 1–4 larger inner teeth anteriorly and about 13–20 smaller outer teeth. Vomerine

Fig. 2.47 *Gymnothorax pictus* (Ahl 1789). (Courtesy of Hamid Badar Osmany,)Pakistan

teeth are biserial and diverging anteriorly in large specimens. The body has variable coloration but is usually light tan or gray. There are fine speckles with small dark spots, often grouped to form irregular blotches, sometimes with pale centers. They sometimes have irregular edges forming a snowflake-like pattern. The iris is yellowish white with an irregular black circle at the margin (Weitkamp and Sullivan 2003).

Distribution

Individuals of this moray eel are widely distributed in the Indo-Pacific region. They are reported from the East African coast to Central America (Weitkamp and Sullivan 2003). They also reported from the Hawaiian and Ryukyu Islands. They are recorded in Australia and the Kermadec Islands (Weitkamp and Sullivan 2003). This species is uncommon in the Red Sea. Randall and Golani (1995) listed only one specimen.

Habitat, Biology, and Ecology

Individuals of this moray eel live solitary in shallow waters on flat or rocky reefs. They feed on small fishes and crustaceans, sometimes leaving the water in search for prey (Lieske and Myers 1994; Kuiter and Tonozuka 2001). This species prefers benthic habitats (Mundy 2005). Concurrent hermaphroditism is known for this species (De Mitcheson and Liu 2008).

Conservation Status

Smith et al. (2019e) evaluated the conservation status of this moray eel species and listed it as Least Concern in the Red List of the IUCN based on the following justifications: It is widespread and common throughout its range. There are no major threats identified to this species, although it is irregularly caught and sold in local markets in several regions in its range and may appear in the aquarium trade. There are no conservation proposals put forward for the protection of this species; nevertheless, its wide range overlaps numerous marine-protected areas.

Human Interaction

As a species of the moray eel group, it has the potential for sever bites to humans.

Gymnothorax porphyreus (Guichenot 1848) (Fig. 2.48).

Common name: Lowfin moray.

Etymology: *Gymnothorax:* Greek, *gymnos* = naked + Greek, thorax, *-akos* = breast (Romero 2002).

Diagnostic Features

This moray eel has a huge and elongated body. The anus is near the midbody. The head is moderate to short in size. The snout is short and broad. The jaws are moderate in length. The eye is located above midgape and slightly closer to rictus. The anterior nostril is in the form of a short tube, while the posterior nostril is located above and before the eye. It is raised and has the shape of a short tube with a flared crenulate margin. Head pores are typical of those of the other moray eel species. The gill opening has a diagonal slit and is located at midside (Bohlke and McCosker 2001).

The teeth are strong, triangular, smooth, partially biserial, and moderate in number. The main row of outer premaxillary teeth usually has six teeth, with 3–12 small outer teeth present in both small and large specimens. There are usually three caniniform teeth on midline. The inner row contains 1–6 long, slender maxillary teeth anteriorly, while the outer row is composed of 10–18 shorter stout teeth changing little in size. The vomerine teeth are short and sharp, and they usually fall in a single row of 6–12 teeth. The lower jaw is with 18–27 teeth, the first 4–7 are large and

Fig. 2.48 *Gymnothorax porphyreus* (Guichenot 1848). (Courtesy of Yuri Hooker)

increasing in size and then abruptly smaller, changing little in size. The outer row is composed of 2–15 short, pointed teeth flanking main row anteriorly, extending back in some specimens (Böhlke and McCosker 2001).

The general body color is dark gray or brown with darker mottling. The head, body, and fins have a similar color. Nostrils and head pores are noticeably dark. The snout and the mouth angle are dark. The gill opening is dusky or dark, but not in a defined black blotch (Böhlke and McCosker 2001).

Distribution

Individuals of this moray eel are distributed in the South Pacific region. It is reported in subtropical waters from Lord Howe Island to South America.

Habitat, Biology, and Ecology

This species of eel grows to a large size; the largest specimen recorded by Böhlke and McCosker (2001) is 1030 mm. The species prefers cool waters at shallow depths recorded to 13 m of the eastern Pacific off Chile and Peru and is found at Easter Island, the Kermadec Islands, Norfolk and Lord Howe Islands, and waters of New Zealand, all south of 25°S (Böhlke and McCosker 2001).

Conservation Status

McCosker and Béarez (2010) evaluated this species of moray eel and listed it as Least Concern in the Red List of IUCN. They based their decision on the following justifications: It is a widespread species in the Eastern Pacific region. No major threats to this species are known, and there is no current indication of population decline.

Human Interaction

As a species of the moray eel group, it has the potential for sever bites to humans.

Gymnothorax prasinus (Richardson 1848) (Fig. 2.49).

Common name: yellow moray.

Etymology: *Gymnothorax:* Greek, *gymnos* = naked + Greek, thorax, *-akos* = breast (Romero 2002).

Fig. 2.49 *Gymnothorax prasinus* (Richardson 1848). (Courtesy of Ian Skipworth)

Diagnostic Features

The body of this moray eel is very large, long, and tapering toward the tail. The anus is usually near or before the midbody point. The head is short, and the snout is moderately short. The upper jaw is moderate in length, and eyes are small and located above midgape. The anterior nostril is of a moderate to long tube in shape. The posterior nostril has a short tube shape and is located above and behind anterior margin of the eye. Head pores have a typical pattern of other moray eel species. The teeth are stout and smooth. The maxillary teeth are biserial anteriorly. The main intermaxillary teeth are 7–10, plus 1–6 small outer teeth. There are three strong median teeth. The vomerine teeth are 9–14, small, and usually uniserial. There are usually four large inner dentary teeth, outer row of 19–25 teeth and anterior of 3–6 small and enclosing the large teeth. Body and fins are dark brown, usually covered with pale (greenish) mucus. The head is pale brown, with median dorsal pale streak that bordered on each side by dark brown streak from the snout to behind the eye. Nostrils and head pores are dark, contrasting with head coloration. The anterior nostril is dark around the base, while posterior nostril is dark brown. Head pores are dark and sometimes with white rims bordered by dark brown pigment. Fins are dark. The tip of the tail is sometimes pale.

Distribution
Individuals of this moray eel are distributed in the Southwest Pacific. They are reported from New Zealand and southern Australia.

Habitat, Biology, and Ecology
This moray eel prefers living in holes and crevices. It feeds mainly on crabs and fish (Russell 1983). It is a well-known species in south temperate waters of Australia and New Zealand, and it is common off New Zealand, southeastern and southwestern (Böhlke and McCosker 2001). Individuals of this moray eel prefer to live at depths of 400 m.

Conservation Status
Smith et al. (2019h) evaluated this moray eel and listed it as Least Concern in the Red List of the IUCN based on the following justifications: It is the most common moray in southeastern Australia and is abundant where suitable rocky habitat is available. This species is occasionally involved in the aquarium trade; however, the take of this species is not of large enough quantities to be considered a major threat. There are no major threats to this species, and its range overlaps several marine-protected areas.

Human Interaction
As a species of the moray eel group, it has the potential for sever bites to humans.

Gymnothorax prionodon (Ogilby 1895) (Fig. 2.50).
Common name: Australian mottled moray.
Etymology: *Gymnothorax:* Greek, *gymnos* = naked + Greek, thorax, *-akos* = breast (Romero 2002).

Diagnostic Features
This is a very large moray eel, with a tapering tail. The anus is located before the midbody point. The head is elongated, and the snout is long and narrow. The jaws are long. The eyes are small and located above midgape. The anterior nostril has a long tube shape, while the posterior nostril has a raised rim and is located above the anterior margin of the eye. There are two branchial pores above and before the gill opening. The dorsal fin originates above or before the first branchial pore. The gill opening is located at midside. The dentition is uniserial. The teeth are smooth (not serrate), few, long, slender, and frequently broken or missing. The intermaxillary teeth are 6–6, 0–3 long median teeth (absent in

Fig. 2.50 *Gymnothorax prionodon* (Ogilby 1895). (Courtesy of Ian Skipworth)

adults). The maxillary teeth are 7–15, decreasing moderately in size posteriorly. The vomerine teeth are 0–8, short and hidden in muscle folds. The teeth on the lower jaw are tall and slender, 11–23, decreasing in size posteriorly. The background color of the body is medium brown, with small-to-moderate pale spots. The pattern of body coloration is highly variable. There are small spots on the dorsal side of the head. The lower jaw is uniformly tan. The thorax and abdomen are with spots in large specimens. Spots on the body are small, ocellated, and moderately separated in small specimens, becoming larger and divided into segments in larger specimens, sometimes fewer, more widely separated, and ocellated on the tail. Both dorsal and anal fins are slightly dark, with 1–2 rows of spots, frequently a row on margin of fins, some of which are joined, but not forming a continuous pale margin on fins. Mouth angle and gill opening are dusky (Böhlke and McCosker 2001).

Distribution

Individuals of this moray eel are distributed in the Western Pacific region. They reported from the southern Japan to the South China Sea (Masuda et al. 1984), Australia, and New Zealand.

Habitat, Biology, and Ecology

Individuals of this moray eel prefer inshore waters (Paxton et al. 1989). If provoked, this species will make a lightning-fast bite before retreating (Ayling and Cox 1982). It reaches up to 1000 mm (Coleman 1983).

Conservation Status

Smith et al. (2019f) evaluated this species of moray eel and listed it as Least Concern in the Red List of the IUCN based on the following justifications: It has no commercial usage, nor is it threatened by fisheries. There are no species-specific conservation mechanisms in place for *G. prionodon*; however, its range may overlap with marine-protected areas throughout its distribution. Information about the biology and ecology of this species is infrequent.

Human Interaction

As a species of the moray eel group, it has the potential for sever bites to humans. If this eel is provoked, it makes a lightning-fast bite before retreating (Ayling and Cox 1982).

Gymnothorax thyrsoideus (Richardson 1845) (Fig. 2.51).

Common name: Greyface moray.

Etymology: *Gymnothorax:* Greek, *gymnos* = naked + Greek, thorax, *-akos* = breast (Romero 2002).

Diagnostic Features

The dorsal fin originates about halfway between the rictus of the mouth and the gill opening. The teeth are conical and form two rows in the upper and lower jaws. There are two conical teeth in the median row of intermaxillary. Vomerine teeth are grouped in two rows, diverging anteriorly in large adults. Head and body are brown, but the head is darker, densely mottled with darker-brown flecking and random marks. The gill opening is surrounded by dark-brown-to-dusky spot. The iris is brilliant white (Trnski et al. 2015).

Distribution

Individuals of this moray eel species are distributed in the Indo-Pacific region. They are reported from the Christmas Island to French Polynesia,

Fig. 2.51 *Gymnothorax thyrsoideus* (Richardson 1845). (Courtesy of Nick Hobgood)

north to the Ryukyu Islands, and south to Tonga, throughout Micronesia. Recently, this species is recorded in New Zealand from the NW corner of North Meyer Island and Kermadec Islands (29°14.48'S, 177°52.72'W) (Trnski et al. 2015).

Habitat, Biology, and Ecology

This species of moray lives in association with coral reefs at depths up to 30 m (Fricke et al. 2011). It feeds on crustaceans, small fishes, and octopi (Bacchet et al. 2006). It is often found in shipwrecks and usually swims in pairs or small aggregations and shares habitat with other morays (Kuiter and Tonozuka 2001). A new record shows it being captured using rotenone and hand nets at 16–18 m depth and found in rocky bottom interspersed with sand and gravel, corals, and seaweeds (Trnski et al. 2015).

Conservation Status

Smith et al. (2019g) evaluated this species of moray eel and listed it as Least Concern in the Red List of IUCN based on the following justifications: It is a widespread moray eel species throughout the Indo-West Pacific. It is common and moderately abundant throughout its range. This species has no commercial value, and there are no major threats known to affect it. Although there are no species-specific conservation measures in place for this species, it is likely that this species is found in marine-protected areas in the Western Pacific.

Human Interaction

As a species of the moray eel group, it has the potential for sever bites to humans.

Gymnothorax ypsilon (Hatooka and Randall 1992) (Fig. 2.52).

Common name: Y-bar moray.

Etymology: *Gymnothorax:* Greek, *gymnos* = naked + Greek, thorax, *-akos* = breast (Romero 2002); *ypsilon: ypsilon* after the twentieth letter of the Greek alphabet, in reference to the Y-shaped dark bars often found on this species (Hatooka and Randall 1992).

Fig. 2.52 *Gymnothorax ypsilon* (Hatooka and Randall 1992). (Courtesy of Chen Hongming)

Diagnostic Features

This species of moray eel has a moderate body length and compressed, with pointed snout. The gill opening is nearly horizontal having a center located slightly below midbody. The anterior nostril has a slender tube shape and is located on both tips of the snout. The posterior nostril has a raised rim and is found over the front edge of the eye. The upper jaw is slightly longer than lower, and the mouth closes completely. Teeth in both jaws are uniserial, stout, pointed, and slightly retrorse. Mandibular teeth are very large and in a single row, 16 on the left and 17 on the right side of the jaw. The dorsal fin is moderately high and originates before gill opening. The anal fin is low and originates just behind the anus. The anus is located before the midbody. Jaw teeth are sharp and uniserial. The body is creamy light brown with 26–35 distinct, narrow, dark crossbars extending onto fins (some bars on most specimens Y-shaped). The anterior part of the head is uniformly light brown. The anal fin has a white margin (Hatooka and Randall 1992).

Distribution

Individuals of this moray eel are distributed in the Pacific Ocean origin and reported from southern Japan and the Hawaiian Islands.

Habitat, Biology, and Ecology

This species of moray eel prefers deep water, at depths more than 100 m (Hatooka and Randall 1992).

Conservation

Smith et al. (2019h) have evaluated this species of moray eel and listed it as Least Concern in the Red List of the IUCN based on the following justifications: It is restricted to deeper waters and is found between 120 and 185 m depth. There are only a handful of specimens known for this species. This species has no commercial value, nor does it have any major threats. Its range overlaps several marine-protected areas. However, there is a need for more research regarding the full range of this species, ecology, population, and the likelihood of threats.

Human Interaction

As a species of the moray eel group, it has the potential for sever bites to humans.

2.2.1.1 Bite by Moray Eel

The family Muraenidae encompasses over 100 species of the moray eel. These fishes differ from the remaining eels in having small gill openings and missing pectoral fins. Their body has no scales, but it is covered with a thick mucus material. Their eyesight is weak, but their sense of smell is outstanding, and they use it to locate food in their environment. Their jaws are equipped with strong muscles and strong and sharp teeth, which give them the ability to hold prey tight (Riordan et al. 2004).

Moray eels are known for their hostile behavior, and several severe cases of attack against humans were reported, the most severe being that reported by Riordan et al. (2004). Such cases might be augmented in the future with the increase of human population visiting seashores (Riordan et al. 2004). Eels frequently hide in crevices and holes during the day and come out during the night to hunt their food. They are so eye-catching as they swim in an attractive way.

The bite of a moray eel is one of the serious injuries that need to be given extreme care and medication. There are two outcomes of the eel's bite (Wet Web Media 2016). If the eel is poisonous, then the victim will have a dose of antitoxin injected into his or her body. The toxic side of the eel's bite will be dealt with later in the section on poisonous and venomous fishes. Another concern is the inflammation caused by the secondary infection, which is more serious and puts health at jeopardy.

The eel has many teeth of different shapes and sizes, some of them packed with no space between them and others with a gap separating them. Food regularly rots in those spaces between the teeth, and germs of different kinds grow in such an environment. Bacteria and other germs can be transferred directly to the body of the victim once the bite happens. Among these horrifying bacteria is *Vibrio*, which can produce septicemia (infection of a blood). Infections with this type of bacteria can lead to death. Another less dangerous but still worth noting possibility is being infected by *Pseudomonas* bacteria. Eels may cause a large wound that can lead to a substantial loss of tissue and blood (Howard et al. 1985; Chang and Pien 1986).

All eel species stated above and present in New Zealand waters can potentially make a severe attack. Those species considered harmless to humans may also suddenly attack, as the behavior of moray eels is not predictable.

Below are some general guidelines posed for people who encounter moray eels such as divers and aquarium keepers (Wet Web Media 2016):

A. Try not to feed the eel by hand. It is still possible for people who have eels as pet for their aquariums at home, aquarium keepers, or divers who feel confident enough to come close to and feed eels to get bitten by eels. The movement of the hand when feeding provokes the eel and puts it in an attack mode. The eel has the tendency to change its mind on how to get its food. This information is very significant for aquarium keepers so they can always keep their guard when feeding eels in the tank by hand.
B. Do not attempt to stay close to or feed large eel species. Eel bites differ in harshness and depth according to the size of the eel. A bite

from a few-centimeter-long eel is completely different and less severe than that by a 5 m eel. The latter can easily tear a large piece of tissue from any part of the body.

C. Use tools to handle eels. New aquarium or fish keepers are usually taught to keep their hands out of the tanks and water as much as possible. Several handy tools currently exist to assist people in daily contact with dangerous aquatic animals. These tools can aid in delivering food, with the human feeder standing a good distance away from the tank.

D. Never get distracted with another activity while feeding. This is the most difficult rule to keep. Divers usually carry their camera and other equipment while diving and, at the same time, trying to feed an eel bulging its head from a hole with the aim of getting it out of its hide. In doing so, the diver often concentrates more on the camera than on the spear that carries the dead fish. There is a good chance the eel might attack the diver's body instead of the lure fish (Auerbach 1984). For aquarium keepers, it is significant to not get in a discussion with another person while feeding the eels or cleaning their tank. It only takes a few seconds for the eel to attack; hence full attention should be given on where you are putting your hand. Injuries caused by moray eels fall in the second group of injuries caused by marine animals. This class is about an animal with the potential to cause a severe injury as it armed with sharp and long teeth. The grasp that the eel has on its prey is very strong, and, in most instances, you need to kill the eel to free the body part. In addition to the long teeth of different shapes, the roof of the mouth is supplied with small sharp teeth that help in holding the prey. The best treatment of a moray eel bite starts at the site, and then the victim should be taken directly to a hospital to determine the tetanus status. At the bite site, direct pressure should be applied to control the bleeding until the patient is received by the hospital. An inclusive check of the bite should take place to know if any teeth remain in the wound, which might cause sepsis later on. Use of oxygen locally on the wound appears to be a good therapy for healing (LaVan and Hunt 1990). Hyperbaric oxygen therapy will inhibit the activity of the anaerobic bacteria and certain toxin-forming pathogenic microorganisms (Ellis and Mandal 1983; Petzold et al. 1999).

An augmented responsiveness among both emergency-treating personnel and ocean users should reduce the threat of encounters. All divers and especially people entering waters for recreation should be alerted to avoid direct contact with moray eels. Continued efforts in spreading awareness as to how moray eels attack are relevant. Such information will help for better understanding of the motivations of these creatures and reduce the hazard of dangerous encounters in the future (Riordan et al. 2004).

2.2.1.2 Incidences of Moray Eel Bites: Examples

Most injuries resulting from eel bites are considered superficial and managed effectively in the emergency department and outpatient setting (Erickson et al. 1992). Nevertheless, with about 80% of the world's population nowadays living in coastal areas and with more people using the ocean for recreational events than ever before, encounters with dangerous marine animals will certainly upsurge (Rosson and Tolle 1989).

Riordan et al. (2004) reported on an unprovoked incidence of moray eel attack on humans that caused a shocking injury to a diver's upper arm. A 37-year-old man was attacked by a moray eel while scuba diving on a coral reef close to the Cuban island of Varodaro. The diver was dressed in a "shortie"-type wetsuit, which exposed his forearms and calves. Before entering the water, the instructor gave instructions to his diving group when a large 10-year-old green moray eel (*Gymnothorax funebris*) came across the reef (estimated size of the eel was 1.5 m). It is not clear if preceding groups of divers had been feeding the eel. The eel did, however, seem to associate the divers with food, for it emerged to meet them soon after they went down to 30 m. Originally, it moved toward the instructor and several other divers before circling the diver.

Without obvious incitement, the eel seized the volar aspect of the diver's right forearm tightly. The diver made an unsuccessful attempt to capture the attention of his diving group before inflating his buoyancy device and making an emergency uncontrolled ascent to the surface. The eel eventually released its grip at a depth of 15 m. On board, initial assessment revealed two large avulsion wounds on the volar aspect of the right forearm. A cloth bandage was placed around the diver's arm to minimize blood loss, and the boat returned to shore 35 min later. The diver was taken to a local hospital, where the extent of the injury was assessed. Despite the extensive tissue loss to the volar aspect of the forearm, the distal vascularity was intact.

In the operation theatre, the victim was examined. Two segmental areas of skin loss were on the volar aspect of the forearm, proximally 7 × 3 cm and distally 5 x 4 cm, with a bridge of intact skin separating them. (For photos of the incidence, please see Riordan et al. (2004). It is free article online.) The median, ulnar, and radial sensory nerves had been divided at two levels. The ulnar artery was divided, but the radial artery was intact. The entire flexor compartment muscles, except flexor pollicis longus, were completely divided at both levels. The muscle bellies to extensor digitorum communis, extensor carpi radialis longus and brevis, and extensor carpi ulnaris were partially divided. After taking wound swabs and performing wound lavage and debridement, the surgeons repaired the structures. The overlying skin was closed primarily with a drain in situ. The patient was placed in a hyperbaric oxygen chamber to optimize wound healing for 1 h each day while recovering in Cuba. One week after the operation, when the patient was stable and was off antibiotics, he was transferred home to his native country. Because the patient reported increasing pain and low-grade pyrexia, we elected to surgically explore his wound on the day of admission. Some residual necrotic muscle was excised, and a split thickness skin graft was applied to close the wound without tension. He started to take intravenous ciprofloxacin empirically for 24 h and orally for 5 days. Postoperatively, the pyrexia settled, and hyperbaric oxygen therapy was recommenced for 1 week to enhance wound healing. After a 3-week period of protected mobilization in static splints, a rigorous hand-therapy program was initiated.

At 1-year follow-up, substantial enduring disability continued. An advancing Tinel's sign could be elicited at the mid-palm level with acceptable deep and superficial pain feeling. Two-point static discrimination in the fingertips was greater than 15 mm. Despite nonstop strapping, a noteworthy claw-hand deformity continued. Broad intrinsic muscle wasting occurred in the hand, along with reduced pinch grip and thumb abduction. The total active range was considerably reduced to 120° in the index finger and 100° in the middle, ring and little fingers. The patient returned to work 9 months after the injury while still undergoing intensive rehabilitation by the hand therapist.

The moray eel bite falls under the second category of injuries and illnesses caused by marine animals. This category consists of animals that are dangerous because of their size or their ability to inflict traumatic injuries.

Historically, moray eels have been dreaded for many reasons. In ancient times, Roman rulers such as Nero were rumored to have thrown disobedient servants into pits filled with ravenous moray eels (Erickson et al. 1992). When they do attack, fanglike teeth and vicelike jaws impose the wound. It is sometimes required to kill an eel or disarticulate its jaws to release its grip (Auerbach 1984). The roof of the moray eel's mouth is equipped with a third row of teeth, which help grasp prey. Most injuries tend to occur when divers insert their hands into holes occupied by eels or when freshly cut fish that divers carry into the water (Auerbach 1984) attract the eels. Consequently, hands tend to be the region most bitten. A rising trend of sport divers that practice hand feeding will certainly lead to an augmented number of incidents.

The ideal treatment of a moray eel bite starts in the prehospital phase. All injuries should be mentioned to a hospital for valuation and determination of tetanus status. Although the risk of developing tetanus is unknown, the usual practice in all hospitals is to administer tetanus toxoid for all marine-inflicted wounds if the victim's tetanus-immune status is indistinct or inadequate. Direct pressure should be applied to any bleeding points, and arrangements should be made to transfer the patient to a receiving hospital. Specialized training in first aid and shock management for lifeguards and sport-diving operators should help optimize the victim before taking him/her to the hospital.

Wound care in the hospital necessitates management in the operating theatre. Valuation of tendon, vessel, and peripheral nerve injury should be recognized. Plain radiographs should be obtained in all injured areas to identify areas of periosteal stripping or particles of teeth, which can continue entrenched in the bone and act as a source of sepsis if not identified and removed. Wounds should be swabbed for culture. The jeopardy of infection from these injuries is unclear. Erickson et al. (1992) reported three superficial attacks that all healed drearily without wound infection. Nevertheless, each victim took prophylactic antibiotics at the initial presentation. Irrespective of whether prophylactic antibiotics are given, any early signs of invasive wound infection should be treated quickly.

Even though little sign implicates moray eels as indiscriminate predators, a bigger awareness among both emergency-treating personnel and ocean users of the potential severity of moray eel bites should minimize the risk of encounters. Recreational divers entering waters populated by moray eels should be alerted to evade direct contact. Because little is known to clarify why moray eels infrequently bite divers, gathered information from these happenstances is a must. This information will allow us to better apprehend the incentives of these creatures and then diminish the threat of dangerous encounters.

Barreiros and Haddad (2008) reported on a case of moray eel attack against a 42-year-old male living in Angra do Heroísmo, Azores islands, Portugal. The victim is a fish biology researcher and a knowledgeable diver and underwater spear fisherman for more than 25 years. He was free diving (spear fishing) around an islet about 9 km off the south coast of Terceira island. At a depth of 27 m, when he was placing his left hand on a rocky shelf, it was aggressively pushed, and a moray eel (*Muraena helena*) was seen soon afterward swimming away quickly (approximate total length 1 m). At the surface, the wound was evaluated, and third and fourth finger flexor tendons were visible. Instant and sharp pain that continued for about 6 h was felt. Bleeding was organized and the wound cautiously washed. The man was transferred, as rapidly as possible, to the local hospital. There surgery showed a deep laceration of the skin and muscle over the dorsum of the hand, total laceration of the third finger flexor tendon, and a partial tear of the fourth. Tendons were stitched, and the wound was closed with seven stitches plus another three on an accessory wound on the basal distal muscle of the hand. Instantly, anti-inflammatory and antibiotic medications were given. Later, assessment of the wound show no sign of inflammation. After 12 days, mobility was partially recovered, although application of manual force was practically impossible. The victim was totally recovered after 4 months since the incidence happened.

According to the four categories of accidents with aquatic animals suggested by some authors (Strauss and Orris 1974; Haddad 2000), the incidence of the victim described above agrees with the classification of Riordan et al. (2004) that moray eels instigate injury by biting. It is known that bites by moray eels mainly involve divers, spear anglers, and traditional fishermen when handling hooked or netted specimens (Haddad 2000).

In the incidence reported by Barreiros and Haddad (2008), because the victim was completely aware of the possible development and onset of serious infection, the wound developed very favorably, particularly owing to the prehospital stage when the lesion was comprehensively washed in seawater, exhaustively cleaned, and disinfected. Nevertheless, moray eel bites can lead to possible serious bacterial infections (Haddad 2004). There is clinical evidence that toxins exist in the saliva of the moray eel because pain persisted after the bite (Haddad 2000). Injuries caused by these animals must be checked until completely healed.

2.2.1.3 The Mechanism of the Moray Eel Bite

Even though morays are recognized to feed on relatively large prey, dietary references in the literature are scarce, and few investigations document prey eaten in relation to the size of the predator (Randall 1967; Yukihira et al. 1994; Young and Winn 2003). From biomechanical, physiological, and ecological perspectives, it is remarkable how a moderately large aquatic predator with a reduced cross-sectional area, a large gape, and a noncircular mouth opening can consume large prey. Recently, Mehta and Wainwright (2007a, b) explored the functional morphology of prey capture and transport in moray eels, noting the significant reduction in moveable cranial elements and the small size of the hyoid bar, both of which are key features in many suction-feeding fishes.

As shown in the study of Mehta and Wainwright (2007a), *Muraena retifera* capture prey by direct biting. After prey capture, morays initiate prey transport by ventrally flexing their head while concurrently advancing the postcranial portion of their body. Such effort, which shortens the distance between the posterior part of the branchial basket and the oral cavity, seems to assist protraction of the pharyngeal jaws between the dentaries and into the buccal cavity. Due to the protraction of pharyngeal teeth, there is obvious ventral displacement. Based on the anatomy of the pharyngeal jaw apparatus, its associates to the dentary, and the extensive reduction of the hyoid apparatus, the anterior displacement of the skull was caused by the lower pharyngeal jaw during protraction. Once the pharyngeal jaws obtained the prey item, the oral jaws released their grip by increasing their gape distance, allowing the pharyngeal jaws to retract with the prey item toward the esophagus. During pharyngeal jaw retraction, eels stays in place while either the prey disappears farther into the pharynx or advances their body over the prey. The pharyngeal jaws drag prey into the esophagus. Mehta and Wainwright (2007a, b) noticed the same prey transport kinematics in three other muraenines: *Echidna nebulosa*, *Echidna rhodochilus*, and *Gymnothorax funebris*. Of the four species of eels for which Mehta and Wainwright (2007a, b) were able to observe capture and transport behaviors, the upper pharyngobranchial was observed contacting prey for only *Muraena retifera* and *Gymnothorax funebris*. During prey transport, anterior movement of the branchial arch region, ventral displacement, and ventral rotation of the neurocranium of the other two morays, *Echidna nebulosa* and *Echidna rhodochilus*, were observed. These kinematic behaviors presumably caused the protraction of the pharyngeal jaw apparatus, although the pharyngeal jaws never made contact with the prey in the oral cavity. This may be because the event occurred posterior to the jaw joint and, therefore, was out of lateral view.

Mehta and Wainwright (2007a, b) studied the kinematics of capture and transport and suggested that prey capture and transport in morays display functionally homologous acts in that both stages of feeding involve biting. The large gape involves elevation of the neurocranium, and depression of the lower jaw serves different functions during prey capture and transport (Mehta and Wainwright 2007a, b). These functional changes are outcomes in dramatic differences in the timing of peak excursion events. During prey capture, peak gape functions to arrange the upper and lower jaws to offer a forceful bite and leads the prey capture event, while during prey transport, growing the oral gape functions to release the prey for swallowing. Angular excursions of the head and lower jaw reached similar values in both behaviors as did gape distance. Generally, prey capture was considerably faster than prey transport. Mehta and Wainwright (2007a, b) in their study recognized three different stages in moray prey transport behavior: pharyngeal protraction, pharyngeal-prey contact, and pharyngeal retraction.

Comparative investigations portraying the branchial arches and their corresponding muscular attachments have proposed that morays have the most specified gill arches among the *Anguilliformes*. Nelson (1966, 1967) determined that the reduction in branchial structures, enlargement of pharyngeal jaws, and their muscular attachments allow morays to transport relatively large prey into the esophagus. Mehta and

Wainwright (2007b) tested Nelson's hypothesis and agreed that moray eels have, indeed, evolved an alternative to hydraulic transport, the most used prey transport mechanism in teleost fishes.

2.2.2 Barracuda Species

Family: Sphyraenidae.
Sphyraena qenie (Klunzinger 1870) (Fig. 2.53).
Common name: Blackfin barracuda.
Etymology: *Sphyraena:* Greek, *sphyraina*, -es = the name of a fish (Romero 2002).

Diagnostic Features
The body is long with tough head and jaws; no gillrakers on gill arches. Eyes are moderate in size. Absence of membranous flap on the corner of the pre-operculum and no cartilaginous bump on the median lower jaw. The maxilla passes the anterior edge of the eye. Teeth are erect, sharp, and close-set on both jaws. The dorsal fin originates slightly anterior to the tip of the pectoral fin. The caudal fin has a small lobe at the inner side of each lobe. The body is gray in color, with several dark bars across the lateral line (Randall 1995; Froese and Pauly 2016).

Distribution
Species of barracuda are distributed in the Indo-Pacific region. They are reported from the Arabian/Persian Gulf, Red Sea, and East Africa to the central Indian Ocean and French Polynesia. They are also recorded from the Eastern Pacific region in Mexico and Panama. The exact range is uncertain because of confusion with *Sphyraena jello* and *Sphyraena putnamae* (Senou 2001). They are also reported from waters around New Zealand (Roberts et al. 2015). There is no record of this species from the Arabian/Persian Gulf, but it is reported in both the Sea of Oman (Randall 1995) and the south coast of the Arabian Peninsula (Randall 1995; Manilo and Bogorodsky 2003; Froese and Pauly 2022).

Fig. 2.53 *Sphyraena qenie* (Klunzinger 1870). (Courtesy of Nosy Iranja Madagascar)

Habitat, Biology, and Ecology
This species of barracuda is a marine species that lives in association with reefs at a depth range of 1–100 m (Bacchet et al. 2006). It reaches a maximum length of 1700 mm TL (Lieske and Myers 1994), with a common length of 800 mm TL (Senou 2001). Individuals of this species are found near current-swept lagoon and seaward reefs (Lieske and Myers 1994). They feed on fishes (Bacchet et al. 2006). They usually aggregate in large semi-stationary schools during the day (Senou 2001), which have a propensity to occupy the same site for months or even years at a time. This species is a night feeder (Myers 1999).

Economic Value
This species has high economic value as the meat is used for human consumption.

Conservation Status
Not evaluated.

Human Interaction
As a species of barracuda, this species has the potential for attacking humans. Some species of barracuda are reputed to be dangerous to swimmers. Barracudas are scavengers and may mistake snorkelers for large predators. These species then follow these snorkelers, hoping to eat the remains of their prey. There have been reports of bites by barracudas, but such incidents are rare and are possibly caused by poor visibility. Large barracudas can be encountered in muddy shallows on rare occasions. Barracudas may mistake things that glint and shine for prey (Myers 1999).

In New Zealand, a man underwent plastic surgery after a barracuda savaged his arm, which left him with serious tendon damage and paralyzed fingers. The victim was swimming in warm water near the mouth of the Waikanae River, Wellington, on January 11, 2008, (New Zealand Herald 2018) when a barracuda bit into his left arm with its razor-sharp teeth. According to the victim, "It was excruciating pain like nothing I've experienced before. I just ran like hell from the water." Due to panic, he grabbed the fish's head and ripped it off his arm while it was still under the water. "It probably would have let go—I don't know if that caused more damage or not," he said. The victim added there were children swimming nearby, so it was fortunate the barracuda chose him to chew on. Once he managed to get out of the water, he was taken to the hospital where his wound was examined. "I didn't know what had happened to begin with, if it was a barracuda or a stingray. I knew it wasn't a shark because they just shred your arm," he said. Three holes in his arm looked more like stab wounds. "They were more like knife marks. My tendons were sliced, I had nerve damage and had paralysed fingers," he said.

The other incidence of barracuda attack was that of a woman on a kayaking trip in Florida Keys. The barracuda attacked her and sliced her lung. The *Daily Mail* brought this news on the 18th of January 2011 (Daily Mail 2011). The woman, who was saved by her boyfriend, suffered a potentially fatal injury. The victim was sliced open by the fish after it jumped out of the water and struck her, puncturing her lung. The force of the impact was so great that she also suffered broken ribs and was knocked out of the two-seat kayak. Her boyfriend assisted her and applied pressure on her wound with his hand and rang for help. Due to the cut in the lung, the victim could hardly breathe; with the assistance of her boyfriend who applied pressure on the wound, the victim managed to continue breathing. The couple was kayaking in the thick mangroves off the Florida coast when the incident occurred. US Coast Guard vessels could not reach the couple because they were in such shallow waters, so a local tugboat volunteered to go out into the mangroves looking for them. After a long search, with the victim losing a lot of blood, the tugboat managed to find them and bring them out into the open where she was airlifted to a hospital. The victim, 46 years old, said she was thinking of her daughter during the agonizing 30-minute wait. She said, "One of the first things I thought of is, she's getting married June 30 and I need to be able to walk her down the aisle." The victim spent 9 days recovering in intensive care, with an impressive scar to show for her ordeal.

The case of Thomas Goreau from Cambridge, Massachusetts, USA, represents an outstanding barracuda attack on humans. He was surface snorkeling with his family when a very big barracuda attacked him from below, cut his finger, and caused severe injuries to his hand and other fingers. This incident was completely unprovoked as the snorkeler was neither spearfishing nor offering food for the barracuda or other fish in the vicinity. It also added new information to the barracuda attacks on humans as it was previously thought that barracudas usually attack humans only after being provoked (Goreau 2015).

2.2.2.1 Barracuda Bites

The species of barracuda fish are among apex predators roaming the seas globally. It mainly feeds on fish (97% of stomach contents), and maximum reported is over 2 m in length (Gudger 1918; de Sylva 1963; Randall 1967; Blaber 1982; Schmidt 1989; Barreiros et al. 2002). A barracuda is a swift piscivore that uses its strong eyesight and sense of smell to find its prey (Sinha 1987). Rapid swimming speed is the technique barracudas follow to catch their prey (Walters 1966) and hold them with its long-serrated jaws, slicing into the flesh with an assortment of sharp caniniform teeth. Whereas attacks on humans are rare (Wright 1948; de Sylva 1963), de Sylva (1963) reported on 29 barracuda attacks on humans between 1873 and 1962, including fatal or limb amputation incidences. In contrary to most shark attacks, these vicious fish attacks are committed by considerably smaller fish (Wright, 1948), a fact that highlights the possibly tremendous cutting forces in their bite. Certainly, large barracudas can rapidly slaughter large prey by cutting them in two (Yasuda 1960; de Sylva 1963; Randall 1967; Porter and Motta 2004). Inspection of the stomach contents and prey items of a barracuda proposes that its feeding habits may be exclusive in comparison to other fishes: there are numerous reports of only "back-halves" of prey, items found in barracuda stomachs (de Sylva 1963; Randall 1967). Nevertheless, the dynamics and mechanisms of the feeding of barracuda have received little consideration. Porter and Motta (2004), for example, have described the kinematics of the strike of the great barracuda in small juvenile fish, and Gudger (1918), Gregory (1933), and de Sylva (1963) have explored their feeding morphology. Nevertheless, no study has dealt with the functional morphology and biomechanics of their powerful cutting jaws (Grubich et al. 2008).

Biomechanical simulations can offer vital information and the reality of how the mechanical phase of feeding and the force and motion involved (e.g., Alexander 2003; Koehl 2003; Westneat 2003). The latest theoretical simulations have integrated muscle contraction kinetics into the mechanics of the lower jaw, revealing jaw mechanisms as dynamic (Westneat 2003). Such models in fish skulls have a pronounced capability for testing theories of mechanical design across a diversity of fishes (Westneat 2004; Alfaro et al. 2005).

Barracudas can attack fish larger than the gape of their jaws, and to gulp large prey, it can slice the prey into pieces with its powerful jaws supplied with sharp cutting teeth. Grubich et al. (2008) examined the functional morphology and biomechanics of "ram-biting" behavior in great barracudas where the posterior portions of the oral jaws are used to slice through prey. They used fresh fish and preserved museum specimens to examine the jaw mechanism of an ontogenetic series of barracuda ranging from 20 g to 8.2 kg. Their results revealed that prey is impacted at the corner of the mouth during capture in an orthogonal position where rapid recurring bites and short lateral headshakes result in cutting the prey into pieces. A robust palatine bone entrenched with large dagger-like teeth opposes the mandible at the rear of the jaws offering for a scissor-like bite accomplished of shearing through the flesh and bone of the prey.

Investigation on feeding capability in fishes has revealed that the diameter of a fish's mouth is a usually good judge of the maximum prey size a fish can effectively seize and devour (Yasuda 1960; Werner 1974; Wainwright and Richard 1995). Usually, the optimal prey size (i.e., body depth) for suction-feeding fishes where the greatest net energy return is achieved ranges from 40% to 70% of the predator's mouth diameter (Werner 1974, 1977; Kislalioglu and Gibson 1976a, b; Hoyle and Keast 1987; Wainwright and Richard 1995). However, barracuda species such the great barracudas are distinguished for attacking and eating prey much larger than the gape or width of their jaws (Gudger 1918; de Sylva 1963; Randall 1967). Indeed, large adult barracudas (about 2 m total length) can sever 1 m-long amberjack, *Seriola dumerili*, in half (Grubich et al. 2008). This great biting aptitude is not limited to large individuals, as the juvenile barracuda (30 cm) could also decapitate large goldfish (Grubich et al. 2008). Nevertheless, the bite force scales with positive allometry across the lower jaw, suggesting that the importance of prey slicing behavior may increase with increasing predator size (Grubich et al. 2008). Such positive allometry of bite force scaling is comparable to that seen in sharks (Huber et al. 2006), lizards and turtles (Herrel and O'Reilly 2006), and finches (van der Meij and Bout 2004). Being able to devour larger prey may offer greater energy returns per feeding bout for barracudas.

2.2.3 Species of Triggerfish

Family: Balistidae.

Canthidermis maculata (Bloch 1786) (Fig. 2.54).

Common name: Rough triggerfish.

Etymology: *Canthidermis:* Greek, *kanthos* = the outer or inner corner of the eye, where the lids meet, 1646 + Greek, *derma* = skin (Romero 2002).

Fig. 2.54 *Canthidermis maculata* (Bloch 1786). (Courtesy of Moazzam Khan, Pakistan)

Diagnostic Features

The body is elongated, with convex head profile and a prominent chin. A deep groove is behind the eye, below nostrils, and no grooves on the cheek. Lips are broad at the center and narrow at the edges. The interorbital area is convex, having a shape of groove equal in length to the eye diameter, narrow, deep anteriorly, broader posteriorly, and connected to the nasal depression by narrow groove. The first dorsal spine is pointed, with large spinules at the anterior end. Nostrils are placed in a shallow depression with blunt round protuberances bordering the depression. Teeth are white and rectangular with conical edges. No scales are present above the base of pectoral fin. Gill rakers are short and do not project above the edge of the branchial arch. The second dorsal and anal fins are highly elevated anteriorly and short posteriorly. The pectoral fin is rounded. The scales on the cheek have round protuberances and ridges and arranged in vertical rows. Body scales are with a large ridge at the anterior middle and with many ridges and round protuberances. The scales on the abdomen are rectangular to rhomboid in shape with many ridges and round protuberances arranged in 3–6 oblique rows. Scales on caudal peduncle are with ridge at the anterior, middle, and round protuberances and arranged in horizontal rows. The ventral flap is reduced. Ventral spines are modified into a single row of modified scales present at the ventral side, and each scale has many spines directed backward, giving a comb-like appearance. Body and head are dark, lighter below, with many elongated white spots; these spots less prominent or missing from large adult specimens; all fins are dark.

Habitat, Biology, and Ecology

This is a marine species that lives in association with reefs at a depth range of 1–110 m (Lieske and Myers 1994). It reaches a maximum size of 500 mm TL (Bussing 1995), with a common length of 350 mm TL (Jiménez Prado and Béarez 2004). The length of specimen collected from the Kermadec Islands and the New Zealand EEZ is 54 mm SL (Trnski et al. 2015).

Individuals of this species are found in shallow coastal and offshore waters (Bussing 1995; Yamada et al. 1995). They spend most of their lives in the epipelagic area and are frequently linked with drifting objects (Myers 1999; Fedoryako 1980). Sometimes, they are found on deep rocky slopes (Robins and Ray 1986). They feed on zooplanktons (Bacchet et al. 2006).

Distribution

Individuals of this species have **a** circumglobal distribution in warm and tropical seas (not in the Mediterranean). Recently, it has been recorded from the Kermadec Islands and the New Zealand EEZ (Trnski et al. 2015).

Economic Value

This species has commercial value as it is used for human consumption in certain regions of its distribution range (Matsuura 2001).

Conservation Status

Not evaluated.

Human Interaction

Reported as a source of ciguatera poisoning and may inflict serious bites to humans

Rhinecanthus rectangulus (Bloch and Schneider 1801) (Fig. 2.55).

Common name: Wedgetail triggerfish.

Etymology: *Rhinecanthus:* Greek, *rhinos* = nose + Greek, *akantha* = thorn (Romero 2002).

Fig. 2.55 *Rhinecanthus rectangulus* (Bloch and Schneider 1801). (Courtesy of Reyes, Rodolfo Philippines)

Diagnostic Features

The body is orange-brown above. Head and belly are white. Soft dorsal, anal, and pectoral fins are pale. The caudal fin is dusky (Smith and Heemstra 1986). This species is characterized by the presence of a black band through the eye, connecting expansive black area between pectoral and anal fin bases. The caudal peduncle/posterior body has a distinctive black triangular mark. There is no groove in front of the eye. Head profile is nearly straight dorsal and ventral. The caudal peduncle and adjacent posterior body area are with 4–5 horizontal rows of small, anterior projecting spines. The caudal fin is rounded in shape (Allen and Erdmann 2012).

Habitat, Biology, and Ecology

Individuals of this species prefer the shallow outer reefs subject to surge (Matsuura 2001, Kuiter and Tonozuka 2001). They are normally found over barren rocks or the spur-and-groove zone where there is a mixture of bare rock, rubble, and coral. Juveniles live on algae reef (Kuiter and Tonozuka 2001). They are benthopelagic (Mundy 2005). This species feeds on algae, detritus, mollusks, crustaceans, worms, echinoderms, fishes, sponges, foraminiferans, and eggs. Individuals of this species have a territorial behavior. Females are oviparous (Breder and Rosen 1966).

Distribution

Individuals of this species are distributed in the Indo-Pacific region. They are found from the Red Sea south to East Africa through Indonesia and northern Australia up to Hawaii and Pitcairn islands and north to southern Japan. They are also reported from waters around New Zealand (Roberts et al. 2015).

Economic Value

This species has a minor commercial value as food for humans, but it has a substantial value in the aquarium trade.

Conservation Status

Not evaluated.

Human Interaction

This species can cause severe bites to humans.

2.2.4 Reported Cases of Triggerfish Bites

There are always hazards of different kinds facing divers whenever they go down in the water. These threats are not limited to certain aquatic organisms such as sharks and poisonous aquatic animals of various species and can come from unexpected organism. The piece of information that they are not regularly aware of is the risk that might come from a triggerfish. No previous books that gave information about the dangerous organisms living in the sea have shed a light on the fact that a triggerfish might cause risk to humans and give harsh bites (Phillips and Brady 1953; Halstead 1974).

The distinguished characteristic of triggerfishes is cruising in water instead of swimming fast. Such movement is performed by moving the second dorsal and anal fins, but when they need to swim fast, they move the caudal fin, which gives them a good push onward. Under no circumstances do triggerfishes fight imminent divers, but they go in their territory not realizing where they are going or not noticing the presence of adult fish protecting the nest. The technique that triggerfishes use while they are in reef holes to keep themselves immovable is to erect the first dorsal spine and depress their pelvic fin bone. For the diver to remove the fish from its position, the diver should catch the fish, push the second dorsal spine down to release the lock, and then pull the fish out.

Randall and Millington (1990) have reported seven cases of unprovoked attack by different species of triggerfish. The species stated are not present in the waters around New Zealand, but it is worth mentioning here to give an idea about how the triggerfish might be aggressive and to stress that divers should take extra care.

In most of these seven incidents, divers advanced or were near the triggerfish territory and got slight bites, except for one case where the diver had severe bites from a large adult triggerfish. The area in front of the ear and the parotid area were pursued. The fish persisted sucking the skin and soft tissue, and the fish was removed by a strong knock from the diver. The repeated bites caused lacerations and a slightly deep wound on the left cheek.

Randall and Millington (1990) have resolved that care should be taken when divers come face to face with triggerfish as they have the potential for severe and unanticipated bites. They also advise divers to be vigilant when they dive in any area outside New Zealand waters, where the two species *Pseudobalistes fuscus* and *Balistoides viridescens* live, as these do not move away when approached. In addition, care should be taken to stop feeding triggerfish by hand.

Lanelli et al. (2007) reported on an incidence that happened in Borneo in 2007, saying, "Triggerfish nest during periods of the full moon. The females usually guard the eggs and will attack anything that swims near to it. The victim didn't do anything to provoke the attack; he just happened to be in the wrong place." The triggerfish, with its mouth wide open, followed the victim and lunged at his bare right lower leg. It bit the victim savagely on the calf. Blood gushed from a wound the size of a dime. The victim grabbed his leg and kicked diagonally to the surface and onto the waiting boat.

So, what should divers and snorkelers do? First, watch the moon. If it is full, give triggerfish a wide berth, especially if they exhibit unusual behavior. This fish's usual behavior is to nibble on coral, totally ignoring divers and snorkelers. If instead of eating and ignoring you they are facing you or your friends and darting back and forth, then swim away, keeping your fins between you and them. "The locals in Maldives told me that triggerfish defend a cone-shaped area around and above their nests. Therefore, if attacked, they said, dive either down—away from the nest, or swim away on the same level as the nest. Never swim straight up because that's still 'in their area within the cone." Having already rid his body of accumulated nitrogen, the diver should swim away at the same level as the nest and gradually rise to the surface.

2.3 Species of Ribbonfish

Family: Trichiuridae.

Aphanopus mikhailini (Parin 1983).

Common name: Mikhailin's scabbardfish.

Etymology: *Aphanopus:* Greek, *aphanes* = hidden + Greek, *pous* = foot (Romero 2002).

Diagnostic Features

Body is elongated, with a distinct caudal fin. The anus is situated below penultimate or last dorsal fin spine. The distance from the snout to the anus ranges between 1.8 and 1.9 times in standard length; dorsal fin with XLIII to XLV spines and 61 to 65 soft rays (total 104 to 109 fin elements). The anal fin has two spines, its origin situated below third to fourth dorsal fin soft ray, and 51 to 54 soft rays.

Habitat, Biology, and Ecology

A marine species prefers living in the benthopelagic region at a depth range of 905–2000 m (Nakamura and Parin 1993). It reaches a maximum length of 900 mm TL (Bianchi et al. 1999). Individuals of this species are found on the continental slope or underwater rises (Parin 1990).

Distribution

Individuals of this species are distributed in the Southeast Atlantic from Namibia and at Walvis Ridge. They are also reported from the Southwest Atlantic at areas off Argentina. They are recorded from the Indian Ocean from the seamounts off

Mozambique and the West Australian Ridges. Their presence is also reported from the Western Pacific at the Southeast Australia, North Island in New Zealand, and the Challenger Plateau (Parin 1995).

Economic Value
No economic value is known for this species.

Conservation Status
Collen et al. (2010) assessed *Aphanopus mikhailini* as Least Concern in the Red List of the IUCN based on the following information: This species has wide-ranging distribution and has no commercial value to fisheries; however, it may be taken as bycatch by fisheries operating within its range. However, there are a number of unexploited seamounts throughout the southern ocean where this species is impacted by major threats. Harvesting as bycatch is only present in a small portion of this species range.

Human Interaction
As a trichiurid species, it may pose severe bites to humans.

Benthodesmus elongatus (Clarke 1879) (Fig. 2.56).
Common name: Elongate frostfish.
Etymology: *Benthodesmus:* Greek, *benthos* = depth of the sea + Greek, *desmos* = bond, chain (Romero 2002).

Diagnostic Features
The body is extremely long and compressed. The head profile is smooth, gently rising from the tip of the snout to the origin of the dorsal fin. Frontal ridges are not elevated, with nape flattened and without sagittal crest. The lower hind margin of gill cover is convex. The lower jaw extends anterior to the upper jaw. Jaw dentition includes anterior fangs and slightly compressed lateral teeth. The dorsal fin is long, partially divided by a shallow notch, with base of spinous part about half as long as base of soft part. The anus is under the first to fourth dorsal fin soft ray. The anal fin has two spines close to each other, the first rudimentary and completely concealed in adults and the second with a delicate cardiform shape, well detached from the rest of the fin. Pectoral fins have 12 soft rays. Pelvic fins are tiny, composed of a scale-like spine and a rudimentary soft ray, inserted before, below, or behind the base of pectoral fins. The caudal fin is small and forked. The body has a silvery color. Jaws and opercle are blackish. Area inside the mouth and gill cavities is black.

Habitat, Biology, and Ecology
This is a marine species inhabiting the benthopelagic region at a depth range of 178–950 m (Nakamura and Parin 1993). Individuals of this species reach maturity at 570 mm TL and reach a maximum total length of 1000 mm (Nakamura 1995). Juveniles of this species are mesopelagic (Nakamura and Parin 1993), and in some cases, they are benthopelagic (Scott and Scott 1988). This species feeds on crustaceans (mainly prawns and euphausiids), small fishes (e.g., *Maurolicus*), and squid (Parin et al. 1990).

Distribution
This species is reported from the Atlantic, Indian, and Pacific Oceans. It is present off south of Brazil and Argentina, Rio Grande Rise, south-

Fig. 2.56 *Benthodesmus elongatus* (Clarke 1879). (Courtesy of Luciano Gomes, Brazil)

eastern Africa, Madagascar Ridge, southeastern Australia, New Zealand, and Sala y Gomez Ridge.

Economic Value

No commercial information is available about this species.

Conservation Status

This species is not evaluated for the Red List of the IUCN.

Human Interaction

As a trichiurid species, it may pose severe bites to humans.

Benthodesmus tenuis (Günther 1877) (Figs. 2.57 and 2.58).

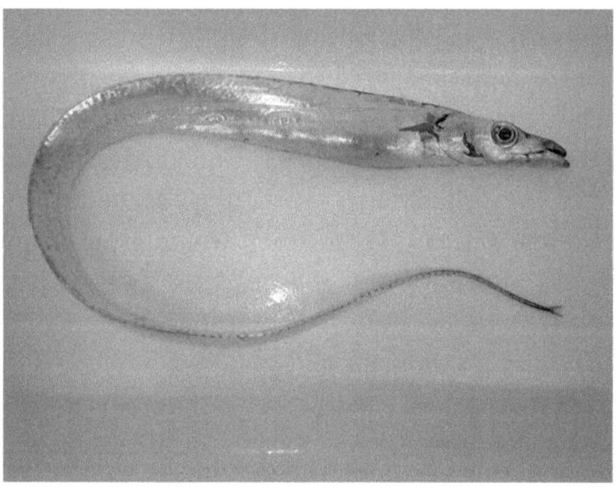

Fig. 2.57 *Benthodesmus tenuis* (Günther 1877). (Courtesy of Yoko Ishikawa through Yukio Ueta)

Fig. 2.58 Head of *Benthodesmus tenuis* (Günther 1877). (Courtesy of Yoko Ishikawa through Yukio Ueta)

Common name: Slender frostfish.

Etymology: *Benthodesmus:* Greek, *benthos* = depth of the sea + Greek, *desmos* = bond, chain (Romero 2002).

Diagnostic Features

The body is elongated and laterally compressed. The anus is located below fourth to seventh dorsal fin soft ray. The distance from the snout to the anus is 2.2 to 2.4 times in standard length. The anal fin has two spines located below the sixth to 11th soft dorsal fin ray and has 69–76 soft rays. The pelvic fins are inserted before or below pectoral fin base.

Habitat, Biology, and Ecology

Benthodesmus tenuis is a marine species living in the benthopelagic region at a depth range of 200–850 m (Nakamura and Parin 1993). It reaches a maximum total length of 2300 mm (Parin 1990). Individuals of this species are mainly found over the continental slope (Bianchi et al. 1999).

Distribution

This species is distributed in the Western Atlantic region. It is reported from Canada (Scott and Scott 1988), off Cape Hatteras, Gulf of Mexico, Suriname, and southern Brazil. In the Eastern Atlantic, it is reported from the Gulf of Guinea to Angola, while in the Western Indian Ocean, it is recorded off Natal, South Africa (Heemstra 1995). It is also reported from the Western Pacific at Emperor Seamounts, Japan, Ryukyu Islands, Vietnam, and the Sulu Sea. In the Indian Ocean, it is recorded from south of Java, Indonesia. It is also recorded from New Zealand (Roberts et al. 2015).

Economic Value

No commercial information is available about this species.

Conservation Status

Iwamoto (2015) evaluated this species as Least Concern in the Red List of the IUCN based on the following facts: it has a circumglobal distribution from the tropical through warm temperate seas. It prefers to live in deep areas of the mesopelagic region of the continental slope. It has no major threats.

Human Interaction
As a trichiurid species, it may pose severe bites to humans.

Benthodesmus tuckeri (Parin and Becker 1970).
Common name: Tucker's frostfish.
Etymology: *Benthodesmus:* Greek, *benthos* = depth of the sea + Greek, *desmos* = bond, chain (Romero 2002).

Diagnostic Features
The body is elongated and laterally compressed. The anus is located below fifth or seventh dorsal fin soft ray. The dorsal fin has 39 to 44 spines and 88 to 96 soft rays (total 130 to 137 fin elements). The anal fin has two spines, located below the eighth to 11th dorsal fin soft ray, and 76 to 83 soft external rays developed throughout its base. Pelvic fins are inserted before or below anterior edge of pectoral-fin base.

Habitat, Biology, and Ecology
This species lives in a totally marine environment and prefers benthopelagic areas at a depth range of 500–790 m (Nakamura and Parin 1993). It reaches a maximum total length of 770 mm (Nakamura and Parin 1993). The juveniles are mesopelagic and living at depth of 500 m.

Distribution
Individuals of this species are distributed in the Indo-West Pacific region. They are reported from the Philippines, Vietnam, Molucca Islands and southeastern Australia to Socotra Island, Saya de Malha Bank, Mozambique Channel, off Natal (Heemstra 1995), and south of Java. It is recorded from waters around New Zealand (Roberts et al. 2015).

Economic Value
No commercial information is available about this species.

Conservation Status
This species is not evaluated for the Red List of the IUCN.

Human Interaction
As a trichiurid species, it may pose severe bites to humans.

Lepidopus caudatus (Euphrasen 1788) (Figs. 2.59 and 2.60).
Common name: Silver scabbardfish.
Etymology: *Lepidopus:* Greek, *lepis* = scale + Greek, *pous* = foot (Romero 2002).

Diagnostic Features
The body is elongated and compressed. The anus is located below the 36th to 40th dorsal fin soft ray. The upper head profile is oblique and con-

Fig. 2.59 *Lepidopus caudatus* (Euphrasen 1788). (Courtesy of Peter Wirtz)

Fig. 2.60 Jaws of *Lepidopus caudatus* (Euphrasen 1788). (Courtesy of Costas Giannakenas)

cave, rising gently from the tip of the snout to the middle of orbits and more steeply to the dorsal origin. The posterior convergence of frontal ridges is behind the rear margin of the orbits. The sagittal crest is restricted to the nape area. Orbits are nearly touching the dorsal profile, with interorbital space flat or slightly concave. Gill rakers are 17 to 20. Dorsal fin elements are 98 to 110. The anal fin has two spines, with second plate-like or triangular, shorter than the pupil. Pelvic fins are inserted below eighth to ninth dorsal-fin soft ray. The body has a uniform silvery coloration. The dorsal fin is blackish gray (in North Atlantic populations) or with black margin of membrane between the first three soft dorsal fin rays and seventh to ninth soft rays (in South Hemisphere populations).

Habitat, Biology, and Ecology

It is a marine species that prefers the benthopelagic environment and has an oceanodromous habit (Reide 2004), living at a depth range of 42–620 m (Mytilineou et al. 2005). This species matures at 1110 mm TL and reaches a maximum total length of 2100 mm (Mytilineou et al. 2005).

Individuals of this species live on continental shelf, along its edge and upper slope, down to 400 m (600 m in Australia), usually over sandy and muddy bottoms from 100 to 250 m (over 300 m in Australia). They migrate into midwater at night. They form schools, occasionally found inshore in upwelling of deep water when it appears at surface, and feed on crustaceans, small squid, and fish, and their eggs and larvae are pelagic (Parin 1986).

Distribution

This trichiurid species is distributed in the Eastern Atlantic. It is reported from France and western Mediterranean to Senegal, including Azores, Madeira, the Canary Islands, and offshore seamounts, Cape Fria, Namibia, to Agulhas Bank, South Africa, including northern Walvis Ridge. It is also found in the Southern Indian Ocean in the seamounts area (30 to 35°S). Reports of this species from the Southwest Pacific include Australia (New South Wales to southern West Australia) and New Zealand, while the Southeast Pacific contain Peru.

Economic Value

This species has an economic value in the eastern North Atlantic, mainly off Portugal and Morocco. Trawls off Namibia and New Zealand (Nakamura and Parin 1993) also catch it.

Conservation Status

Iwamoto (2015) evaluated this species as Data Deficient based on the following facts: This species is widely distributed throughout the world. In addition to its geographically correlated differences in morphology, there is a significant intra-specific genetic distance between northern (Atlantic and Mediterranean) and southern (Australasian, Namibian, and South African) subpopulations of *L. caudatus*, and this species have been suggested to need taxonomic revision. It is found in depths of 400 m in the eastern Atlantic and 600 m off New Zealand. It is a benthopelagic species that is found on the continental shelf and slope, over offshore pinnacles, and sea mounts. It is fast-growing and reproduces early and is considered a moderate-productivity species by FAO. This fishery is characterized by vessels such as trawlers, gillnetters, and longliners. It is not considered an explicit target of any one fishery; however, it is taken in deep-sea industrial fisheries operating at depths of 200–1000 m.

Human Interaction

As a trichiurid species, it may pose severe bites to humans.

2.4 Bites of Ribbonfishes

All ribbonfish species mentioned above have long fangs in either the upper or lower jaw or both. These fangs are long and very sharp and are prepared to cut and cause severe wounds. In different parts of the world, many anglers encounter these fishes either on fishing boats or through processing their daily catch. There have been several incidents of attacks by these species on fishermen and people handling fresh fish specimens. Cases of attacks originate from Fao City, southern Iraq (Salem 2010), Iran (Mehdi 2008), and south Oman (Ali 2008). Regarding the incidence from Iraqi marine waters, two anglers pulled their fishing net from the water, and when they started to clear the catch from the net, two large ribbonfishes were freed, which bit the right hand of one of the fishermen and the arm of the other before the two men were able to control the angry fishes. The cuts were very deep, and the two men were evacuated to a nearby hospital for treatment. The Iranian incidence is like that of the Iraqi, but it happened to one fisherman working on the deck of a small trawler. He started to sort out the catch from the net when a large ribbonfish jumped and bit his forearm. The wound was very deep, and the fisherman required treatment immediately. The Omani incidence happened on the shore when the fish catch was brought back from the fishing boat and put on a plastic mat for sorting; suddenly, a large ribbonfish jumped and had the young angler by his arm. The wound was deep, and the affected victim was taken to the hospital for treatment. In all these cases, the fishermen or the people handling the fresh fish thought that the fishes were dead, and they felt free to handle them from any part of the fish's body. Such routine is not accurate particularly with freshly caught fishes as some large specimens stay alive for longer periods and get aggressive once they have been touched.

At this point, it is important to have some information about the mechanism that ribbonfishes have in their jaws and mouth to accomplish such strong and severe bites.

Ribbonfishes have a long jaw, armed with large and sharp teeth. In such type of mouth, the fish needs to have maximal speed to snap and maximal force to press against the prey (De Schepper et al. 2008). With the mechanism of the mouth closing, in trichiurid fishes, as seen in *T. lepturus,* studied by De Schepper et al. (2008), there are several different muscles involved. Some of these muscles are responsible for closing the lower jaw and at the same time dragging the food toward the esophagus.

Trichiurids have the aptitude to close their mouth very quickly and leave very little chance for the prey to escape. Such feature is underway with the development of the feeding habit and goes along with the type of food that these fishes feed on and with the development of a powerful bite (Wojciechowski 1972; Martins and Haimovici 1997; Friel and Wainwright 1998; Costa et al. 2000; Swan et al. 2003). These fundamentals are shown in the morphology of the jaws of the trichiurid species, which are displayed to have elongated jaws. Furthermore, long jaws give the fish the ability to swallow large prey (Norton and Brainerd 1993; Porter and Motta 2004). Trichiurids have a streamlined head, which allows them to come very close to their prey before attacking. The streamlined head does not generate momentum on water in front of the head, which diminishes prey recognition of the predator (Porter and Motta 2004). These species move toward their prey slowly holding their body rigidly (Bone 1971). The less undulation there is, the less visual recognition there is by the prey for the predator (Porter and Motta 2004).

References

Adams DH, McMichael RH. Mercury levels in four species of sharks from the Atlantic coast of Florida. Fish Bull. 1999;97:372–9.

Afonso AS, Hazin FHV. Vertical movement patterns and ontogenetic niche expansion in the Tiger Shark, *Galeocerdo cuvier*. PLoS One. 2015;10(1):e0116720.

Afonso AS, Hazin FHV, Barreto RR, Santana FM, Lessa RP. Extraordinary growth in tiger sharks *Galeocerdo cuvier* from the South Atlantic Ocean. J Fish Biol. 2012;81:2080–5.

Alexander RM. Modelling approaches in biomechanics. Philos Trans R Soc B. 2003;358:1429–35.

Alfaro ME, Bolnick DI, Wainwright PC. Evolutionary consequences of many-to-one mapping of jaw morphology to mechanics in labrid fishes. Am Nat. 2005;165:E140–54.

Ali FG. Fishes from the Gulf of Oman. Muscat: Oman Publishing House; 2008. 87 p.

Allen GR, Erdmann MV. Reef fishes of the East Indies, Volumes I–III. Tropical reef research. Perth: University of Hawai'i Press; 2012.

Andrews AH, Kerr LA. Validated age estimates for large white sharks of the northeastern Pacific Ocean: altered perceptions of vertebral growth shed light on complicated bomb Δ14C results. Environ Biol Fish. 2015;98(3):971–8.

Ardizzone D, Cailliet GM, Natanson LJ, Andrews AH, Kerr LA, Brown TA. Application of bomb radiocarbon chronologies to shortfin mako (Isurus oxyrinchus) age validation. Environ Biol Fish. 2006;77:355–66.

Auerbach P. Hazardous marine animals. Emerg Med Clin North Am. 1984;2:531–44.

Auerbach PS, Burgess GH. Injuries from nonvenomous aquatic animals. In: Auerbach PS, editor. Wilderness medicine. 5th ed. New York: Elsevier; 2007. p. 1654–91.

Ayling T, Cox GJ. Collins guide to the sea fishes of New Zealand. International Specialized Book Service Incorporated; 1982.

Bacchet P, Zysman T, Lefèvre Y. Guide des poissons de Tahiti et ses îles. Tahiti (Polynésie Francaise): Éditions Au Vent des Îles; 2006. 608 p.

Baerez P. Lista de los peces marinos del Ecuador Continental. Rev Biol Trop. 1996;44:731–41.

Baldridge HD. Shark aggression against man: beginnings of an understanding. Calif Fish Game. 1988;74(4):208–17.

Barlow GW. Modal action patterns. In: Sebeok TA, editor. How animals communicate. Bloomington: Indiana University Press; 1977. p. 98–134.

Barnett A, Stevens JD, Frusher SD, Semmens JM. Seasonal occurrence and population structure of the broadnose sevengill shark *Notorynchus cepedianus* in coastal habitats of South-East Tasmania. J Fish Biol. 2010;77(7):1688–701.

Barreiros JP, Haddad V Jr. Bite by moray eel. J Venom Anim Toxins Incl Trop Dis. 2008;14(3):541–5.

Barreiros JP, Santos RS, Borba AESD. Food habits, schooling and predatory behaviour of the yellowmouth barracuda, *Sphyraena viridensis* Cuvier, 1829 (Perciformes: Sphyraenidae) in the Azores. Cybium-Int J Ichthyol. 2002;26(2):83–8.

Barreiros JP, Gadig OB, Haddad V. An unprovoked attack by a blue shark Prionace glauca (Chondrichthyes: Carcharhinidae) on a spear fisherman in Terceira Island, Azores, Northeast Atlantic. Wilderness Environ Med. 2014;25(3):371–2.

Bass AJ, D'Aubrey JD, Kistnasamy N. Sharks of the east coast of southern Africa. l. The genus Carcharhinus (Carcharhinidae). Investigational Report. Oceanographic Research Institute, Durban No. 33; 1973. 168 pp.

Bass AJ, D'Aubrey JD, Kistnasamy N. Sharks of the east coast of southern Africa. IV. The families Odontaspididae, Scapanorhynchidae, Isuridae, Cetorhinidae, Alopiidae, Orectolobidae and Rhiniodontidae. Investigational Report. Oceanographic Research Institute, Durban No. 39; 1975. 102 pp.

Bass AJ, Compagno LJV, Heemstra PC. Squalidae. In: Smith MM, Heemstra PC, editors. Smiths' sea fishes. Berlin: Springer; 1986a. p. 49–62.

Bass AJ, Heemstra PC, Compagno LJV. Carcharhinidae. In: Smith MM, Heemstra PC, editors. Smiths' sea fishes. Berlin: Springer; 1986b. p. 67–87.

Bass NC, Day J, Guttridge TL, Mourier J, Knott NA, Vila Pouca C, Brown C. Residency and movement patterns of adult Port Jackson sharks (Heterodontus portusjacksoni) at a breeding aggregation site. J Fish Biol. 2021;99(4):1455–66.

Benchley P. Great white sharks. The unmistakable jaws of the great white have long been synonymous with terror. Photographs by David Doubilet. Natl Geogr. 2000;197(4):2–29.

Bennett MB, Gordon I, Kyne PM. (SSG Australia & Oceania Regional Workshop, March 2003). Carcharhinus galapagensis. The IUCN Red List of Threatened Species 2003: e.T41736A10550977. 2003. https://doi.org/10.2305/IUCN.UK.2003.RLTS.T41736A10550977. en. Downloaded 26 Feb 2016.

Bernal D, Sepulveda C, Graham JB. Water-tunnel studies of heat balance in swimming mako sharks. J Exp Biol. 2001;204(23):4043–54.

Berteaux H, Prindle B, May D. The deep sea moorings fish bite problem. In: OCEANS'87. IEEE; 1987. p. 661–73.

Bester C. Biological profiles: Great Hammerhead. Florida Museum of Natural History Ichthyology Department. 2008. Retrieved 18 Oct 2008.

Bianchi G, Carpenter KE, Roux J-P, Molloy FJ, Boyer D, Boyer HJ. FAO species identification guide for fishery purposes. Field guide to the living marine resources of Namibia. Rome: FAO; 1999. 265p.

Biggs RD. Pre-Sargonic riddles from Lagash. JNES. 1973;32:26–33.

Bird J. Jonathan Bird's Blue World. Webisode 15: the shark and the lamprey. 2009. http://www.blueworldtv.com/s1_e15.html (video).

Bishop SDH, Francis MP, Duffy C, Montgomery JC. Age, growth, maturity, longevity and natural mortality of the shortfin mako shark (*Isurus oxyrinchus*) in New Zealand waters. Mar Freshwat Res. 2006;57:143–54.

Blaber SJM. The ecology of Sphyraena barracuda (Osteichthyes, Perciformes) in the Kosi system with notes on the Sphyraenidae of other Natal estuaries. S Afr J Zool. 1982;17:171–6.

Boehmer RM. Vom Hassek Höyük bis zum Buch Tobias: Von Sägefischen und Haien im Altertum. Baghdader Mitteilungen. 2002;33:7–43.

Bohlke EB, McCosker JE. The moray eels of Australia and New Zealand, with the description of two new species (Anguilliformes: Muraenidae). Rec-Aust Mus. 2001;53(1):71–102.

Bone Q. On the scabbard fish *Aphanopus carbo*. J Mar Biol Assoc UK. 1971;51:219–25.

Bonfil R. The biology and ecology of the silky shark, Carcharhinus falciformis. In: Camhi M, Pikitch EK, Babcock EA, editors. Sharks of the open ocean: biology, fisheries and conservation. Oxford: Blackwell Science; 2008. p. 114–27. ISBN: 0-632-05995-8.

Bonfil R, Meyër MA, Scholl MC, Johnson RL, O'Brian S, Oosthuizen WH, Swanson S, Kotze D, Patterson M. Transoceanic migration, spatial dynamics, and population linkages of white sharks. Science. 2005;310:100–3. https://doi.org/10.1126/science.1114898.

Boustany AM, Davis SF, Pyle P, Anderson SD, Le Boeuf BJ, Block BA. Expanded niche for white sharks. Nature. 2002;415(6867):35–6.

Bovcon ND, Cochia PD, Navoa X, Ledesma P, Caille GM, Baigun CR. First report on a pupping area of the tope shark Galeorhinus galeus (Carcharhiniformes, Triakidae) in the south-west Atlantic. J Fish Biol. 2018;93(6):1229–32.

Braccini M, Waltrick D. Species-specific at-vessel mortality of sharks and rays captured by demersal longlines. Mar Policy. 2019;99:94–8.

Brading JJ. Behavioral and environmental conditions associated with shark attacks on humans. Degree of Honors Bachelor of Science in Biology. Dalhousie University Halifax, Nova Scotia April 2006

Branco V, Canario J, Vale C, Raimundo J, Reis C. Total and organic mercury concentrations in muscle tissue of the blue shark (Prionace glauca L. 1758) from the Northeast Atlantic. Mar Pollut Bull. 2004;49:854e874.

Branco V, Vale C, Canário J, Dos Santos MN. Mercury and selenium in blue shark (Prionace glauca, L. 1758) and swordfish (Xiphias gladius, L. 1758) from two areas of the Atlantic Ocean. Environ Pollut. 2007;150(3):373–80.

Branstetter S, Burgess GH. Commercial shark fishery observer program. Characterization and comparisons of the directed commercial shark fishery in the eastern Gulf of Mexico and off North Carolina through an observer program. Final Report, Marfin Award NA47FF0008, 1996. 33pp.

Branstetter S, Musick JA, Colvocoresses JA. A comparison of age and growth of the tiger shark, *Galeocerdo cuvieri*, from off Virginia and from the northwestern Gulf of Mexico. Fish Bull. 1987;85:269–79.

Breder CM, Rosen DE. Modes of reproduction in fishes. Neptune City: T. F. H. Publications; 1966, 941 p.

Brewer DT, Morello EB, Griffiths S, Fry G, Heales D, Apte SC, Venables WN, Rothlisberg PC, Moeseneder C, Lansdell M, Pendrey R. Impacts of gold mine waste disposal on a tropical pelagic ecosystem. Mar Pollut Bull. 2012;64(12):2790–806.

Bruce BD. The biology and ecology of the white shark, *Carcharodon carcharias*. In: Camhi MD, Pikitch EK, Babcock EA, editors. Sharks of the open ocean. Oxford: Blackwell; 2008. p. 69–76.

Buck JD, Spotte S, Gadbaw JJ. Bacteriology of the teeth from a great white shark: potential medical implications for shark bite victims. J Clin Microbiol. 1984 Nov;20(5):849–51.

Buglass S, Nagy S, Ebert D, Sepa P, Turchik A, Bell KL, Rivera F, Giddens J. First records of the seven-gilled *Notorynchus cepedianus* and six-gilled *Hexanchus griseus* sharks (Chondrichthyes: Hexanchiformes: Hexanchidae) found in the Galápagos marine reserve. J Fish Biol. 2020;97(3):926–9.

Burgess GH. Shark attack and the international shark attack file. In: Gruber SH, editor. Discovering sharks. Sandy Hook: American Littoral Society; 1990. p. 101–5.

Burgess GH. International shark attack Wle. Florida Museum of Natural History. 2008. http://www.flmnh.ufl.edu/fish/sharks/ISAF/ISAF.htm

Bussing WA. Balistidae. Pejepuercos, calafates, gatillos. In: Fischer W, Krupp F, Schneider W, Sommer C, Carpenter KE, Niem V, editors. Guia FAO para Identification de Especies para lo Fines de la Pesca. Pacifico Centro-Oriental. 3 Vols. Rome: FAO; 1995. p. 905–9.

Buttler M. Man bitten by shark at Elwood beach. Herald Sun. 2011. Retrieved 03 June 2021.

Byard RW, Gilbert JD, Brown K. Pathologic features of fatal shark attacks. Am J Forensic Med Pathol. 2000;21(3):225–9.

Byard RW, Wick R, Simpson E, Gilbert JD. The pathological features and circumstances of death of lethal crush/traumatic asphyxia in adults—a 25-year study. Forensic Sci Int. 2006;159(2):200–5.

Cacutt L. The big-game fishing handbook. Stackpole Books; 2000.

Cailliet GM, Martin LK, Harvey JT, Kusher D, Welden BA. Preliminary studies on the age and growth of blue (Prionace glauca), common thresher (Alopias vulpinus), and shortfin mako (Isurus oxyrinchus) sharks from California waters. In: Prince ED, Pulos LM, editors. Proceedings international workshop on age determination of oceanic pelagic fishes: tunas, billfishes, sharks. NOAA Tech. Rep. NMFS, vol. 8, 1983. pp. 179–188.

Caldicott DG, Mahajani R, Kuhn M. The anatomy of a shark attack: a case report and review of the literature. Injury. 2001;32(6):445–53.

Camhi M, Pikitch EK, Babcock EA, editors. Sharks of the open ocean: biology, fisheries and conservation. Oxford: Blackwell Science; 2008. 231p. ISBN: 0-632-05995-8.

Campana SE, Joyce WN. Temperature and depth associations of porbeagle shark (*Lamna nasus*) in the Northwest Atlantic. Fish Oceanogr. 2004;13(1):52–64.

Campana SE, Natanson LJ, Myklevoll S. Bomb dating and age determination of large pelagic sharks. Can J Fish Aquat Sci. 2002;59(3):450–5.

Campana SE, Joyce W, Manning MJ. Bycatch and discard mortality in commercially caught blue sharks Prionace glauca assessed using archival satellite pop-up tags. Mar Ecol Prog Ser. 2009;387:241–53.

Campana SE, Joyce W, Fowler M, Showell M. Discards, hooking, and post-release mortality of porbeagle (Lamna nasus), shortfin mako (Isurus oxyrinchus), and blue shark (Prionace glauca) in the Canadian pelagic longline fishery. ICES J Mar Sci. 2016;73(2):520–8.

Capapé C, Mellinger J. Nouvelles donnees sur la biologie de la reproduction du milandre, Galeorhinus galeus (Linne, 1758), (Pisces, Triakidae) des cotes tunisiennes. Cahiers Biol Mar. 1988;29:135–46.

Carey FG, Teal JM, Kanwisher JW. The visceral temperatures of mackerel sharks (Lamnidae). Physiol Zool. 1981;54(3):334–44.

Casey JG, Kohler NE. Tagging studies on the shortfin mako shark (*Isurus oxyrinchus*) in the western North Atlantic. Aust J Mar Freshwat Res. 1992;43:45–60.

Castle PHJ, McCosker JE. Muraenidae. In: Smith MM, Heemstra PC, editors. Smiths' sea fishes. Berlin: Springer; 1986. p. 165–76.

Castro JI, Woodley CM, Brudek RL. A preliminary evaluation of the status of shark species. FAO Fish. Tech. P. 380, Rome: FAO; 1999, 72 p.

Cavigneaux A. A scholar's library in Meturan? With an edition of the tablet H 72 (Textes de tell Haddad VII). In: Mesopotamian magic: textual, historical and interpretative perspectives. Brill; 2000. p. 251–73.

Cedrola PV, Caille GM, Chiaramonte GE, Pettovello AD. Demographic structure of broadnose seven-gill shark, Notorynchus cepedianus, caught by anglers in southern Patagonia. Argentina: Marine Biodiversity Records; 2009. p. 2.

Chang W, Pien F. Marine acquired infections: hazards of the ocean environment. Postgrad Med. 1986;80:30–41.

Chapman BK, McPhee D. Global shark attack hotspots: identifying underlying factors behind increased unprovoked shark bite incidence. Ocean Coast Manage. 2016;133:72–84.

Chen HM, Shao KT, Chen CT. A review of the muraenid eels (family Muraenidae) from Taiwan with descriptions of twelve new records. Zool Stud. 1994;33(1):44–64.

Christiansen HM, Campana SE, Fisk AT, Cliff G, Wintner SP, Dudley SF, Kerr LA, Hussey NE. Using bomb radiocarbon to estimate age and growth of the white shark, Carcharodon carcharias, from the southwestern Indian Ocean. Mar Biol. 2016;163(6):1–13.

Clarke S, Magnusson JE, Abercrombie DL, McAllister M, Shivji MS. Identification of shark species composition and proportion in the Hong Kong shark fin market using molecular genetics and trade records. Conserv Biol. 2006a;20:201–11. Clarke et al., 2007

Clarke SC, McAllister MK, Milner-Gulland EJ, Kirkwood GP, Michielsens CGJ, Agnew DJ, Pikitch EK, Nakano H, Shivji MS. Global estimates of shark catches using trade records from commercial markets. Ecol Lett. 2006b;9:1115–26.

Claro R. Caracterı'sticas generales de la ictiofauna. In: Claro R, editor. Ecologı'a de los peces marinos de Cuba. Instituto de Oceanologı'a Academia de Ciencias de Cuba and Centro de Investigaciones de Quintana Roo; 1994. p. 55–70.

Cliff G. Shark attacks on the south African coast between 1960 and 1990. S Afr J Sci. 1991;87(10):513–8.

Cliff G, Dudley SFJ. Sharks caught in the protective gill nets off Natal, South Africa. 6. The copper shark Carcharhinus brachyurus (Günther). S Afr J Mar Sci. 1992;12(1):663–74.

Cliff G, Dudley SFJ, Davis B. Sharks caught in the protective gill nets off Natal, South Africa. 3. The shortfin mako shark Isurus oxyrinchus (Rafinesque). S Afr J Mar Sci. 1990;9(1):115–26.

Clua E, Reid D. Features and motivation of a fatal attack by a juvenile white shark, Carcharodon carcharias, on a young male surfer in New Caledonia (South Pacific). J Forensic Leg Med. 2013;20(5):551–4. https://doi.org/10.1016/j.jflm.2013.03.009.

Clua E, Séret B. Unprovoked fatal shark attack in Lifou Island (Loyalty Islands, New Caledonia, South Pacific) by a great white shark, Carcharodon carcharias. Am J Forensic Med Pathol. 2010;31(3):281–6.

Clua E, Bescond PM, Reid D. Fatal attack by a juvenile tiger shark, Galeocerdo cuvier, on a kitesurfer in New Caledonia (South Pacific). J Forensic Leg Med. 2014;25:67–70.

Coelho R, Infante P, Santos MN. Application of generalized linear models and generalized estimation equations to model at-haulback mortality of blue sharks captured in a pelagic longline fishery in the Atlantic Ocean. Fish Res. 2013;145:66–75.

Coleman N. Australian Sea fishes south of 30°S. Lane Cove (Sydney): Doubleday Australia; 1983. p. 302.

Collen B, Richman N, Ram M. (Sampled Red List Index Coordinating Team). Aphanopus mikhailini. The IUCN Red List of Threatened Species 2010: e.T154813A4640067. 2010. https://doi.org/10.2305/IUCN.UK.2010-4.RLTS.T154813A4640067.en.

Colloca F, Scannella D, Geraci M, Falsone F, Battista G, Vitale S, Di Lorenzo M, Bono G. British sharks in Sicily: records of long distance migration of tope shark (Galeorhinus galeus) from Northeastern Atlantic to Mediterranean Sea. Mediterr Mar Sci. 2019;20(2):309–13.

Compagno LJV. FAO Species Catalogue. Vol. 4. Sharks of the world. An annotated and illustrated catalogue of shark species known to date. Part 1—Hexanchiformes to Lamniformes. FAO Fish. Synop. 125(4/1):1–249. Rome: FAO; 1984a.

Compagno LJV. Sharks of the world. An annotated and illustrated catalogue of shark species known to date. Part 2, Carcharhiniformes. Rome: Food and Agriculture Organization of the United Nations; 1984b.

Compagno LJV. Lamnidae. Mackerel sharks, makos, white sharks, porbeagles. In: Carpenter KE, Niem VH, editors. FAO identification guide for fishery purposes. The Living Marine Resources of the Western Central Pacific. Rome: FAO; 1998. p. 1274–8.

Compagno LJV. Checklist of living elasmobranchs. In: Hamlett WC, editor. Sharks, skates, and rays: the biology of elasmobranch fishes. Maryland: Johns Hopkins University Press; 1999. p. 471–98.

Compagno LJV. Sharks of the world. An annotated and illustrated catalogue of shark species known to date. Vol. 2. Bullhead, mackerel and carpet

sharks (Heterodontiformes, Lamniformes and Orectolobiformes). FAO Spec. Cat. Fish. Purp. 1(2):269 p. Rome: FAO; 2001.
Compagno LJ V. Sharks of the world. An annotated and illustrated catalogue of shark species known to date. Vol. 2. Bullhead, mackerel and carpet sharks (Heterodontiformes, Lamniformes and Orectolobiformes). Rome: FAO; 2002.
Compagno LJV. Notorynchus cepedianus. In The IUCN Red List of Threatened Species 2009: e.T39324A10200310. 2009
Compagno LJV, Niem VH. Triakidae. Houndsharks, smoothhounds, topes. In: Carpenter KE, Niem VH, editors. FAO identification guide for fishery purposes. The living marine resources of the Western Central Pacific. Rome: FAO; 1998. p. 1297–304.
Compagno LJV, Niem VH. Carcharhinidae. Requiem sharks. In: Carpenter KE, Niem VH, editors. FAO identification guide for fishery purposes. The living marine resources of the Western Central Pacific. Rome: FAO; 1998a. p. 1312–60.
Compagno LJV, Niem VH. Odontaspididae. Sand tiger sharks. In: Carpenter KE, Niem VH, editors. FAO identification guide for fishery purposes. The living marine resources of the Western Central Pacific. Rome: FAO; 1998b. p. 1264–7.
Compagno LJV, Dando D, Fowler S. Sharks of the world. Princeton field guides. London: Harper Collins; 2005. 368 p.
Cooper JS, Kong EL, Murphy-Lavoie HM. Shark trauma. 2018. Study Guide from Stat Pearls Publishing, Treasure Island. 2018. PMID: 29939632
Cooper JS, Kong EL, Murphy-Lavoie HM. Shark trauma. StatPearls; 2021.
Cortés E. Standardized diet compositions and trophic levels of sharks. ICES J Mar Sci. 1999;56:707–17.
Cortés E, Domingo A, Miller P, Forselledo R, Mas F, Arocha F, Campana S, Coelho R, Da Silva C, Hazin FHV, Holtzhausen H, Keene K, Lucena F, Ramirez K, Santos MN, Semba-Murakami Y, Yokawa K. Expanded ecological risk assessment of pelagic sharks caught in Atlantic pelagic longline fisheries. Collect Vol Sci Pap ICCAT. 2015;71(6):2637–88.
Costa G, Chubb JC, Veltkamp CJ. Cystacanths of *Bolbosoma vasculosum* in the black scabbard fish *Aphanopus carbo*, oceanic horse mackerel *Trachurus* and common dolphin *Delphinus delphis* from Madeira, Portugal. J Helminthol. 2000;74:113–20.
Cousteau J-Y, Cousteau P. The shark: splendid savage of the sea. Garden City: Doubleday; 1970.
Cox G, Francis M. Sharks and rays of New Zealand. Canterbury University Press/University of Canterbury; 1997. 68 p.
Crespi-Abril AC, Garcia NA, Crespo EA, Coscarella MA. Consumption of marine mammals by broadnose sevengill shark Notorynchus cepedianus in the northern and central Patagonian shelf. Lat Am J Aquat Mamm. 2003:101–7.
D'Alberto BM, Chin A, Smart JA, Baje L, White WT, Simpfendorfer CA. Age, growth and maturity of oceanic whitetip shark (Carcharhinus longimanus) from Papua New Guinea. Mar Freshw Res. 2017;68(6):1118–29.
Da Silva C, Bürgener M. South Africa's demersal shark meat harvest. Traffic Bull. 2007;21(2):55–65.
Daily Mail. Barracuda attack breaks women's ribs. 2011. https://www.dailymail.co.uk/news/article-1348152/Barracuda-attack-breaks-womans-ribs-forces-boyfriend-seal-punctured-lung-hand.html
Damian R. Girl continues to surf after possible shark bite at Oreti Beach. stuff.co.nz. 2020. Retrieved 26 November, 2021.
Davies DH, Campbell GD. The aetiology, clinical pathology and treatment of shark attack. J R Nav Med Serv. 1961;48(110–136):74.
De Crosta MA, Taylor LR Jr., Parrish JD. Age determination, growth, and energetics of three species of carcharhinid sharks in Hawai'i. In: Proceedings of the second symposium on resource investigations of the North-western Hawaiian Islands, vol. 2. University of Hawai'i Sea Grant, UNIHI-SEAGRANT-MR-84-01; 1984. pp. 75–95.
De Mitcheson YS, Liu M. Functional hermaphroditism in teleosts. Fish Fish. 2008;9(1):1–43.
De Schepper N, Van Wassenbergh S, Adriaens D. Morphology of the jaw system in trichiurids: trade-offs between mouth closing and biting performance. Zool J Linnean Soc. 2008;152(4):717–36.
de Sylva DP. Systematics and life history of the great barracuda Sphyraena barracuda (Walbaum). Stud Trop Ocean. 1963;3:1–179.
Dent F, Clarke S (2015) State of the global market for shark products. FAO Fisheries and Aquaculture technical paper no. 590. FAO, Rome.
Domeier ML. A new life-history hypothesis for white sharks, Carcharodon carcharias, in the Northeastern Pacific. In: Global perspectives on the biology and life history of the white shark; 2012. p. 199–224.
Domeier ML, Nasby-Lucas N. Migration patterns of white sharks Carcharodon carcharias tagged at Guadalupe Island, Mexico, and identification of an eastern Pacific shared offshore foraging area. Mar Ecol Prog Ser. 2008;370:221–37.
Domi N, Bouquegneau JM, Das K. Feeding ecology of five commercial shark species of the Celtic Sea through stable isotope and trace metal analysis. Mar Environ Res. 2005;60(5):551–69.
Dudley SFJ, Simpfendorfer CA. Population status of 14 shark species caught in the protective gillnets off KwaZulu-Natal beaches, South Africa, 1978-2003. Mar Freshw Res. 2006;57:225–40.
Dudley SJF, Cliff G, Zungu MP, Smale MJ. Sharks caught in the protective gillnets off KwaZulu-Natal, South Africa. 10. The dusky shark Carcharhinus obscurus (LeSueur, 1818). S Afr J Mar Sci. 2005;27(1):107–27.

Duffy C, Gordon I. Carcharhinus brachyurus. In: IUCN 2006. 2006 IUCN red list of threatened species. 2003. Available: http://www.iucnredlist.org. Accessed 19 Aug 2007.

Duffy CMF, Dunn M, Finucci B, Ford R, Hitchmough R, Rolfe J. Conservation status of New Zealand chondrichthyans (chimaeras, sharks and rays). New Zealand Threat Classification Series, Department of Conservation Te Papa Atawhai; 2018. 13p.

Dulvy NK, Reynolds JD. Evolutionary transitions among egg-laying, live-bearing and maternal inputs in sharks and rays. Proc R Soc Lond Ser B Biol Sci. 1997;264:1309–15.

Dulvy NK, Baum JK, Clarke S, Compagno LJ, Cortés E, Domingo A, Fordham S, Fowler S, Francis MP, Gibson C, Martínez J. You can swim but you can't hide: the global status and conservation of oceanic pelagic sharks and rays. Aquat Conserv Mar Freshwat Ecosyst. 2008;18(5):459–82.

DVG. Shark. depi.vic.gov.au. *Archived from the original on June 26, 2015.* 2015. Retrieved November 26, 2021.

Dyer BS, Westneat MW. Taxonomy and biogeography of the coastal fishes of Juan Fernández archipelago and Desventuradas Islands, Chile. Rev Biol Mar Oceanogr. 2010;45(S1):589–617.

Ebert DA. Sharks, rays, and chimaeras of California. Berkeley: University of California Press; 2003. p. 160–2.

Ebert DA, Mostarda E. Guía Para La Identificación de Peces Cartilaginosos de Aguas Profundas Del Océano Pacífico Sudoriental. Roma: Organización de las Naciones Unidas para la Alimentación y la Agricultura; 2016.

Ebert DA, Fowler S, Compagno L. Sharks of the world. A fully illustrated guide. Plymouth: Wild Nature Press; 2013.

Eddy C, Brill R, Bernal D. Rates of at-vessel mortality and post-release survival of pelagic sharks captured with tuna purse seines around drifting fish aggregating devices (FADs) in the equatorial eastern Pacific Ocean. Fish Res. 2016;174:109–17.

Edwards A, Lubbock HR. The shark population of Saint Paul's Rocks. Copeia. 1982;1982(1):223–5.

Ellis ME, Mandal BK. Hyperbaric oxygen treatment: 10 years' experience of a regional infectious diseases unit. J Infect. 1983;6:17–28.

Ellis JR, McCully Philips SR, Poisson F. A review of capture and post-release mortality of elasmobranchs. J Fish Biol. 2017;90(3):653–722.

Enax J, Prymak O, Raabe D, Epple M. Structure, composition, and mechanical properties of shark teeth. J Struct Biol. 2012;178(3):290–9.

Endo T, Hisamichi Y, Haraguchi K, Kato Y, Ohta C, Koga N. Hg, Zn and Cu levels in the muscle and liver of tiger sharks (Galeocerdo cuvier) from the coast of Ishigaki Island, Japan: relationship between metal concentrations and body length. Mar Pollut Bull. 2008;56(10):1774–80.

Erickson T, Vanden Hoek TL, Kuritza A, Leiken JB. The emergency management of moray eel bites. Ann Emerg Med. 1992;21:212–6.

Fahmi, White WT. First record of the basking shark *Cetorhinus maximus* (Lamniformes: Cetorhinidae) in Indonesia. Mar Biodivers Rec. 2015;8:e18.

FAO, 1994. FAO Yearbook. Fishery statistics: catches and landings. Vol. 74: 1992. FAO Fish. Ser. 43:677 p.

FAO. Report of the FAO/CECAF Technical Workshop on Deep-sea Fisheries and Vulnerable Marine Ecosystems in the Eastern Central Atlantic. FAO Fisheries and Aquaculture Report No. 1184, Senegal: Dakar; 2016.

FAO, Fisheries Department. World review of highly migratory species and straddling stocks. FAO Fish. Tech. Pap. No. 337. Rome: FAO; 1994. 70 p.

Fedoryako BI. The Ichthyofauna of the surface waters of Sargasso Sea south-west of Bermuda. J Ichthyol. 1980;20(4):1–9.

Fernandez L, Salmeron F, Ramos A. Change in elasmobranchs and other incidental species in the Spanish Deepwater Black hake trawl fishery off Mauritania (1992–2001). J Northwest Atl Fish Sci. 2005;35(art. 30)

Ferreira LC, Simpfendorfer C. Galeocerdo cuvier. The IUCN Red List of Threatened Species 2019: e.T39378A2913541. 2019. https://doi.org/10.2305/IUCN.UK.2019-1.RLTS.T39378A2913541.en

Ferreira BP, Vooren CM. Age, growth, and structure of vertebra in the school shark *Galeorhinus galeus* (Linnaeus, 1758) from southern Brazil. Fish Bull. 1991;89:19–31.

Fields AT, Fischer GA, Shea SKH, Zhang H, Abercrombie DL, Feldheim KA, Babcock EA, Chapman DD. Species composition of the international shark fin trade assessed through retail market survey in Hong Kong. Conserv Biol. 2017;32(2):376–89.

Fields AT, Fischer GA, Shea SKH, Zhang H, Abercrombie DL, Feldheim KA, Babcock EA, Chapman DD. Species composition of the international shark fin trade assessed through retail market survey in Hong Kong. Conserv Biol. 2018;32(2):376–89.

Finkelstein R, Oren I. Soft tissue infections caused by marine bacterial pathogens: epidemiology, diagnosis, and management. Curr Infect Dis Rep. 2011;13(5):470–7.

Finucci B, Edwards CTT, Anderson OF, Ballara SL. Fish and invertebrate bycatch in New Zealand deepwater fisheries from 1990–91 until 2016–17. New Zealand Aquatic Environment and Biodiversity Report No. 210; 2019.

Finucci B, Barnett A, Cheok J, Cotton CF, Kulka DW, Neat FC, Pacoureau N, Rigby CL, Tanaka S, Walker TI. Notorynchus cepedianus. The IUCN Red List of Threatened Species 2020: e.T39324A2896914. 2020. https://doi.org/10.2305/IUCN.UK.2020-3.RLTS.T39324A2896914.en.

Finucci B, Cheok J, Cotton CF, Kulka DW, Neat FC, Pacoureau N, Rigby CL, Tanaka S, Walker TI. Scymnodon ringens. The IUCN Red List of Threatened Species 2020: e.T161717A116739828.

2020a. https://doi.org/10.2305/IUCN.UK.2020-3.RLTS.T161717A116739828.en

Fishes of Australia. 2021. https://fishesofaustralia.net.au/home/species/2002. Retrieved 26 November 2021.

FLMNH. Sevengill shark. flmnh.ufl.edu. 2013. Archived from the original on January 26, 2013. Retrieved November 26, 2021.

Florida Museum of Natural History. International shark attack file; 2021.

Francis MP. Checklist of the coastal fishes of Lord Howe, Norfolk, and Kermadec Islands, Southwest Pacific Ocean. Pac Sci. 1993;47(2):136–70.

Francis MP. Observations on a pregnant white shark with a review of reproductive biology. In: Great White sharks: the biology of *Carcharodon carcharias*. New York: Academic Press; 1996. p. 157–72.

Francis MP, Duffy C. Length at maturity in three pelagic sharks (*Lamna nasus*, *Isurus oxyrinchus* and *Prionace glauca*) from New Zealand. Fish Bull. 2005;103:489–500.

Francis MP, Mulligan KP. Age and growth of New Zealand school shark *Galeorhinus galeus*. NZ J Mar Freshwat Res. 1998;32(3):427–40.

Francis MP, Stevens JD. Reproduction, embryonic development, and growth of the porbeagle shark, Lamna nasus, in the Southwest Pacific Ocean. Fish Bull. 2000;98:41–63.

Francis MP, Duffy CAJ, Bonfil R, Manning MJ. The third dimension—vertical habitat use by white sharks, Carcharodon carcharias, in New Zealand and in oceanic and tropical waters of the Southwest Pacific Ocean. In: Domeier ML, editor. Global perspectives on the biology and life history of the great White shark. Boca Raton: CRC Press; 2012. p. 319–42.

Francis MP, Roberts J, MacGibbon DJ. Indicator based analysis of the status of eight shark and chimaera species in New Zealand waters. New Zealand Fisheries Assessment Report 2016/65. Wellington: Ministry for Primary Industry; 2016.

Frazzetta TH. The mechanics of cutting and the form of shark teeth (Chondrichthyes, Elasmobranchii). Zoomorphology. 1988;108(2):93–107.

Freer DWL. The commercial fishery for sharks in the South-Western Cape, with an analysis of the biology of the two principal target species, Callorhinchus Capensis Dumeril and Galeorhinus Galeus Linn (Master's thesis, University of Cape Town); 1992.

Fricke R. Fishes of the Mascarene Islands (Réunion, Mauritius, Rodriguez): an annotated checklist, with descriptions of new species. Theses Zoologicae, 31. Koenigstein: Koeltz Scientific Books; 1999. 759 p.

Fricke R, Bilecenoglu M, Sari HM. Annotated checklist of fish and lamprey species (Gnathostoma and Petromyzontomorphi) of Turkey, including a red list of threatened and declining species. Stuttgarter Beitr Naturk Sea A. 2007;706:1–172.

Fricke R, Kulbicki M, Wantiez L. Checklist of the fishes of New Caledonia, and their distribution in the Southwest Pacific Ocean (Pisces). Stuttgarter Beiträge zur Naturkunde A Neue Serie. 2011;4:341–463.

Friel JP, Wainwright PC. Evolution of motor patterns in Teteraodontiform fishes: does muscle duplication lead to functional diversification? Brain Behav Evol. 1998;52:159–70.

Frimodt C. Multilingual illustrated guide to the world's commercial warmwater fish. Oxford: Fishing News Books; 1995. 215 p.

Froese R, Pauly D, editors. FishBase. World Wide Web Electronic Publication; 2016. www.fishbase.org, version (01/2022).

Froese R, Pauly D, editors. FishBase. World Wide Web Electronic Publication; 2021. www.fishbase.org, version (01/2022).

Froese R, Pauly D, editors. FishBase. World Wide Web Electronic Publication; 2022. www.fishbase.org, version (01/2022).

FSANZ. The 20th Australian Total Diet Study. Supplementary Material: Part 1A. Canberra; 2002

Gallo-Reynoso J-P, Figueroa-Carranza A-L. A cookiecutter shark wound on a Guadalupe fur seal male. Mar Mamm Sci. 1992;8:428–30.

Garrick JA. Revision of sharks of genus *Isurus* with description of a new species (Galeoidea, Lamnidae). Proc US Natl Mus. 1967;118(3537):663–90.

Gasparini JL, Sazima I. A stranded melon-headed whale, Peponocephala electra, in southeastern Brazil, with comments on wounds from the cookiecutter shark, Isistius brasiliensis. Mar Mamm Sci. 1996;12:308–12.

Gauld JA. Records of porbeagles landed in Scotland, with observations on the biology, distribution and exploitation of the species. Scott Fish Res Rep. 1989;45. Scotland (Department of Agriculture and Fisheries for Scotland):15.

George JC, Philo LM, Hazard K, Withrow D, Carroll GM, Suydam R. Frequency of killer whale (Orcinus orca) attacks and ship collisions based on scarring on bowhead whales (Balaena mysticetus) of the Bering Chukchi-Beaufort Seas stock. Arctic. 1994;47:247–55.

Gilbert JM, Reichelt-Brushett AJ, Butcher PA, McGrath SP, Peddemors VM, Bowling AC, Christidis L. Metal and metalloid concentrations in the tissues of dusky Carcharhinus obscurus, sandbar C. Plumbeus and white Carcharodon carcharias sharks from south-eastern Australian waters, and the implications for human consumption. Mar Pollut Bull. 2015;92(1–2):186–94.

Gliwicz ZM. A lunar cycle in zooplankton. Ecology. 1986;67:882–97.

Goff BL. Symbols of prehistoric Mesopotamia. New Haven: Yale University Press; 1963.

Goldman KJ. Thresher shark Alopias vulpinus. In: Fowler SL, Cavanagh RD, Camhi M, Burgess GH, Cailliet GM, Fordham SV, Simpfendorfer CA, Musick JA, editors. Sharks, rays and chimaeras: the status of the Chondrichthyan fishes. Status survey. Gland/Cambridge: IUCN SSC Shark Specialist Group/IUCN; 2005. p. 250–2.

Gomon MF, Glover CJM, Kuiter RH editors. The fishes of Australia's south coast. Adelaide: State Print; 1994. 992 p.

Gomon MF, Bray DJ, Kuiter RH. Fishes of Australia's Southern Coast. Reed New Holland, Museum Victoria, Chatswood; 2008.

Gonzalez-Pestana A, Kouri JC, Velez-Zuazo X. Shark fisheries in the Southeast Pacific: a 61- year analysis from Peru. F1000Research. 2014;3(164):1.

Gore MA, Rowat D, Hall J, Gell FR, Ormond RF. Transatlantic migration and deep midocean diving by basking shark. Biol Lett. 2008;4(4):395–8.

Goreau TJ. Undercurrent Diver's Blog; 2015. http://www.undercurrent.org/blog/2011/01/04/barracudaattack/

Gray J. Biomagnification in marine systems: the perspective of an ecologist. Mar Pollut Bull. 2002;45:46–52.

Green A, Black J. Gods, demons and symbols. In: Ancient Mesopotamia: an Illustrated Dictionary, *London*; 1992 190p.

Gregory WK. Fish skulls: a study of the evolution of natural mechanisms. Trans Am Philos Soc. 1933;23:75–468.

Griffin LT. Descriptions of New Zealand Fishes. Trans N Z Inst. 1926;56:538–46. Pls. 93–98.

Grubich JR, Rice AN, Westneat MW. Functional morphology of bite mechanics in the great barracuda (Sphyraena barracuda). Zoology. 2008;111(1):16–29.

Guardian News and Media Limited. New Zealand: woman dies after suspected shark attack near Auckland. 2021. https://www.theguardian.com/world/2021/jan/08/new-zealand-woman-dies-after-suspected-shark-attack-near-auckland

Gudger EW. Sphyræna Barracuda; its morpholog Y. Papers from the Department of Marine Biology of the Carnegie Institution of Washington, 1918. pp. 211–214, 53.

Gudger EW, Smith BG. The natural history of the frilled shark Chlamydoselachus anguineus. In: Gudger EW, editor. Bashford dean memorial volume on archaic fishes. New York: American Museum of Natural History; 1933. p. 245–319.

Günther A. VIII.—A contribution to the fauna of the river Tigris. J Nat Hist. 1874;14(79):36–8.

Haddad JRV. Atlas de animais aquáticos perigosos do Brasil: guia médico de identificação e tratamento (Atlas of dangerous aquatic animals of Brazil: a medical guide to identification and treatment). São Paulo: Editora Roca; 2000. p. 60p.

Haddad JRV. Infecções cutâneas e acidentes por animais traumatizantes e venenosos ocorridos em aquários comerciais e domésticos no Brasil: descrição de 18 casos e revisão sobre o tema. An Bras Dermatol. 2004;79:157–67.

Hajimoradi M, et al. The effect of shark liver oil on the tumour infiltrating lymphocytes and cytokine pattern in mice. J Ethnopharmacol. 2009;126(3):565–70.

Halstead BW. Marine biotoxicology. In: Environmental quality and safety. Academic Press; 1974. p. 212–39.

Hamady LL, et al. Vertebral bomb radiocarbon suggests extreme longevity in white sharks. PloS One. 2014;9(1):e84006.

Hardiman N, Burgin S, Shao J. How sharks and shark–human interactions are reported in major Australian newspapers. Sustain For. 2020;12(7):2683.

Hart JL. Pacific fishes of Canada. Bull. Fish. Res. Board Can. 1973;180:740.

Harvey-Clark CJ, Stobo WT, Helle E, Mattson M. Putative mating behavior in basking sharks off the Nova Scotia coast. Copeia. 1999;1999(3):780–2.

Hatooka K, Randall JE. A new moray eel (*Gymnothorax*: Muraenidae) from Japan and Hawaii. Jap J Ichthyol. 1992;39(3):183–90.

Haynes D, Johnson JE. Organochlorine, heavy metal and polyaromatic hydrocarbon pollutant concentrations in the great barrier reef (Australia) environment: a review. Mar Pollut Bull. 2000;41:267–78.

Hazin F, Burgess GH, Carvalho F. Shark attack outbreak of Recife, Pernambuco, Brazil: 1992–2006. Bull Mar Sci. 2008;89:199–212.

Heemstra PC. Additions and corrections for the 1995 impression. In: Smith MM, Heemstra PC, editors. Revised edition of Smiths' sea fishes. Berlin: Springer; 1995. p. v–xv.

Heessen HJL. Development of Elasmobranch Assessments DELASS DG Fish Study Contract 99/055. Final Report. January 2003.

Heithaus MR. The biology of tiger sharks, Galeocerdo cuvier, in Shark Bay, Western Australia: sex ratio, size distribution, diet, and seasonal changes in catch rates. Environ Biol Fish. 2001;61(1):25–36.

Heithaus M, Frid A, Dill L. Shark-inflicted injury frequencies, escape ability, and habitat use of green and loggerhead turtles. Mar Biol. 2002;140(2):229–36.

HELCOM. HELCOM red list of threatened and declining species of lampreys and fishes of the Baltic Sea. Balt Sea Environ Proc. 2007;109:40.

Henderson AC, Reeve AJ. Noteworthy elasmobranch records from Oman. Afr J Mar Sci. 2011;33(1):171–5.

Herrel A, O'Reilly JC. Ontogenetic scaling of bite force in lizards and turtles. Physiol Biochem Zool. 2006;79:31–42.

Hoese DF, Bray DJ, Paxton JR, Allen GR. Fishes. ABRS & CSIRO Publishing/IUCN; 2006. 2019

Hoffmayer ER, Franks JS, Driggers WB III, Grace MA. Movements and habitat preferences of dusky (Carcharhinus obscurus) and silky (Carcharhinus falciformis) sharks in the Northern Gulf of Mexico: Preliminary Results. 2009 MTI Bird and Fish Tracking Conference Proceedings; 26 Mar 2009.

Holmes BJ, Pepperell JG, Griffiths SP, Jaine FR, Tibbetts IR, Bennett MB. Tiger shark (Galeocerdo cuvier) movement patterns and habitat use determined by satellite tagging in eastern Australian waters. Mar Biol. 2014;161:2645–58.

Holmes BJ, Peddemors CM, Gutteridge AN, Geraghty PT, Chan RWK, Tibbetts IR, Bennett MB. Age and growth of the tiger shark Galeocerdo cuvier off the east coast of Australia. J Fish Biol. 2015;87:422–48.

Holts DB, Bedford DW. Horizontal and vertical movements of the shortfin mako shark, Isurus oxyrinchus, in the Southern California Bight. Mar Freshw Res. 1993;44(6):901–9.

Honebrink R, Buch R, Galpin P, Burgess GH. First documented attack on a live human by a cookiecutter shark (Squaliformes, Dalatiidae: Isistius sp.). Pac Sci. 2011;65(3):365–74.

Howard RJ, Pessa ME, Brennaman BH, Ramphal R. Necrotizing soft-tissue infections caused by marine vibrios. Surgery. 1985;98:126–30.

Hoyle JA, Keast A. The effect of prey morphology and size on handling time in a piscivore, the largemouth bass (Micropterus salmoides). Can J Zool. 1987;65:1972–7.

Huber DR, Eason TG, Hueter RE, Motta PJ. Analysis of the bite force and mechanical design of the feeding mechanism of the durophagous horn shark Heterodontus francisci. J Exp Biol. 2005;208(18):3553–71.

Huber DR, Weggelaar CL, Motta PJ. Scaling of bite force in the blacktip shark Carcharhinus limbatus. Zoology. 2006;109:109–19.

Hussain NA, Rasen AK, Al BY, Coad BW. Bull shark occurrence Carcharhinus leucas (Valenciennes, 1839) at the inland waters of southern Iraq. J Univ Duhok. 2012;15(1):140–3.

Hussey NE, Olin JA, Kinney MJ, McMeans BC, Fisk AT. Lipid extraction effects on stable isotope values (δ13C and δ15N) of elasmobranch muscle tissue. J Exp Mar Biol Ecol. 2012;434:7–15.

Huveneers C, Simpfendorfer C. Heterodontus portusjacksoni. The IUCN Red List of Threatened Species 2015: e.T39334A68625721. 2015. https://doi.org/10.2305/IUCN.UK.2015-4.RLTS.T39334A68625721.en.

ISAF Statistics on Attacking Species of Shark. International Shark Attack File, Florida Museum of Natural History, University of Florida. Retrieved on May 18, 2009.

ISAF Statistics on Attacking Species of Shark. International Shark Attack File, Florida Museum of Natural History, University of Florida. Retrieved on May 18, 2012.

ISAF Statistics on Attacking Species of Shark. International Shark Attack File, Florida Museum of Natural History, University of Florida. Retrieved on May 18, 2018.

ISAF Statistics on Attacking Species of Shark. International Shark Attack File, Florida Museum of Natural History, University of Florida. Retrieved on May 18, 2021.

Islam MS, Khan S, Tanaka M. Waste loading in shrimp and fish processing effluents: potential source of hazards to the coastal and nearshore environments. Mar Pollut Bull. 2004;49(1–2):103–10.

Iwamoto T. Lepidopus caudatus. The IUCN Red List of Threatened Species 2015: e.T198721A42691759. 2015. https://doi.org/10.2305/IUCN.UK.2015-4.RLTS.T198721A42691759.en

Izzo C, Rodda KR. Comparative rates of growth of the Port Jackson shark throughout its southern Australian range. Mar Freshw Res. 2012;63:687–94.

Jabado RW, Al-Ghais SM, Hamza W, Henderson AC, Ahmad MA. First record of the sand tiger shark, *Carcharias taurus*, from United Arab Emirates. Mar Biodivers Rec. 2013;2013(6):e27.

Jabado RW, Al Ghais SM, Hamza W, Henderson AC, Spaet JLY, Shivji MS, Hanner RH. The trade in sharks and their products in The United Arab Emirates. Biol Conserv. 2015;181:190–8.

Jalili SH. Quality of gelatin extracted from skin of dominant ray fishes of the Persian Gulf. Iran Sci Fish J. 2004;13:55–68.

Jawad LA. Dangerous fishes of the eastern and southern Arabian Peninsula. Springer; 2017. 322pp.

Jiménez Prado P, Béarez P. Peces Marinos del Ecuador continental. Tomo 2: Guía de Especies/Marine fishes of continental Ecuador. Volume 2: Species Guide. SIMBIOE/NAZCA/IFEA; 2004.

Jones EC. Isistius-brasiliensis, a squaloid shark, probable cause of crater wounds on fishes and cetaceans. Fish Bull Nat Oceanic Atmos Adm. 1971;69(4):791.

Jorgensen SJ, Reeb CA, Chapple TK, Anderson SD, Perle C, Van Sommeran SR, Fritz-Cope C, Brown AC, Klimley AP, Block BA. Philopatry and migration of Pacific white sharks. Proc R Soc B Biol Sci. 2010;277:679–88. https://doi.org/10.1098/rspb.2009.1155.

Joung SJ, Chen NF, Hsu HH, Liu KM. Estimates of life history parameters of the oceanic whitetip shark, Carcharhinus longimanus, in the Western North Pacific Ocean. Mar Biol Res. 2016;12(7):758–68.

Kabasakal H, De Madalena A. A huge shortfin mako shark *Isurus oxyrinchus* Rafinesque, 1810 (Chondrichthyes: Lamnidae) from the waters of Marmaris, Turkey. Ann Ser Hist Nat. 2011;21:21–4.

Kennedy WP. Some additions to the fauna of Iraq. J Bombay Nat Hist Soc. 1937;39:745–9.

Khan N. An investigation of the bacterial profile recovered from the oral cavity of sharks, on the coast of KwaZulu-Natal, South Africa. Doctoral dissertation; 2016.

Kislalioglu M, Gibson RN. Prey handling time and its importance in food selection by 15-spined stickleback, *Spinachia spinachia* (L.). J Exp Mar Biol Ecol. 1976a;25:115–58.

Kislalioglu M, Gibson RN. Some factors governing prey selection by 15-spined stickleback, *Spinachia spinachia* (L.). J Exp Mar Biol Ecol. 1976b;25:159–69.

Klimley AP, Beavers SC, Curtis TH, Jorgensen SJ. Move ments and swimming behavior of three species of sharks in La Jolla Canyon, California. Environ Biol Fish. 2002;63:117–35.

Kneebone J, Natanson LJ, Andrews AH, Howell WH. Using bomb radiocarbon analyses to validate age and growth estimates for the tiger shark, Galeocerdo cuvier, in the western North Atlantic. Mar Biol. 2008;60:1–87.

Koehl MAR. Physical modelling in biomechanics. Philos Trans R Soc B. 2003;358:1589–96.

Kojadinovic J, Potier M, Le Corre M, Cosson RP, Bustamante P. Mercury content in commercial pelagic fish and its risk assessment in the Western Indian Ocean. Sci Total Environ. 2006;366(2–3):688–700.

KPBS. Sharks attracting attention in San Diego Waters. kpbs.org. June 4, 2013. Retrieved November 26, 2021.

Kuiter RH, Tonozuka T. Pictorial guide to Indonesian reef fishes. Part 3. Jawfishes – Sunfishes, Opistognathidae – Molidae. Zoonetics, Australia; 2001. pp. 623–893.

Kunzlik PA. The basking shark. Dept. Agric. Fish. Scotland, Scottish Fisheries Information Pamphlet (14); 1988. 21 p.

Kyne PM. Isistius brasiliensis. The IUCN Red List of Threatened Species 2018: e.T41830A2956761; 2018. https://doi.org/10.2305/IUCN.UK.2018-2.RLTS.T41830A2956761.en

Kyne PM, Barreto R, Carlson J, Fernando D, Fordham S, Francis MP, Herman K, Jabado RW, Liu KM, Pacoureau N, Romanov E, Sherley RB. Carcharhinus galapagensis. The IUCN Red List of Threatened Species 2019: e.T41736A2954286;2019. https://doi.org/10.2305/IUCN.UK.2019-3.RLTS.T41736A2954286.en

Lamberth SJ. White shark and other chondrichthyan interactions with the beach-seine (treknet) fishery in False Bay, South Africa. Afr J Mar Sci. 2006;28(3–4):723–7.

Lamberth SJ, Bennett BA, Clark BM. Catch composition of the commercial beach-seine fishery in False Bay, South Africa. S Afr J Mar Sci. 1994;14:69–78.

Lanelli Y, Schimelpfenig S, Shekher M, McBride D, Schoen R, Wilderness C. Beware the vicious triggerfish!; 2007.

Langer R. Isolation of bioactive compounds from sharks. Rep. Mass. Inst. Technol. Sea Grant Program; 1989. 34 p.

Last PR, Stevens JD. Sharks and rays of Australia. Hobart: CSIRO; 1994. 513 p.

Last PR, Stevens JD. Sharks and rays of Australia. 2nd ed. Melbourne: CSIRO; 2009.

LaVan F, Hunt T. Oxygen and wound healing. Clin Plast Surg. 1990;17:463–72.

Lavenberg RJ, Grove JS, Seigel JA. Conservation alert: status of fisheries for Galapagos sharks, with a checklist of known species. Natural History Museum of Los Angeles County; 1994.

Lea JSE, Wetherbee BM, Queiroz N, Burnie N, Aming C, Sousa LL, Mucientes GR, Humphries NE, Harvey GM, Sims SW, Shivji MS. Repeated, long-distance migrations by a philopatric predator targeting highly contrasting ecosystems. Sci Rep. 2015;5:11202.

Lee AK, van Beuzekom M, Glowacki J, Langer R. Inhibitors, enzymes and growth factors from shark cartilages. Comp Biochem Physiol. 1984;788(3):609–16.

Lentz AK, Burgess GH, Perrin K, Brown JA, Mozingo DW, Lottenberg L. Mortality and management of 96 shark attacks and development of a shark bite severity-scoring system. Am Surg. 2010;76(1):101–6.

Lessa R, Santana FM, Paglerani R. Age, growth and stock structure of the oceanic whitetip shark, (*Carcharhinus longimanus*), from the southwestern equatorial Atlantic. Mar Freshw Res. 1999;50:383–8.

Lieske E, Myers R. Collins Pocket Guide. Coral reef fishes. Indo-Pacific & Caribbean including the Red Sea: Haper Collins; 1994. p. 400.

Limbaugh C. Field notes on sharks. In sharks and survival (P. W. Gilbert, ed.), Lubbock, R. (1980). The shore fishes of Ascension Island. J Fish Biol. 1963;17:283–303.

Liu KM, Tsai W. Catch and life history parameters of pelagic sharks in the Northwestern Pacific. ISC/11/SHARKWG-1/06;2011.

Long DJ. Apparent predation by a white shark *Carcharodon carcharias* on a pygmy sperm whale *Kogia breviceps*. Fish Bull. 1991;89:538–40.

Lopez S, Meléndez R, Barría P. Preliminary diet analysis of the blue shark Prionace glauca in the eastern South Pacific. Rev Biol Mar Oceanogr. 2010;45(S1):745–9.

Lowe CG. Ontogenetic dietary shifts and feeding behavior of the tiger shark, Galeocerdo cuvier, in Hawaiian waters. Environ Biol Fishes. 1996;47:203–11.

Lowry M, William D, Metti Y. Lunar landings: relationship between lunar phase and catch rates for an Australian gamefish-tournament fishery. Fish Res. 2007;88:15–23.

Lowry D, de Castro ALF, Mara K, Whitenack LB, Delius B, Burgess GH, Motta P. Determining shark size from forensic analysis of bite damage. Mar Biol. 2009;156(12):2483–92.

Lyle J. Mercury concentrations in four Carcharhinid and three hammerhead sharks from coastal waters of the Northern Territory. Aust J Mar Freshw Res. 1984;35:441–51.

Mallefet J, Stevens DW, Duchatelet L. Bioluminescence of the largest luminous vertebrate, the kitefin shark, Dalatias licha: first insights and comparative aspects. Front Mar Sci. 2021;8:153.

Mamtani R, Stern P, Dawood I, Cheema S. Metals and disease: a global primary health care perspective. J Toxicol. 2011;2011.1:319136.

Manilo LG, Bogorodsky SV. Taxonomic composition, diversity and distribution of coastal fishes of the Arabian Sea. J Ichthyol. 2003;43(1):S75.

Marinebio. Tiger shark, Galeocerdo cuvier at marinebio.org; 2021. Accessed 20 Nov 2021.

Martin RA. Elasmo research. Reef Quest; 2006. Retrieved 6 Feb.

Martins AS, Haimovici M. Distribution, abundance and biological interactions of the cutlassfish *Trichiurus lepturus* in the southern Brazil subtropical convergence ecosystem. Fish Res. 1997;30:217–27.

Martins R, Coasta FO, Murta AG, Carniero M, Landi M. First record of *Zenion hololepis* (Zenionidae) in Portuguese continental waters: the northernmost occurrence in the eastern Atlantic. Mar Biodiv Rec. 2012;5:1–3.

Martínez-Ortiz J, Aires-da-Silva AM, Lennert-Cody CE, Maunder MN. The Ecuadorian artisanal fishery for large pelagics: species composition and spatio-temporal dynamics. PLoS One. 2015;10(8):e0135136.

Masuda H, Amaoka K, Araga C, Uyeno T, Yoshino T. The fishes of the Japanese Archipelago. Vol. 1. Tokyo: Tokai University Press; 1984. 437 p. (text).

Matsuura K. Balistidae. Triggerfishes. In: Carpenter KE, Niem V, editors. FAO species identification guide for fishery purposes. The living marine resources of the Western Central Pacific. Vol. 6. Bony fishes part 4 (Labridae to Latimeriidae), estuarine crocodiles. Rome: FAO; 2001. pp. 3911–3928

McAuley R, Simpfendorfer C, Hall N. A method for evaluating the impacts of fishing mortality and stochastic influences on the demography of two longlived shark stocks. ICES J Mar Sci. 2007;64:1710–22.

McCandless CT, Conn P, Cooper P, Cortés E, Laporte SW, Nammack M. Status review report: northwest Atlantic dusky shark (Carcharhinus obscurus). Report to National Marine Fisheries Service, Office of Protected Resources; 2014.

McCosker J, Béarez P. Gymnothorax porphyreus. The IUCN Red List of Threatened Species 2010: e.T183859A8189838; 2010

McCosker JE, Stewart AL. Additions to the New Zealand marine eel fauna with the description of a new moray, *Anarchias supremus* (Teleostei: Muraenidae), and comments on the identity of *Gymnothorax griffini* Whitley & Phillips. J R Soc N Z. 2006;36(2):83–95.

Mclaughln RH, O'Gower AK. Life history and underwater studies of a heterodont shark. Ecol Monogr. 1971;41(4):271–89.

McMillan PJ, Francis MP, James GD, Paul LJ, Marriott PJ, Mackay E, Wood BA, Griggs LH, Sui H, Wei F. New Zealand fishes. Volume 1: A field guide to common species caught by bottom and midwater fishing. New Zealand Aquatic Environment and Biodiversity Report No. 68; 2011. 329 p.

Mecklenburg CW, Lynghammar A, Johannesen E, Byrkjedal I, Christiansen JS, Dolgov AV, Karamushko OV, Mecklenburg TA, Møller PR, Steinke D, Wienerroither PR. Marine fishes of the Arctic Region. Volume I. Conservation of Arctic Flora and Fauna, Akureyri, Iceland. CAFF Monitoring Series Report 28:1–454; 2018

Mehdi KL. Ribbonfishes. Tehran: Persian Gulf News; 2008. 34 p.

Mehta RS, Wainwright PC. Raptorial pharyngeal jaws help moray eels swallow large prey. Nature. 2007a;449:79–82.

Mehta RS, Wainwright PC. Biting releases constraints on moray eel feeding kinematics. J Exp Biol. 2007b;210:495–505.

Menni RC, Lucifora LO. Condrictios de la Argentina y Uruguay. ProBiota, FCNyM, UNLP, Serie Te'cnicaDida'ctica, La Plata, Argentina 11:1–15; 2007.

Michael SW. Reef sharks and rays of the world. A guide to their identification, behavior, and ecology. Monterey: Sea Challengers; 1993. 107 p.

Midway SR, Wagner T, Burgess GH. Trends in global shark attacks. PLoS One. 2019;14(2):e0211049.

Mollet HF, Cailliet GM. Comparative population demography of elasmobranchs using life history tables, Leslie matrices and stage-based matrix models. Mar Freshw Res. 2002;53(8):503–16.

Momigliano P, Harcourt R. Shark conservation, governance and management: the science-law disconnect. In: Techera EJ, Klein N, editors. Sharks: conservation, governance and management. New York: Routledge; 2014.

Moore AB, White WT, Ward RD, Naylor GJ, Peirce R. Rediscovery and redescription of the smooth tooth blacktip shark, Carcharhinus leiodon (Carcharhinidae), from Kuwait, with notes on its possible conservation status. Mar Freshw Res. 2011;62(6):528–39.

Moore ABM, Ward RD, Peirce R. Sharks of the Persian (Arabian) Gulf: a first annotated checklist (Chondrichthyes: Elasmobranchii). Zootaxa. 2012;3167:1–16.

Moss SA. Feeding mechanism in sharks. Am Zool. 1977;17:355–64.

Moss SA. Shark feeding mechanisms. Oceanus 1981;24(4):23–29.

Motta PJ, Wilga CD. Advances in the study of feeding behaviours, mechanisms, and mechanics of sharks. Environ Biol Fish. 2001;60:131–56.

Moura T, Fernandes A, Figueiredo I, Alpoim R, Azevedo M. Management of deep-water sharks' bycatch in the Portuguese anglerfish fishery: from EU regulations to practice. Mar Policy. 2018;90:55–67.

Mundy BC. Checklist of the fishes of the Hawaiian archipelago. Bishop Mus Bull Zool. 2005;6:1–704.

Murua H, et al. Updated Ecological Risk Assessment (ERA) for shark species caught in fisheries managed by the Indian Ocean Tuna Commission (IOTC). Mahé, Seychelles: Indian Ocean Tuna Commission; 2018.

Myers RF. Micronesian reef fishes: a comprehensive guide to the coral reef fishes of Micronesia, 3rd revised and expanded edition. Barrigada: Coral Graphics; 1999. 330 p.

Myers RA, Baum JK, Shepherd TD, Powers SP, Peterson CH. Cascading effects of the loss of apex predatory sharks from a coastal ocean. Science. 2007;315(5820):1846–50.

Mytilineou C, Politou C-Y, Papaconstantinou C, Kavadas S, D'Onghia G, Sion L. Deep-water fish fauna in the eastern Ionian Sea. Belg J Zool. 2005;135(2):229–33.

Nakamura I. Trichiuridae. Peces sables, cintillas. In: Fischer W, Krupp F, Schneider W, Sommer C, Carpenter KE, Niem V, editors. Guia FAO para Identification de Especies para lo Fines de la Pesca. Pacifico Centro-Oriental, vol. 3 Vols. Rome: FAO; 1995. p. 1638–42.

Nakamura I, Parin NV. FAO Species Catalogue. Vol. 15. Snake mackerels and cutlassfishes of the world (families Gempylidae and Trichiuridae). An annotated and illustrated catalogue of the snake mackerels, snoeks, escolars, gemfishes, sackfishes, domine, oilfish, cutlassfishes. scabbardfishes, hairtails, and frostfishes known to date. FAO Fish Synop 1993;125(15):136 p.

Nakano H, Stevens JD. The biology and ecology of blue shark, *Prionace glauca*. In: Canhi MD, Pikitch EK, Babcock EA, editors. Sharks of the open ocean; biology, fisheries and conservation. Blackwell; 2008. p. 140–51.

Narberhaus I, Krause J, Bernitt U, editors. Threatened biodiversity in the German north and Baltic seas, Naturschutz und Biologische Vielfalt, Heft, vol. 117. Bonn: Federal Agency for Nature Conservation; 2012.

Natanson LJ, Skomal GB. Age and growth of the white shark, *Carcharodon carcharias*, in the western North Atlantic Ocean. Mar Freshw Res. 2015;66(5):387–98.

Natanson LJ, Casey JG, Kohler NE, Colket T. Growth of the tiger shark, Galeocerdo cuvier, in the western North Atlantic based on tag returns and length frequencies; and a note on the effects of tagging. Fish Bull. 1999;97:944–53.

Natanson LJ, Kohler NE, Ardizzone D, Cailliet GM, Wintner SP, Mollet HF. Validated age and growth estimates for the shortfin mako, Isurus oxyrinchus, in the North Atlantic Ocean. Environ Biol Fish. 2006;77:367–83.

Natanson LJ, Hamady LL, Gervelis BJ. Analysis of bomb radiocarbon data for common thresher sharks, *Alopias vulpinus*, in the northwestern Atlantic Ocean with revised growth curves. Environ Biol Fish. 2016;99(1):39–47.

Naylor E. Marine animal behaviour in relation to lunar phase. Earth Moon Planet. 1999;85–86:291–302.

Neff C, Hueter R. Science, policy, and the public discourse of shark "attack": a proposal for reclassifying human–shark interactions. J Environ Stud Sci. 2013;3(1):65–73.

Nelson GJ. Gill arches of teleostean fishes of the order Anguilliformes. Pac Sci. 1966;20:391–408.

Nelson GJ. Branchial muscles in representatives of five eel families. Pac Sci. 1967;21:348–63.

New Zealand Herald. New Zealand surfer airlifted to hospital after shark attack. 2018. https://www.independent.co.uk/news/world/australasia/shark-attack-new-zealand-surfer-baylys-beach-bitten-survives-a8591681.html

NIWA. Summer Series 6: Broadnose Sevengill Shark. National Institute of Water and Atmospheric Research. January 31, 2012. Retrieved June 22, 2015.

Norton SF, Brainerd EL. Convergence in the feeding mechanics of ecomorphologically similar species in the Centrarchidae and Cichlidae. J Exp Biol. 1993;176:11–29.

O'Gower AK. Speculations on a spatial memory for the Port Jackson shark (Heterodontus portusjacksoni) (Meyer) (Heterodontidae). Mar Freshw Res. 1995;46:861–71.

O'Gower AK, Nash AR. Dispersion of the Port Jackson shark in Australian waters. Arlington: Office of Naval Research. Department of Navy; 1978.

Papastamatiou YP, Wetherbee BM, O'Sullivan J, Goodmanlowe GD, Lowe CG. Foraging ecology of Cookiecutter sharks (Isistius brasiliensis) on pelagic fishes in Hawaii, inferred from prey bite wounds. Environ Biol Fish. 2000;88:361–8.

Pardini AT, Jones CS, Noble LR, Kreiser B, Malcolm H, Bruce BD, Stevens JD, Cliff G, Scholl MC, Francis M, Duffy C, Martin AP. Sex-biased dispersal of great white sharks. Nature. 2001;412:139–40.

Parin NV. Trichiuridae. In: Whitehead PJP, Bauchot M-L, Hureau J-C, Nielsen J, Tortonese E, editors. Fishes of the North-Eastern Atlantic and the Mediterranean, vol. 2. Paris: UNESCO; 1986. p. 976–80.

Parin NV. Trichiuridae. In: Quéro JC, Hureau JC, Karrer C, Post A, Saldanha L, editors. Check-list of the fishes of the eastern tropical Atlantic (CLOFETA). Vol. 2. Lisbon/Paris: JNICT/SEI/UNESCO; 1990. pp. 974–977.

Parin NV. Three new species and new records of cutlass fishes of the genus *Aphanopus* (Trichiuridae). J Ichthyol. 1995;35(2):128–38.

Parin NV, Gorelova TA, Borodulina OD. Feeding and trophic relationships of fishes inhabiting the Nazca and Sala y Gomez submarine ridges. Trudy Inst Okenol. 1990;125:37–57.

Parker HW, Stott FC. Age, size and vertebral calcification in the basking shark, *Cetorhinus maximus* (Gunnerus). Zoologische Mededelingen. 1965;40(34):305–19.

Paulin C, Roberts C. The rockpool fishes of New Zealand. Museum of New Zealand Te Papa Tongarewa; 1992. p. 177p.

Paulin C, Stewart A, Roberts C, McMillan P. New Zealand fish: a complete guide. National Museum of New Zealand Miscellaneous Series No. 19;1989. 279 p.

Pauly D. A critique of some literature data on the growth, reproduction and mortality of the lamnid shark *Cetorhinus maximus* (Gunnerus). ICES CM 1978/H: 17. Copenhagen: International Council for the Exploration of the Sea; 1978.10 p.

Pauly D. Growth and mortality of the basking shark *Cetorhinus maximus* and their implications for management of whale sharks *Rhincodon typus*. IUCN SSC SHARK SPECIALIST GROUP. In: Elasmobranch biodiversity, conservation and management: Proceedings of the International Seminar and Workshop, Sabah, Malaysia, 1997. Gland/Cambridge: IUCN: 2002. pp. 199–208.

Paxton JR, Hoese DF, Allen GR, Hanley JE. Pisces. Petromyzontidae to Carangidae. Zoological Catalogue of Australia, Vol. 7. Canberra: Australian Government Publishing; 1989. 665 p.

Pequeño G. Peces de Chile. Lista sistematica revisada y comentada. Rev Biol Mar Valparaiso. 1989;24(2):1–132.

Peres MB, Vooren CM. Sexual development, reproductive cycle, and fecundity of the school shark Galeorhinus galeus off southern Brazil. Fish Bull. 1991;89:655–67.

Perry B. Cookie-cutter sharks 'sort of a mosquito of the sea'. The Maui News. 2009. Retrieved on January 26, 2010.

Pethybridge H, Cossa D, Butler ECV. Mercury in 16 demersal sharks from Southeast Australia: biotic and abiotic sources of variation and consumer health implications. Mar Environ Res. 2010;69:18–26.

Petzold T, Feindt PR, Carl UM, Gams E. Hyperbaric oxygen therapy in deep sternal wound infection after heart transplantation. Chest. 1999;115:1455–8.

Phillips C, Brady WH. Sea pests- poisonous and harmful sea life of Florida and West Indies. Coral Gables: University of Miami Press; 1953.

Pike GC. Lamprey marks on whales. J Fish Res Board Can. 1951;8:275–80.

Porter HT, Motta PJ. A comparison of strike and prey capture kinematics of three species of piscivorous fishes: Florida gar (Lepisosteus platyrhincus), redfin needlefish (Strongylura notata), and great barracuda (Sphyraena barracuda). Mar Biol. 2004;145:989–1000.

Powter D. Conservation biology of the port Jackson shark, Heterodontus portusjacksoni, in New South Wales. University of Newcastle; 2007.

Powter DM, Gladstone W. Embryonic mortality and predation on egg capsules of the Port Jackson shark *Heterodontus portusjacksoni* (Meyer). J Fish Biol. 2008;72:573–84.

Pratt HL, Casey JG. Age and growth of the shortfin mako, Isurus oxyrinchus. In: Prince ED, Pulos LM, editors. Proceedings of the international workshop on age determination of oceanic pelagic fishes: Tunas, billfishes, and sharks NOAA Tech. Rep. NMFS 8; 1983. pp. 175–177.

Pratt HL, Carrier JC. A review of elasmobranch reproductive behavior with a case study on the nurse shark, Ginglymostoma cirratum. Environ Biol Fish. 2001;60:157–88.

Pratt HL Jr. Reproduction in the blue shark, Prionace glauca. Fish Bull. 1979;77:445–69.

Pratt HL Jr. Reproduction in the male white shark (*Carcharodon carcharias*). In: Klimley AP, Ainley DG, editors. Great white sharks: ecology and behaviour. Orlando: Academic Press; 1996.

Quigley TG. Shar attack in Irish water? Sherkin comment, Issue No 44; 2007. 18.

Rabbani A, Abdosamadi S, Sari-Saraf N. Affinity of anticancer drug daunomycin to core histones in solution: comparison of free and cross-liked proteins. Acta Pharmacol Sin. 2007;28:731–7.

Rabbani-Chadegani A, Abdossamadi S, Bargahi A, Yousef-Masboogh M. Identification of low-molecular- weight protein (SCP1) from shark cartilage with anti-angiogenesis activity and sequence similarity to parvalbumin. J Pharm Biomed Anal. 2008;46(3):563–7.

Ramos RO. Age and growth estimates for the Port Jackson shark, *Heterodontus portusjacksoni* (Mayer, 1793) from New South Wales. University of Newcastle; 2007.

Ramos SGAC. Diet and trophic position of deep-sea sharks in the southwest coast of Portugal: using stable isotopes analysis and nucleic acids ratios (RNA/DNA). Doctoral dissertation; 2018.

Rancurel P, Intes A. Le requin tigre, Galeocerdo cuvier, Lacepède, des eaux Néo-calédoniennes: examen des contenus stomacaux. Tethys. 1982;10:195–9.

Randall JE. Food habits of reef fishes of the West Indies. Stud Trop Ocean. 1967;5:665–847.

Randall JE. Review of the biology of the tiger shark (*Galeocerdo cuvier*). Aust J Mar Freshwat Res. 1992;43:21–31.

Randall JE. Coastal fishes of Oman. Honolulu: University of Hawaii Press; 1995.

Randall JE, Golani D. Review of the moray eels (Anguilliformes: Muraenidae) of the Red Sea. Bull Mar Sci. 1995;56(3):849–80.

Randall JE, Levy MF. A near-fatal shark attack by a mako in the northern Red Sea. Isr J Ecol Evol. 1976;25(1–2):61–70.

Randall JE, Millington JT. Triggerfish bite–a little-known marine hazard. J Wilderness Med. 1990;1(2):79–85.

Randall JE, van Egmond J. Marine fishes from the Seychelles: 108 new records. Zool Verh Leiden. 1994;297:43–83.

Randall IE, Smith CL, Feinberg MN. Report on fish collections from Rapa, French Polynesia. Am Mus Nov. 1990;296:1–42.

Ratkowsky DA, Dix TG, Wilson KC. Mercury in fish in the Derwent estuary, Tasmania, and its relation to the position of the fish in the food chain. Aust J Mar Freshw Res. 1975;26:223–31.

Razmi NA, Kalalian Moghadam H, Tabandeh MR, Babakhani AAF. The therapeutic effect of ingestion of shark cartilage (Carcharhinus dussumieri) on the growth of dimethylbenzanthracene induced mammary tumour in Sprague Dawley rats. Iran South Med J. 2008;10:104–11.

Reid OS. Sharks and New Zealand News Media. M.Sc. Thesis submitted to the University of Otago, New Zealand; 2019. 107p.

Reid OS. Sharks and New Zealand News Media (Doctoral dissertation, University of Otago). 2020. 107p.

Reid DD, Robbins WD, Peddemors VM. Decadal trends in shark catches and effort from the New South Wales, Australia, shark meshing program 1950-2010. Mar Freshw Res. 2011;62:676–93.

Reide K. Global register of migratory species—from global to regional scales. Final Report of the R&D Projekt 808 05 081. Federal Agency for Nature Conservation, Bonn; 2004. 329 p.

Ribéreau-Gayon A, Rando C, Schuliar Y, Chapenoire S, Crema ER, Claes J, Seret B, Maleret V, Morgan RM. Extensive unusual lesions on a large number of immersed human victims found to be from cookiecutter sharks (Isistius spp.): an examination of the Yemenia plane crash. Int J Legal Med. 2017, March 1;131(2):423–32.

Ribot-Carballal MC, Galvan Magaña F, Velazquez Q. Age and growth of the shortfin mako shark Isurus oxyrinchus from the western coast of Baja California Sur, Mexico. Fish Res. 2005;76:14–21.

Ricci JA, Vargas CR, Singhal D, Lee BT. Shark attack-related injuries: epidemiology and implications for plastic surgeons. J Plast Reconstr Aesthet Surg. 2016;69(1):108–14.

Rigby CL, Barreto R, Carlson J, Fernando D, Fordham S, Francis MP, Jabado RW, Liu KM, Marshall A,

Pacoureau N, Romanov E, Sherley RB, Winker H. Isurus oxyrinchus. The IUCN Red List of Threatened Species 2019: e.T39341A2903170. 2019a. https://doi.org/10.2305/IUCN.UK.2019-1.RLTS.T39341A2903170.en

Rigby CL, Barreto R, Carlson J, Fernando D, Fordham S, Francis MP, Herman K, Jabado RW, Liu KM, Marshall A, Romanov E, Kyne PM. Cetorhinus maximus (errata version published in 2020). The IUCN Red List of Threatened Species 2019: e.T4292A166822294. 2019b. https://doi.org/10.2305/IUCN.UK.2019-3.RLTS.T4292A166822294.en. Downloaded on 04 June 2021.

Rigby CL, Barreto R, Carlson J, Fernando D, Fordham S, Francis MP, Herman K, Jabado RW, Liu KM, Lowe CG, Marshall A, Pacoureau N, Romanov E, Sherley RB, Winker H. Carcharodon carcharias. The IUCN Red List of Threatened Species 2019: e.T3855A2878674. 2019c. https://doi.org/10.2305/IUCN.UK.2019-3.RLTS.T3855A2878674.en. Downloaded on 06 June 2021.

Riordan C, Hussain M, McCann J. Moray eel attack in the tropics: a case report and review of the literature. Wilderness Environ Med. 2004;15(3):194–7.

Ritter E, Levine M. Use of forensic analysis to better understand shark attack behaviour. J Forensic Odontostomatol. 2004;22(2):40–6.

Ritter EK, Levine M. Bite motivation of sharks reflected by the wound structure on humans. Am J Forensic Med Pathol. 2005;26:136–40.

RNZ. Broadnose Sevengill shark thought behind Curio Bay attack. 2017. https://www.Rnz.co.nz/national/programmes/morningreport/audio/201841814/broadnose-sevengill-shark-thought-behind-curio-bay-attack

Roberts CD, Stewart AL, Struthers CD editors. The Fishes of New Zealand. Te Papa Press, Wellington. 4 volumes, S250 + 1748 pp. [Vol. 1: S1–S250; vol. 2: 1–574; vol. 3: 581–1152; vol. 4: 1157–1748]; 2015.

Robins CR, Ray GC. A field guide to Atlantic coast fishes of North America. Boston: Houghton Mifflin; 1986. 354 p.

Rodda K. Development in the Port Jackson shark embryo. PhD. Thesis presented to the University of Adelaide, Australia; 2000. 238p.

Rodda KR, Seymour RS. Functional morphology of embryonic development in the Port Jackson shark Heterodontus portusjacksoni (Meyer). J Fish Biol. 2008;72:961–84.

Roman B. Porbeagle (On-line). 2010. http://www.flmh.ufl.edu/fish/Gallery/Descript/Porbeagle/Porbeagle.html.Accessed 5 Nov 2021.

Romero P. An etymological dictionary of taxonomy. Madrid, unpublished. 2002.

Romine J. Status, ecology and demographic analysis of the dusky shark, Carcharhinus obscurus, and the sandbar shark, Carcharhinus plumbeus, in the northwest Atlantic. M.Sc. thesis. College of William & Mary;2004.

Rosson C, Tolle S. Management of marine stings and envenomations. West J Med. 1989;150:97–100.

Ruiz-Campos G, Castro-Aguirre JL, Balart EF, Campos-Dávila L, Vélez-Marín R. New specimens and records of chondrichthyan fishes (vertebrata: Chondrichthyes) off the Mexican Pacific coast. Revista Mexicana de Biodiversidad. 2010;81:363–71.

Russell BC. The food and feeding habits of rocky reef fish of northeastern New Zealand. NZ J Mar Freshwat Res. 1983;17(2):121–45.

ISAF Statistics on Attacking Species of Shark. International Shark Attack File, Florida Museum of Natural History, University of Florida. Retrieved on May 18, 2016.

Sahrhage D. Fischfang und Fischkult im alten Mesopotamien. Frankfurt am Main: Peter Lang; 1999. 354pp.

Salem AB. Fishes of the sea. Baghdad: Report to the Ministry of Agriculture; 2010. 120 p.

Sato K, Nakamura M, Tomita T, Toda M, Miyamoto K, Nozu R. How great White sharks nourish their embryos to a large size: evidence of lipid histotrophy in lamnoid shark reproduction. Biol Open. 2016;5:1211–5.

Schmidt TW. Food habits, length-weight relationship and condition factor of young great barracuda, Sphyraena barracuda (Walbaum), from Florida Bay, Everglades National Park, Florida. Bull Mar Sci. 1989;44:163–70.

Schultz K. Fishing encyclopaedia: worldwide angling guide. Foster City: IDG Books Worldwide; 2000. 1916 p.

Scott WB, Scott MG. Atlantic fishes of Canada. Can Bull Fish Aquat Sci 1988; 219:731 p.

Seki T, Taniuchi T, Nakano H, Shimizu M. Age, growth, and reproduction of the oceanic whitetip shark from the Pacific Ocean. Fish Sci Tokyo. 1998;64:14–20.

Senou, H. Sphyraenidae. Barracudas. In: Carpenter KE, Niem V, editors. FAO species identification guide for fishery purposes. The living marine resources of the Western Central Pacific. Vol. 6. Bony fishes part 4 (Labridae to Latimeriidae), estuarine crocodiles. Rome: FAO, 2001. pp. 3685–3697

Sepulveda CA, Kohin S, Chan C, Vetter R, Graham JB. Movement patterns, depth preferences, and stomach temperatures of free-swimming juvenile mako sharks, Isurus oxyrinchus, in the Southern California Bight. Mar Biol. 2004;145(1):191–9.

Shahrokhi N, Hadad MK, Keshavarzi Z, Shabani M. Effects of aqueous extract of water cress on glucose and lipid plasma in streptozotocin induced diabetic rats. Pak J Physiol. 2009;5(2):6–10.

Shark Facts. Carcharhinus longimanus; 2021. http://www.arkive.org/oceanic-whitetip-shark/carcharhinus-longimanus/

Shimada K. The relationship between the tooth size and total body length in the white shark, Carcharodon carcharias (Lamniformes: Lamnidae). J Fossil Res. 2002;35:28–33.

Shimada K. The relationship between the tooth size and total body length in the sandtiger shark, Carcharias taurus (Lamniformes: Odontaspidae). J Fossil Res. 2004;37:76–81.

Shimada K, Seigel JA. The relationship between the tooth size and total body length in the goblin shark,

Mitsukurina owstoni (Lamnidae: Mitsukurinidae). J Fossil Res. 2005;38:49–56.

Shivji MS, Chapman DD, Pikitch EK, Raymond PW. Genetic profiling reveals illegal international trade in fins of the great white shark, Carcharodon carcharias. Conserv Genet. 2005;66(6):1035–9.

Simpfendorfer C. Biology of tiger sharks (*Galeocerdo cuvier*) caught by the Queensland shark meshing program off Townsville, Australia. Aust J Mar Freshwat Res. 1992;43:3–43.

Simpfendorfer CA, McAuley RB, Chidlow J, Unsworth P. Validated age and growth of the dusky shark, *Carcharhinus obscurus*, from Western Australian waters. Mar Freshw Res. 2002;53:567–73.

Simpfendorfer CA, de Jong SK, Sumpton W. Long-term trends in large shark populations from inshore areas of the Great Barrier Reef World Heritage Area: results from the Queensland Shark Control Program. Marine and Tropical Sciences Research Facility Transition Program Report; 2010.

Simpfendorfer CA, Heupel MR, White WT, Dulvy NK. The importance of research and public opinion to conservation management of sharks and rays: a synthesis. Mar Freshw Res. 2011;62:518–27.

Sims DW, Quayle VA. Selective foraging behaviour of basking sharks on zooplankton in a small-scale front. Nature. 1998;393:460–4.

Sims DW, Southall EJ, Richardson AJ, Reid PC, Metcalfe JD. Seasonal movements and behaviour of basking sharks from archival tagging: no evidence of winter hibernation. Mar Ecol Prog Ser. 2003;248:187–96.

Sinha RK. Morphology and anatomy of the olfactory organs of a marine fish Sphyraena jello (C.). Folia Morphol (Praha). 1987;35:202–5.

Skomal GB, Zeeman SI, Chisholm JH, Summers EL, Walsh HJ, McMahon KW, Thorrold SR. Transequatorial migrations by basking sharks in the western Atlantic Ocean. Curr Biol. 2009;19(12):1019–22.

Smale MJ, Goosen AJJ. Reproduction and feeding of spotted gully shark, Triakis megalopterus, off the Eastern Cape, South Africa. Fish Bull. 1999;97:987–98.

Smart JJ, Paul LJ, Fowler SL. Chlamydoselachus anguineus. The IUCN Red List of Threatened Species 2016: e.T41794A68617785. 2016. https://doi.org/10.2305/IUCN.UK.2016-1.RLTS.T41794A68617785.en. Downloaded on 25 November 2021.

Smith CL. National Audubon Society field guide to tropical marine fishes of the Caribbean, the Gulf of Mexico, Florida, the Bahamas, and Bermuda. New York: Alfred A. Knopf; 1997. 720 p.

Smith PJ, Benson PG. Biochemical identification of shark fins and fillets from the coastal fisheries in New Zealand. Fish Bull. 2001;99:351–5.

Smith, Heemstra PC, editors. Smiths' sea fishes. Berlin: Springer; 1986.

Smith JLB, Smith MM. he fishes of Seychelles. Grahamstown: Rhodes University; 1963. 215 p.

Smith SW, Au DW, Show C. Intrinsic rebound potential of 26 species of Pacific sharks. Mar Freshwat Res. 1998;49:663–78.

Smith DG, Brokovich E, Einbinder S. Gymnothorax baranesi, a new moray eel (Anguilliformes: Muraenidae) from the Red Sea. Zootaxa. 2008;1678:63–8.

Smith DG, McCosker J, Tighe K. Gymnothorax eurostus. The IUCN Red List of Threatened Species 2019: e.T195723A2408174; 2019a.

Smith DG, Tighe K, McCosker J. Anarchias supremus. The IUCN Red List of Threatened Species 2019: e.T195675A2400689; 2019b.

Smith DG, Tighe K, McCosker J. Enchelycore ramosa. The IUCN Red List of Threatened Species 2019: e.T195697A2404477; 2019c.

Smith DG, Tighe K, McCosker J. Gymnothorax obesus. The IUCN Red List of Threatened pecies 2019: e.T195770A2415507; 2019d.

Smith DG, McCosker J, Tighe K. Gymnothorax pictus. The IUCN Red List of Threatened Species 2019: e.T195775A2416162; 2019e.

Smith DG, McCosker J, Tighe K. Gymnothorax prionodon. The IUCN Red List of Threatened Species 2019: e.T195782A2417174; 2019f.

Smith DG, Tighe K, McCosker J. Gymnothorax ypsilon. The IUCN Red List of Threatened Species 2019: e.T195807A2420691; 2019g.

Smith DG, McCosker J, Tighe K. Gymnothorax thyrsoideus. The IUCN Red List of Threatened Species 2019: e.T195801A2419428. 2019h.

Snodgrass RE, Heller E. Papers from the Hopkins-Standford Galapagos Expedition, 1898–1899. XVII. Shore fishes of the Revillagigedo, Clipperton, Cocos, and Galapagos Islands. Proc. Wash. Acad. Sci. 1905;6:333–427.

Snyder JO. A catalogue of the shore fishes collected by the steamer "Albatross" about the Hawaiian Islands in 1902. Bull U. S. Fish Comm. 1904;22:513–38. Pls. 1–13. [For a biography see Brittan & Jennings 2010:24].

Southcott RV. Notes on stings of some venomous Australian fishes. Med J Aust. 1970;2(16):722–5.

Southcott RV. Australian venomous and poisonous fishes. Clin Toxicol. 1977;10(3):291–325.

Springer S. Squalidae. In: Quero JC, Hureau JC, Karrer C, Post A, Saldanha L, editors. Check-list of the fishes of the eastern tropical Atlantic (CLOFETA), vol. 1. Lisbon/Paris: JNICT/SEI/UNESCO; 1990. p. 7–19.

Stevens JD. Observations on reproduction in the shortfin mako Isurus oxyrinchus. Copeia. 1983;1983(1):126–30.

Stevens JD. Biological observations on sharks caught by sports fishermen off New South Wales. Aust J Mar Freshwat Res. 1984;35:573–90.

Stevens JD, Brown BA. Occurrence of heavy metals in the blue shark *Prionace glauca* and selected pelagic fish in the NE Atlantic Ocean. Mar Biol. 1974;26:287–93.

Stevens JD, West GJ. Investigation of school and gummy shark nursery areas in Southeastern Australia. FRDC Project 93/061 July 1997. Hobart: CSIRO Marine Research; 1997.

Stillwell CE, Kohler NE. Food, feeding habits, and daily ration of the shortfin mako (Isurus oxyrinchus) in the Northwest Atlantic. Can J Fish Aquat Sci. 1982;39:407–14.

Strauss MB, Orris WL. Injuries to divers by marine animals: a simplified approach to recognition and management. Mil Med. 1974;1974(139):129–30.

Summers AP, Ketcham R, Rowe T. Structure and function of the horn shark (*Heterodontus francisci*) cranium through ontogeny: the development of a hard prey crusher. J Morphol. 2004;260:1–12.

Swan SC, Gordon JDM, Shimmield T. Preliminary investigations on the uses of otolith microchemistry for stock discrimination of the deep-water Black Scabbardfish (*Aphanopus carbo*) I the NorthEast Atlantic. J Northwest Atl Fish Sci 2003;31:221–231.

Tanaka S. Age, growth and genetic status of the white shark (*Carcharodon carcharias*) from Kashima-nada, Japan. Mar Freshw Res. 2011;62:548–56.

Tanaka S, Shiobara Y, Hioki S, Abe H, Nishi G, Yano K, Suzuki K. The reproductive biology of the frilled shark, *Chlamydoselachus anguineus*, from Suruga Bay, Japan. Jpn J Ichthyol. 1990;37(3):273–91.

Tang W-C. Chinese medicinal materials from the sea. Abstr Chin Med. 1987;1(4):571–600.

Taylor VB, Holts DB. Shortfin Mako shark. In: Leet WS, Dewees CM, Klingbeil R, Larson EJ, editors. California's living marine resources: a status report. The Resources Agency, California Department of Fish and Game; 2001. p. 336–7.

Tester AL. Cooperative shark research and control program, University of Hawaii, Honolulu. Final Report; 1969.

Thorburn J, Neat F, Burrett I, Henry L-A, Bailey DM, Jones CS, Noble LR. Ontogenetic variation in movements and depth use, and evidence of partial migration in a Benthopelagic elasmobranch. Front Ecol Evol. 2019;7:353.

Thorson TB, Langhammer JK, Oetinger MI. Reproduction and development of the south American freshwater stingrays, *Potamotrygon circularis* and P. motoro. Environ Biol Fish. 1983;9(1):3–24.

Tirard P, Maillaud C, Borsa P. Fatal tiger shark, *Galeocerdo cuvier* attack in New Caledonia erroneously ascribed to great white shark, *Carcharodon carcharias*. J Forensic Leg Med. 2015;3:68–70.

Tomberg RJ, Cachaper GA, Weingart GS. Shark related injuries: a case series of emergency department patients. Am J Emerg Med. 2018 Sep;36(9):1645–9.

Tovar-Avila J. Reproduction, age validation, growth determination and effects of fishing on the port Jackson shark (Heterodontus portusjacksoni) in south-eastern Australia. University of Melbourne; 2006.

Tovar-Avila J, Troynikov VS, Walker TI, Day RW. Use of stochastic models to estimate the growth of the Port Jackson shark, Heterodontus portusjacksoni, off eastern Victoria, Australia. Fish Res. 2009;95:230–5.

Tracking shark. Rare Cookiecutter Shark Bite in Hawaii. 2019a. https://www.Trackingsharks.com/rare-cookiecutter-shark-bite-in-hawaii/

Tracking shark. Second cookiecutter shark bite reported in Hawaii. 2019b. https://www.Trackingsharks.com/rare-cookiecutter-shark-bite-in-hawaii/

Tracking shark. In rare third incident for year, swimmer attacked by cookiecutter shark. 2019c. https://www.Trackingsharks.com/rare-cookiecutter-shark-bite-in-hawaii/

Tricas TC, McCosker JE. Predatory behaviour of the white shark (Carcharodon carcharias), with notes on its biology (PDF). Proc Calif Acad Sci. 1984;43(14):221–38.

Trnski T, Duffy CA, Francis MP, McGrouther MA, Stewart AL, Struthers CD, Zintzen V. Recent collections of fishes at the Kermadec Islands and new records for the region. Bull Auckl Mus. 2015;20:463–80.

Turner A. Shark attacks diver in Fiordland. stuff.co.nz. 2013. Retrieved 26 November 2021.

Turoczy NJ, Laurenson LJB, Allinson G, Nishikawa M, Lambert DF, Smith C, Cottier JPE, Irvine SB, Stagnitti F. Observations on metal concentrations in three species of shark (Deania calcea, Centroscymnus crepidater and Centroscymnus owstoni) from southeastern Australian waters. J Agric Food Chem. 2000;48:4357–64.

Uchida S, Toda M, Teshima K, Yano K. Preliminary report on near-term foetuses from a white shark caught in Kochi, Japan. In: Klimley AP, Ainley DG, editors. Great white sharks: ecology and behaviour. Orlando: Academic Press; 1996.

Unger NR, Ritter E, Borrego R, Goodman J, Osiyemi OO. Antibiotic susceptibilities of bacteria isolated within the oral flora of Florida blacktip sharks: guidance for empiric antibiotic therapy. PLoS One. 2014;9(8):e104577.

US EPA. Mercury: fish consumption advice. United States Environmental Protection Agency; 2014.

Uyeno T, Matsuura K, Fujii E, editors. Fishes trawled off Suriname and French Guiana. Tokyo: Japan Marine Fishery Resource Research Center; 1983. 519 p.

van Dam JW, Negri AP, Uthicke S, Mueller JF. Chemical pollution on coral reefs: exposure and ecological effects. In: Sanchez-Bayo F, van den Brink P, Mann RM, editors. Ecological impacts of toxic chemicals. Oak Park: Bentham Science; 2011. p. 187–211.

van der Elst R. A guide to the common sea fishes of southern Africa. 3rd ed. Cape Town: Struik; 1993. 398 p.

van der Meij MAA, Bout RG. Scaling of jaw muscle size and maximal bite force in finches. J Exp Biol. 2004;207:2745–53.

Vannuccini, S.. shark utilization, marketing, and trade. Food and Agriculture Organization; 1999. p. 282. ISBN 9251043612.

Verderame, L. The moon and the power of time reckoning in ancient Mesopotamia. In: The Construction of Time in Antiquity; 2017. pp. 124–41.

Vidthayanon C. Elasmobranch diversity and status in Thailand. In: Fowler SL, Reed TM, Dipper FA, editors. Elasmobranch biodiversity, conservation and management: Proceedings of the International Seminar and Workshop, Sabah, Malaysia, July 1997. IUCN SSC Shark Specialist Group. Gland/Cambridge:IUCN; 2002. pp. 104–13.

Wade, A. Shark attacks: Savagery in the ocean haunts victims for life. 2013, March 6. https://www.nzherald.co.nz/nz/news/article.cfm?c_id=1&objectid=10869472

Wainwright PC, Richard BA. Predicting patterns of prey use from morphology of fishes. Environ Biol Fish. 1995;44:97–113.
Walker TI. Effects of species, sex, length and locality on the mercury content of school shark, Galeorhinus australis (Macleay), and gummy shark, *Mustelus antarcticus* (Guenther), from south-eastern Australian waters. Mar Freshw Res. 1976;27:603–16.
Walker TI. Mercury concentrations in edible tissues of elasmobranchs, teleosts, crustaceans and molluscs from southeastern Australian waters. Aust J Mar Freshw Res. 1988;39:39–49.
Walker TI. Can shark resources be harvested sustainably? A question revisited with a review of shark fisheries. Mar Freshw Res. 1998;49:553–72.
Walker TI. Chapter 24. Galeorhinus galeus fisheries of the world. In: Shotton R, editor. Case studies of management of elasmobranch fisheries, FAO fisheries technical paper 378/2. Rome: Food and Agriculture Organization of the United Nations; 1999. p. 728–73.
Walker TI. Chapter 4: Reproduction in fisheries science. In: Hamlett WC, editor. Reproductive biology and phylogeny of Chondrichthyes: sharks, batoids, and chimaeras. Enfield: Science; 2005. p. 81–127.
Walker P, Cavanagh RD, Ducrocq M, Fowler SL. Northeast Atlantic In: Sharks, rays and chimaeras: the status of the chondrichthyan fishes. 2005.
Walker TI, Cavanagh RD, Stevens JD, Carlisle AB, Chiaramonte GE, Domingo A, Ebert DA, Mancusi CM, Massa A, McCord M, Morey G, Paul LJ, Serena F, Vooren CM. Galeorhinus galeus. The IUCN Red List of Threatened Species 2006: e.T39352A10212764. 2006. https://doi.org/10.2305/IUCN.UK.2006.RLTS.T39352A10212764.en. Accessed 2 Nov 2021.
Walker TI, Taylor BL, Brown LP, Punt AE. Chapter 32. Embracing movement and stock structure for assessment of Galeorhinus galeus harvested off southern Australia. In: Camhi MD, Pikitch EK, Babcock EA, editors. Sharks of the open ocean: biology, fisheries and conservation. Oxford: Blackwell; 2008. p. 369–92.
Walker TI, Rigby CL, Pacoureau N, Ellis J, Kulka DW, Chiaramonte GE, Herman K. Galeorhinus galeus. The IUCN Red List of Threatened Species 2020: e.T39352A2907336. 2020. https://doi.org/10.2305/IUCN.UK.2020-2.RLTS.T39352A2907336.en. Downloaded on 05 November 2021.
Walters V. The "problematic" hydrodynamic performance of Gero's great barracuda. Nature. 1966;212:215–6.
WDFW. Bottomfish – Broadnose sevengill shark. wdfw.wa.gov. 2021. Archived from the original on June 23, 2015. Retrieved November 26, 2021.
Weigmann S. Annotated checklist of the living sharks, batoids and chimaeras (Chondrichthyes) of the world, with a focus on biogeographical diversity. J Fish Biol. 2016;88(3):837–1037.
Weitkamp DE, Sullivan RD. Gas bubble disease in resident fish of the lower Clark Fork River. Trans Am Fish Soc. 2003;132(5):865–76.
Werner EE. Fish size, prey size, handling time relation in several sunfishes and some implications. J Fish Res Board Can. 1974;31:1531–6.
Werner EE. Species packing and niche complementarity in three sunfishes. Am Nat. 1977;111:553–78.
Werry JM, Planes S, Berumen ML, Lee KA, Braun CD, Clua E. Reef-fidelity and migration of tiger sharks, Galeocerdo Cuvier, across the Coral Sea. PLoS One. 2014;9:e83249.
West JG. Changing patterns of shark attacks in Australian waters. Mar Freshw Res. 2011;62(6):744–54.
Westneat MW. A biomechanical model for analysis of muscle force, power output and lower jaw motion in fishes. J Theor Biol. 2003;223:269–81.
Westneat MW. Evolution of levers and linkages in the feeding mechanisms of fishes. Int Comp Biol. 2004;44:378–89.
Wet Web Media. Gymnothorax flavimarginatus. 2016. http://www.wetwebmedia.com/
Wetherbee BM, Crow GL, Lowe CG. Distribution, reproduction and diet of the gray reef shark, Carcharhinus amblyrhynchos in Hawaii. Mar Ecol Prog Ser. 1997;151:181–9.
White WT, Last PR, Stevens JD, Yearsley GK, Fahmi and Dharmadi. Economically important sharks and rays of Indonesia. Canberra: ACIAR; 2006.
Whitenack LB, Motta PJ. Performance of shark teeth during puncture and draw: implications for the mechanics of cutting. Biol J Linn Soc. 2010;100(2):271–86.
Whitley GP. Studies in ichthyology. No. 6. Rec. Aust. Mus. 1932;18(6):321–48. Pls. 36–39.
Whitney NM, Crow GL. Reproductive biology of the tiger shark (Galeocerdo cuvier) in Hawaii. Mar Biol. 2007;151:63–70.
Widder EA. A predatory use of counter illumination by the squaloid shark, Isistius brasiliensis. Environ Biol Fish. 1998, November;53(3):267–73.
Wikipedia. 2015. http://en.wikipedia.org/wiki/Barracuda?action¼history
Wikipedia. 2016. http://en.wikipedia.org/wiki/Barracuda?action¼history
Wikipedia 2022. Cookie cutter shark. https://en.wikipedia.org/wiki/Cookiecutter_shark
Windom H, Stickney R, Smith R, White D, Taylor F. Arsenic, cadmium, copper, mercury and zinc in some species of North Atlantic finfish. J. Fish. Res Board Can. 1973;30:275–9.
Wintner SP, Dudley SFJ. Age and growth estimated for the tiger shark, Galeocerdo cuvier, from the east coast of South Africa. Mar Freshw Res. 2000;51:43–53.
Wojciechowski J. Observations on the biology of cutlassfish *Trichiurus lepturus* L. (Trichiuridae) of Mauritania shelf. Acta Ichthyol Piscat. 1972;II:67–75.
Woolgar JD, Cliff G, Nair R, Hafez H, Robbs JV. Shark attack: review of 86 consecutive cases. J Trauma. 2001 May;50(5):887–91.
Worm B, Davis B, Kettemer L, Ward-Paige CA, Chapman D, Heithaus MR, Kessel ST, Gruber SH. Global catches, exploitation rates, and rebuilding options for sharks. Mar Policy. 2013;40:194–204.

WPRFMC, Western Pacific Regional Fishery Management Council. Federal Fishery Managers Vote To Prohibit Shark Feeding (PDF). Western Pacific Regional Fishery Management Council. Archived (PDF) from the original on 2013-06-08. 2013. Retrieved 20 November 2021.

Wright BS. Releasers of attack behavior pattern in shark and barracuda. J Wildl Manag. 1948;12:117–23.

Wright AJ. Lunar cycles and sperm whales (Physeter macrocephalus) stranding on the North Atlantic coastlines of the British Isles and eastern Canada. Mar Mamm Sci. 2005;2:145–9.

Wroe S, Huber DR, Lowry M, McHenry C, Moreno K, Clausen P, Summers AP. Three-dimensional computer analysis of white shark jaw mechanics: how hard can a great white bite? J Zool. 2008;276(4):336–42.

Yamada U, Shirai S, Irie T, Tokimura M, Deng S, Zheng Y, Li C, Kim YU, Kim YS. Names and illustrations of fishes from the East China Sea and the Yellow Sea. Tokyo: Overseas Fishery Cooperation Foundation; 1995. 288 p.

Yasuda F. The feeding mechanisms in some carnivorous fish. Rec Oceanogr Work Jpn. 1960;5:153–60.

Young RF, Winn HE. Activity patterns, diet, and shelter site use for two species of moray eels. Gymnothorax moringa and Gymnothorax vi*cinus*, in Belize. Copeia. 2003;2003:44–55.

Young CN, Carlson J, Hutchinson M, Kobayashi D, McCandless C, Miller MH, Teo S, Warren T. Status review report: common thresher shark (*Alopias vulpinus*) and bigeye thresher shark (Alopias superciliosus). Final Report to National Marine Fisheries Service, Office of Protected Resources; 2016.

Young CN, Carlson J, Hutchinson M, Hutt C, Kobayashi D, McCandless CT, Wraith J. Status review report: oceanic whitetip shark (Carcharhinius longimanus). Final Report to the National Marine Fisheries Service, Office of Protected Resources; 2017.

Yukihira H, Shibuno T, Hashimoto H, Gushima K. Feeding habits of moray eels (Pisces: Muraenidae) at Kuchierabujima. J Fac Appl Biol Sci. 1994;33:159–66.

3 Harmful Fish Group

3.1 Species of Needlefish

Family: Belonidae.
Ablennes hians (Valenciennes 1846) (Fig. 3.1).
Common name: Flat needlefish.
Etymology: *Ablennes:* Greek, a = without + Greek, blennos = mucus (Romero 2002).

Diagnostic Features
The body is elongated and greatly compressed laterally. The upper and lower jaws are greatly elongated and equipped with small sharp teeth. Gill rakers are absent. Anterior parts of dorsal and anal fins are with high falcate lobes. The posterior part of the dorsal fin has a prominent dark lobe. Pectoral fins are falcate in shape. The caudal peduncle is without lateral keels. The caudal fin is deeply forked, with the lower lobe much longer than the upper. The right gonad is absent in females, while it is absent or greatly reduced in males. The body has a bluish green coloration from above and silvery white from below. The sides of the body have broad dark-blue stripes in addition to 12–14 prominent dark. The tip of lower jaw is red. Scales and bones are green in color.

Habitat, Biology, and Ecology
This is a marine species and found in brackish water and reef areas (Riede 2004). It lives at a depth range of 0–3 m (Feitoza et al. 2003). Individuals of this species are usually found in neritic and oceanic waters but more often near islands (Fischer et al. 1990). They are also reported to live in estuaries (Claro 1994), and sometimes, they form large schools (Cervigón et al. 1992). Small fishes are the main food item of this species (Collette 1995a, b). Females are oviparous, and eggs are usually found attached to objects in the water by filaments on the egg's surface (Breder and Rosen 1966). It reaches a maximum total length of 1200 mm.

Distribution
This species has a worldwide distribution in tropical and warm temperate waters. It is reported to come from the Eastern Atlantic at Cape Verde and Dakar to Moçamedes, Angola (Collette and Parin 1990), reaching further south to Namibia (van der Elst 1993). It is also recorded from the Western Atlantic at the Chesapeake Bay (USA), Bermuda, and the northern Gulf of Mexico to Brazil (Robins and Ray 1986). Some of this species are also recorded from the Indian Ocean and the Western Pacific, from off coasts and islands of southern Japan through the East Indies to Australia (Collette 1999), and also from Tuvalu (Chapman and Cusack 1990). In the Eastern Pacific, it is reported to have been found in Mexico to Peru and the entire Caribbean, Antilles, and South American coast to Rio de Janeiro (Smith 1997).

Economic Value
Specimens of this species are usually marketed fresh and salted, smoked, or frozen, but it is a

L. A. Jawad, *Dangerous Fishes of New Zealand*, https://doi.org/10.1007/978-3-031-69698-5_3

Fig. 3.1 *Ablennes hians* (Valenciennes 1846). (Courtesy of Hamid Osmany, Pakistan)

limited market, owing to its green-colored flesh (Cervigón et al. 1992).

Conservation Status

Collette et al. (2015) evaluated this species as Least Concern in the Red List of the IUCN for the following reasons. This species has a worldwide distribution, and it is common in its geographical distribution range. It has a lower commercial value and faces no major threat.

Human Interaction

As a belonid species, it may give severe bites to humans.

Platybelone argalus (Lesueur 1821) (Fig. 3.2).
Common name: Keeltail needlefish.
Etymology: *Platybelone:* Greek, platys = flat + Greek, belone = needle; any fish with sharp pointed snout (Romero 2002).

Diagnostic Features

The body is elongated and equipped with fine teeth. Gill rakers are present. Anterior parts of dorsal and anal fins do not form prominent lobes. The pectoral fin is not falcate in shape. The caudal peduncle is greatly depressed with very large lateral keels. The caudal fin forked, with the upper and lower lobes of about equal length. Predorsal scales (in front of dorsal fin) are moderately few and large, 108–120. Both right and left gonads are present, where the right is longer than the left. The body has a bluish-green coloration on the dorsal side and silvery below. A dark-blue stripe is present along the sides. Fins are pigmented. Scales and bones are green.

Habitat, Biology, and Ecology

Platybelone argalus is a marine species and lives in association with the reef (Gasparini and Floeter 2001). It attains a maximum total length of 500 mm (Claro 1994), with a common total length of 350 cm (Cervigón et al. 1992). Individuals of this species are found offshore, mainly with a large number around islands (Cervigón et al. 1992). Schools of individuals are usually occurring in sheltered parts of the reef (Kuiter and Tonozuka 2001). This species is carnivorous, and small fishes are its main food item. Females are oviparous, and eggs may be found attached to objects in the water by tendrils on the egg's surface (Breder and Rosen 1966).

Distribution

This species is reported from the Western Atlantic at North Carolina, USA, to Brazil (Robins and Ray 1986), including the Bahamas, the Gulf of Mexico, and the Caribbean Sea. It is also recorded from the Western Pacific Ocean, from many island groups extending into the Central and South Pacific. Other subspecies are found worldwide in tropical and warm temperate waters. It is also reported from waters around New Zealand (Roberts et al. 2015).

Economic Value

Specimens of this species are usually marketed fresh but with a limited market (Cervigón et al. 1992).

Conservation Status

Collette et al. (2010) evaluated this species as Least Concern in the Red List of the IUCN based

Fig. 3.2 *Platybelone argalus* (Lesueur 1821). (Courtesy of Rafael Estrada)

on the following reasons. The species is widely distributed and faces no major threats.

Human Interaction

As a belonid species, it may give severe bites to human.

3.1.1 Injuries Caused by Needlefishes

Needlefish live near the surface of the water and are capable to jump outside the water at speeds of 60 km h^{-1}. In most cases, they land on the deck of a shallow boat rather than in the water again. The artificial light during night evokes the needlefish to jump. Usually, night fishers are their victims and are hit by very fast fish with strong and sharp beaks (Scott 1996).

The description of the following cases from around the world gives an idea about how dangerous the species of the family Belonidae are.

The number of injuries of humans by needlefish is not many, but they affect many parts of the human body with some of them fatal. The injuries can be in the head region (i.e., in the orbit, maxilla, ear, and brain), in the neck, in the thorax (i.e., in the heart), in the abdomen, in the knee, in the feet, and in the buttocks. The attack of the needlefish can cause facial analysis and concoction.

Fatal Incidents

Among the fatal cases is the attack on the eye and brain. McCabe et al. (1978) reported on a case where a 10-year-old Hawaiian boy faced a needlefish attack while he was fishing with his father on a boat when a 36–48-cm-long fish jumped from the water, hitting the boy in the right eye.

On admission to hospital, he was found to be comatose, with flaccid paralysis involving the right side of the body. The wound caused by the hit of the fish was about 8 mm long. Later, the patient showed evidence of cerebral infarction and a small left subdural hematoma. After 10 days, the boy died because of brain damage.

Barss (1982) reported three fatal accidents due to the attack of needlefish on people fishing on a boat. In all these cases, the attack happened to the thorax. In one of these cases, the fisher caught the fish, and the fish hit him in the chest. In one of the other two cases, the fisher caught one needlefish, and another one pierced his chest. In the third case, the fisher died when a needlefish attacked him in the chest.

Deutsche Press (2007) carried news about a 16-year-old Vietnamese boy who died because of an attack by a 150-mm-long needlefish. The boy was diving for sea cucumber. The boy's diving friends assisted him in pulling the needlefish snout out of the boy's chest. After the stab to his heart, the boy died before reaching the hospital.

Nawasiwatte et al. (2014) described a case of a 41-year-old Maldivian professional diver at one of the hospitals in Sri Lanka. A needlefish hit the man while he was diving. The right ear was penetrated deep by the long jaws of the fish. The beak and any other particles of the fish were removed. At the start, he showed ear bleeding. The patient developed bilateral visual blurring and sudden-onset left face, arm, and leg weakness. The brain showed an acute infarction, and brain-stem reflexes were stopped and the brain confirmed dead after 72 h. Another fatal case happened to a young man swimming at night in the Arabian-Persian Gulf area of Saudi Arabia. The attack was to his neck, and as result, a severe cut occurred from a strong hit by a needlefish. The man died on his way to the hospital (Sabiq 2015).

Nonfatal Incidents

The incidence of a Hawaiian man reported by Barss (1982) was a stabbing case in the right eye. As a result, the medial canthal ligament was injured, there was a laceration to the lacrimal duct, and decreased vision occurred. After comprehensive examination, it seems that the beak of the fish had penetrated deep and damaged the optic nerve causing blindness in the right eye. The other incidence of attack reported by Barss (1982) happened to a man in Hawaii when he was wading in the water driving fish into a net. A big needlefish jumped and pierced his right lower eyelid. He became blind in the right eye with paralysis of the right cranial nerves once he reached the hospital. It seems like the beak passed on into the brain damaging descending motor tracts from the right cerebral cortex. Two months later, he was still severely disabled with no sign of recovery from the hemiplegia.

In the thorax-abdomen area, Barss (1982) has reported three incidences. In these cases, a 9-year-old girl was stabbed in the stomach while she was wading in the water, which resulted in perforating the stomach and the liver. The second case is about an adult fisher that received a peritoneal hemorrhage with prolonged infection. The third incidence is about an adult fisher that had an attack on the left side of his abdomen causing a perforation of the large and small intestines.

Bendet et al. (1995) have given an incidence of needlefish attack against a man, hitting him in his neck. This case caused severe injuries to the cervical area.

In the lower limbs, a needlefish attack on a knee joint of a man in New Caledonia was reported. The attack caused articular disorder that led to acute arthritis (Labbe et al. 1995).

Among the nonfatal cases, facial paralysis occurred to a man diving near the Maldives when a needlefish weighing 30 kg hit his head. Paralysis of the facial nerve resulted 4–5 h after the collision. After examination and surgical operation, part of the fish jaw was removed from the retromandibular fossa. The fish jaw had compressed the facial nerve (Zwisler and Beigel 1997).

Barton and Bond (2007) described a case where the patient was nearly killed. A jumping needlefish that hit her in the heart attacked a young snorkeler in Florida.

Moreno (2005) reported on a 19-year-old diver stabbed in the stomach by a large needlefish while he was attempting to dive in Kahana Bay, Hawaii. Severe damage to his liver occurred.

An incidence happened to a 31-year-old Indian man who was fishing when a large needlefish struck him in his left eye. The examination of the eye revealed bare light perception fixed and dilated pupil, afferent pupillary defect, absent motility, lid edema, proptosis, and vitreous hemorrhage. The patient lost vision with his left eye after 2 days (Thakker and Usha 2006).

The maxilla is another place in the head area that becomes a target for the attack of the needlefish. Ebner et al. (2009) have reported on the case of a 29-year-old man with a penetrating facial wound caused by needlefish. This incident take place in the early hours of darkness, when two amateur divers were spearfishing off the shores of Herzliya (10 km north of Tel Aviv, Israel). The strong jaws penetrated the maxilla transversely and obliquely from the left canine-fossae, through the nasal cavity, and to the right maxillary sinus, with its tip reaching the right medial-inferior orbital wall. Needlefish jaws were completely removed using a combined endoscopic and external approach.

In 2012, a German kitesurfer was attacked by needlefish in his foot while he was surfing before taking off (Kite Magazine 2012).

The case of Ohtsubo et al. (2013) was about a 27-year-old man who was injured by a fish hitting his right lower eyelid while swimming in Okinawa, a Japanese sea resort. After examination, it looks like there was a small laceration in the right lower eyelid, and a foreign body was detected in the right orbit by computed tomography (CT). Visual function tests indicated diplopia in the right vision and restriction of right eye abduction. Visual acuity of the right eye showed no abnormalities. Computed tomographic findings showed calcifications of 25 mm in the upper part of the superior rectus muscle and 7 mm on the outer side of the lateral rectus muscle in the right orbit. Two elongated bodies that were suspected to be needlefish jaws were located.

There is a high jeopardy of secondary infection after needlefish injury, which is most likely to happen as the members of the family Belonidae are carnivorous and their putrid dentition is a source of infection. In this case, intraorbital foreign bodies were removed 10 days after the injury. The bone had begun to undergo decomposition. Reduction of wound infection was attained by vigorous irrigation and debridement. *Vibrio parahaemolyticus* or *Vibrio vulnificus* are the most suspected *Vibrio* species to be present in such attack. Therefore, antibiotic therapy should be given in such cases.

Mild Nonfatal Incidents

A slight attack by needlefish can be seen in the incidences given by Barss (1982). A man was fishing at night in Hawaii when knocked unconscious into the sea. He was rescued by someone else and survived. A similar case happened to a man hit by needlefish on the side of his head and fell in the sea. The man died probably by drowning. Another slight attack happened to a man paddling a canoe at dusk when he felt a strong hit in his buttock. A deep wound resulted from this attack.

Warrell (2013) reported that needlefish can leap out of the sea at speeds of 60 km/h, attracted by light. They have impaled fishermen and, rarely, surfers and divers, sometimes causing fatal injuries (for images of the incidence, see Warrell 2013) (Fig. 3.3).

3.2 The Species of Surgeonfishes

Family: Acanthuridae.

Acanthurus dussumieri Valenciennes, 1835 (Fig. 3.4).

Common name: Eyestripe surgeonfish.

Etymology: *Acanthurus:* Greek, akantha = thorn + Greek, oura = tail (Romero 2002).

Diagnostic Features

The body is oval in shape. Large-size adults have strong convex foreheads. Lips are pronounced. Eye located well above the line passing through the mouth. Pectoral fin triangularly directed upward, with a broad base. Caudal fin is lunate in shape. The overall body coloration is yellowish brown, with irregular longitudinal blue lines on the head and body. A broad yellow band is pres-

Fig. 3.3 *Acanthurus dussumieri* Valenciennes, 1835. (Courtesy of Stan Shebs)

Fig. 3.4 *Acanthurus nigrofuscus* (Forsskål 1775). (Courtesy of Joe De Vroe)

ent across the front of the interorbital and continuing behind the eye as a large irregular spot. The posterior edge of the operculum is black. Dorsal and anal fins in adults are mainly yellow. The caudal fin is deeply blue in color.

Habitat, Biology, and Ecology

Acanthurus dussumieri is a marine species that lives in association with coral reefs at a depth range of 4–131 m (Lieske and Myers 1994) but usually at a range of 9–130 m (Baensch and Debelius 1997). It reaches a maximum total length of 540 mm (Randall 1986), but the common length is 350 mm TL (Randall 1984). Individuals of this species tend to form schools and usually live in seaward reefs at depths greater than 9 m (Myers 1999). Adults live mainly on deep coastal reef slopes and outer reef walls. They frequently found on deep shipwrecks. Juveniles found on algae-rocky reef (Kuiter and Tonozuka 2001). This species feeds on surface film of fine-green and blue-green algae, diatoms, and detritus covering sand (Myers 1999).

Distribution
Individuals of this specie distributed in the Indo-Pacific region. They reported from East Africa (including the Mascarene Islands) to the Hawaiian and Line islands, north to southern Japan, south to Rowley Shoals, southern Great Barrier Reef, and Lord Howe Island. This species are absent from most of the Central Pacific. They reportedly form waters around New Zealand (Roberts et al. 2015).

Economic Value
This species has a limited commercial value as it is used as food for fish (Randall 1984).

Conservation Status
Choat et al. (2012a) evaluated this species as Least Concern in the Red List of the IUCN based on the following reasons. It is widespread throughout the Indo-Pacific, but it is rare in some regions of its geographical distribution range. In Hawaii, this species is heavily fished. There is no sign of worldwide population decrease. It has a wide spectrum of habitat types; therefore, it provides refuge.

Human Interaction
As an acanthurid species, it may inflict a severe cut and injuries to humans and mainly divers, who are attracted to this fish by its colorful look.

Acanthurus nigrofuscus (Forsskål 1775) (Fig. 3.4).
Common name: Brown surgeonfish.
Etymology: *Acanthurus:* Greek, akantha = thorn + Greek, oura = tail (Romero 2002).

Diagnostic Features
The body is oblong in shape. Eyes are located well above the horizontal line passing through the mouth. The dorsal angle of the operculum is reaching the horizontal line passing through the ventral edge of the eye. Lips are produced. Caudal fin is lunate in shape, and pectoral fins are triangular, with a broad base. Body coloration is mainly brown to pale lavender brown. Head, nape, and chest have small orange spots. Lips are black. The edge of the caudal-spine socket is black. The caudal fin has a white posterior margin.

Habitat, Biology, and Ecology
This acanthurid species inhabits marine environment and is found living in association with coral reefs (Fishelson et al. 1987) at a depth range of 0–25 m (Kuiter and Tonozuka 2001), but it is usually found at a depth range of 2–25 m (Baensch and Debelius 1997). The maximum depth reported is at 25 m (Baensch and Debelius 1997). It reaches maturity at 100–140 mm TL, and its maximum total length is 210 mm (Sommer et al. 1996). Individuals of this species usually live on hard substrates of shallow lagoon and seaward reefs from the lower surge zone to a depth of more than 15 m (Domeier and Colin 1997). This species feeds on filamentous algae. Individuals of this species form spawning aggregations (Domeier and Colin 1997; Kuiter and Tonozuka 2001). Adults usually live in small groups, but they form large schools in some oceanic locations. Juveniles are frequently mixed with juveniles of other species (Kuiter and Tonozuka 2001).

Distribution
Individuals of this species are distributed in the Indo-Pacific region. They are reportedly found in the Red Sea south to Transkei, South Africa, (Randall 1986) and east to the Hawaiian and Tuamoto islands, north to southern Japan, south to the southern Great Barrier Reef, New Caledonia, and Rapa (Austral Islands). These species are also found in waters around New Zealand (Roberts et al. 2015).

Economic Value
This species has an economic value and can be eaten both raw and cooked (Titcomb 1972). It is caught with nets (Sommer et al. 1996).

Conservation Status
Choat et al. (2012b) categorized *A. nigrofuscus* as Least Concern in the Red List of the IUCN based on the following reasons: This species is widespread in the Indo-Pacific Region and is one of the utmost common surgeonfishes on coral reefs

(Randall 2002). It is common and dominant on reefs in most of its range. It is caught as a bycatch and used as food and is a major component of the aquarium trade in parts of its range (e.g., West Hawaii). There was no clear trend of decrease in the overall density in fish replenishment areas in West Hawaii (Walsh et al. 2010). Since this species has not much large market, harvest is not considered a chief threat worldwide. This species occurs in several marine reserves in parts of its distribution.

Human Interaction

In addition to the possibility of inflicting severe cuts and injuries to humans, mainly divers, who are attracted to this fish by its colorful look, this species is reported to cause ciguatera poisoning.

Acanthurus triostegus (Linnaeus 1758) (Fig. 3.5).

Common name: Convict surgeonfish.

Etymology: *Acanthurus:* Greek, akantha = thorn + Greek, oura = tail (Romero 2002).

Diagnostic Features

The body is oval, with a produced snout. The eye is well above the imaginary horizontal line passing through the mouth. The upper corner of the operculum reaches the imaginary horizontal line passing through the edge of the eye. Caudal fin is truncate in shape. The body is mainly olivaceous gray from the dorsal side and shady to white from the ventral side. Five narrow black bars are present on the body and one on the head and passing through the eye.

Habitat, Biology, and Ecology

Acanthurus triostegus is a marine species living in association with coral reefs at a depth of 90 m (Lieske and Myers 1994), where it usually lives at (Baensch and Debelius 1997). It reaches a maximum total length of 270 mm (Randall 1986), with a common total length of 170 mm (Krupp 1995). Adult individuals are found in lagoons and seaward reefs with hard substrate, while the young are common in tide pools (Randall 1986; Kuiter and Tonozuka 2001). Individuals of this species frequently feed near freshwater runoffs where certain algae grow on rocks that are grazed (Kuiter and Tonozuka 2001). This species tends to form schools and feeds on filamentous algae in large aggregations. It forms spawning aggregations (Domeier and Colin 1997).

Distribution

This acanthurid species is distributed in the Indo-Pacific region. It is found through the region, excluding the seas around the Arabian Peninsula. It has been reported from the Eastern Pacific at the lower Gulf of California to Panama, including the Revillagigedo, Cocos, Clipperton, and Galapagos islands. It is recorded from waters around New Zealand (Roberts et al. 2015).

Fig. 3.5 *Acanthurus triostegus* (Linnaeus 1758). (Courtesy of Brian Gratwicke)

Economic Value

This species is used as food for humans (Titcomb 1972).

Conservation Status McIlwain et al. (2012a) evaluated this species as Least Concern in the Red List of the IUCN based on the following reasons: It is distributed widely throughout the Indo-Pacific region. There is a strong geographical pattern in genetic structure throughout the Pacific, but this does not permit species recognition. It is very common in some parts of its range. It is a targeted as a food and recreational species. There are no indications of global population decrease, owing to catch. It is reported from several marine-protected areas in parts of its range.

Human Interaction

In addition to the possibility of inflicting severe cuts and injuries to humans, mainly divers, who are attracted to this fish by its colorful look, this species is reported to cause ciguatera poisoning.

Naso unicornis (Forsskål 1775) (Fig. 3.6).
Common name: Bluespine unicornfish.
Etymology: *Naso:* Latin, nasus = nose (Romero 2002).

Diagnostic Features

The body is deep and compressed, with a tapering bony horn on forehead of adults, projecting anteriorly at eye level but not extending in front of the mouth. It has a pointed, bony, dorsal profile from snout to horn straight. There are two peduncular plates with well-developed forward-curving knife-like spines. The mouth is small. The caudal peduncle is cylindrical in shape. The caudal fin is truncated. The body is yellowish to olivaceous gray dorsally, paler ventrally, with blue peduncular plates and keel spines. The lips are whitish or blue. Dorsal and anal fins are yellowish with narrow blue margins. The caudal fin is orangish basally, shading to gray, with a broad, pale-greenish posterior border. Caudal filaments are edged in blue. Males are with longer horn, larger peduncular keels, and longer caudal filaments than females of the same size.

Habitat, Biology, and Ecology

The bluespine unicornfish is a marine species living in association with coral reef at a depth range of 1–180 m (Bacchet et al. 2006). It reaches a maximum total length of 700 mm (Lieske and Myers 1994), with a common length of 500 mm TL (Randall 1984). The adults live in channels, moats, lagoon, and seaward reefs with strong surge (Kuiter and Tonozuka 2001). Individuals of this species live in small schools and sometimes in solitary (Allen and Erdmann 2012).

Distribution

This species is known to be widely distributed in the Indo-Pacific area. It is reported to be found in the Red Sea, East Africa, South Africa, Socotra, Seychelles, Comoros, Madagascar, and Mascarenes east to Hawaiian Islands, Marquesas Islands, and Tuamotu Archipelago, north to central Japan and Ogasawara Islands, south to Rottnest Island (Western Australia), Jervis Bay (A.C.T., Australia), Lord Howe Island, New Zealand, and Rapa (Eschmeyer 2022; Froese and Pauly 2022). Randall (1995) records it to be from India by Kapoor et al. (2002) and Oman (Arabian Sea coast). It is recently reported from Cox's Bazar, coast of Bangladesh (Siddiqueki et al. 2021). Recently, it has been reported from north to Pakistan coast (Osmany and Moazzam 2021).

Economic Value

This species has a commercial value with its geographical range.

Conservation Status

McIlwain et al. (2012b) evaluated this species as Least Concern in the Red List of the IUCN based on the following facts. This species has wide geographical distribution in the Indo-Pacific region. It is common and abundant in several regions along its geographical distribution. It has commercial value in several parts of its distribution.

Fig. 3.6 *Naso unicornis* (Forsskål 1775). (Courtesy of Rainer Kretzberg, The Maldives Islands)

The populations of this species have not shown a decline anywhere in its distributional range (Fig. 3.7).

Human Interaction

In addition to the possibility of inflicting severe cuts and injuries to humans, mainly divers, who are attracted to this fish by its colorful look, this species is reported to cause ciguatera poisoning.

Prionurus maculatus Ogilby, 1887 (Fig. 3.8).
Common name: Yellowspotted sawtail.
Etymology: *Prionurus:* Greek, prion, -onos = saw + Greek, oura = tail (Romero 2002).

Diagnostic Features

The body is oval in shape. The snout is pointed and straight to steeply concave. Eyes are well over the horizontal imaginary line passing through the snout. The posterior dorsal edge of the operculum is just above the dorsal edge of the eye. Its mouth slightly protrudes. Three peduncular plates are each with a low lateral keel. The caudal fin is slightly emarginate to truncate. The body is bluish gray, with numerous small yellow spots on the head and body. Peduncular plates are black, with blue keels. The caudal fin is gray, with yellow spots on the base.

Habitat, Biology, and Ecology

The yellowspotted sawtail is a marine species that lives in association with coral reefs and sometimes enters brackish water at a depth of 20 m (Fricke et al. 2011). It reaches a maximum total length of 450 mm TL (Randall et al. 1990). Individuals of this species are found along rocky shores and juveniles in estuaries and bays, and they feed on benthic algae (Lieske and Myers 1994).

Fig. 3.7 *Prionurus maculatus* Ogilby, 1887. (Courtesy of Andrew J. Green, Reef life Survey, Australia)

Distribution

This species is distributed in the Southwest Pacific region. It has also been reported in New South Wales and Lord Howe Island to Queensland, only Capricorn Group and Swains Reefs on the Great Barrier Reef, as well as in Norfolk, Kermadec, and New Zealand (Francis 1993).

Economic Value

This species has a commercial value along its geographical range.

Conservation Status

Clements et al. (2012) evaluated *P. maculatus* as a Least Concern in the Red List of the IUCN based on the following facts. It distributed in the Western and Southern Pacific in Australia and New Zealand. It has high abundances in parts of its range (Elizabeth and Middleton Reefs). It is found in protected areas, often in very shallow waters in the surge zone. It is infrequently caught in recreational fisheries.

3.2.1 Injuries Caused by Surgeonfishes

Notable external spines and plates situated laterally on the caudal peduncle of all members of the family Acanthuridae are a distinguished character for this family. There are several shapes of these structures. For example, in the members of the subfamily Nasinae, there are one or two immovable bony plates, while in the members of the subfamily Prionurinae, there are up to 10 peduncular keels (Smith and Heemstra 1986). They may be called armature as they are used in inter- and intraspecific interactions (Nelson 1976; Schober 1984, 1986). Beebe (1926) considered its spines as poisonous and to cause severe pain when they sting. In fact, acanthurid spines are not linked to a poisonous gland but are usually covered with a pigmented epithelium membrane, which may have a toxic effect (Randall 1959).

There is variation in the length of the caudal spine in acanthurid species mainly between species. *Ctenochaetus striatus* and *Acanthurus sohal*

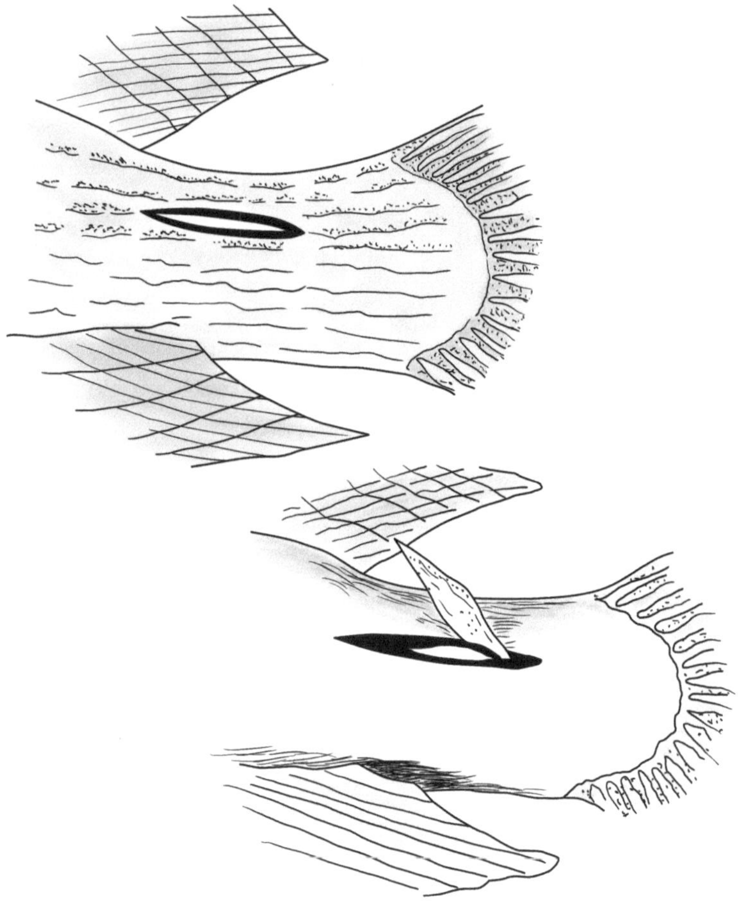

Fig. 3.8 Drawing showing; above, tail of a surgeonfish *Acanthurus* showing the caudal spine in the contracted position; below, tail of *Acanthurus* showing the caudal spine in extended position. After Halstead (1967)

spines are the longest and sharpest caudal spine among the acanthurid species (Randall 1959).

As an example of acanthurid species, the anatomy of the caudal spine apparatus, its connection with the vertebral column, and the locking mechanism are given for the species *Acanthurus leucosternon* (Schober and Ditrich 1992).

In this species, the spine is embedded with its basal plate in the dermal connective tissue and projects anteriorly (Fig. 3.8). It can be erected nearly to an acute angle as a maximum. The erection happens when the muscles connecting the spine to caudal vertebra no. 21 contract and pull the caudal part of the spine, and as a result, the spine erects. The support to the erected spine originates from the process of caudal vertebra no. 22. To reinstate the spine to its resting position, the set of muscles and myosepta present in the region work together to attain this action.

The fish becomes aggressive when an intruder enters its territory. In such behavior, the surgeonfish erects its spine and begins the attack on the intruder. The attack takes the form of beating the tail of the attacker with the body of the intruder to inflict severe injuries and make it leave the territory. A model fish representing an intruder was placed in the territory of a surgeonfish. The latter started its attack by beating its tail with the body of the model intruder fish. Such an attack caused a scratch 20 mm long and a depth of up to 2 mm (Schober and Ditrich 1992). The surgeonfish will continue its attack until the intruder leaves its territory. Such an attack could happen to any divers who might find themselves in the territory of a surgeonfish. During this aggressive attack, the caudal peduncle of the surgeonfish stiffens, and the fish moves rapidly toward the intruder.

The surgeonfish usually causes its injury when another fish or human swipes its body or part of its body against the surgeonfish. This cut is usually deep or sometimes a puncture. The pain that occurs is worse than the wound itself if the membrane that covers the spine is inflicted in the wound (Thomas and Scot 1997).

It is very easy for a surgeonfish to slash divers or snorkelers during a feeding session. It is possible to get a surgeonfish cut when you try to remove surgeonfish from a net, spear, or fish hole as anglers usually do.

To prevent the incidence of being cut by surgeonfish, the following instructions need to be taken into consideration (Thomas and Scot 1997).

1. It is important to throw food away from yourself and others while feeding the surgeonfish underwater.
2. Fishers and anglers should wear thick gloves when handling surgeonfish in nets or on spears.
3. Divers should not chase or corner surgeonfish especially with the territorial species, as they are very aggressive.

The cut caused by the surgeonfish is usually short and deep but bleeds freely. If the membrane covering the spine gets inside the wound, then a severe burning sensation will result. Later, muscle and swelling aches at the cut area will develop. A tendon or nerve damage may cause numbness and inability to move.

Below are first aid techniques that need to be followed in the case of a surgeonfish cut (Thomas and Scot 1997):

1. Gently pull the skin apart, and remove any foreign materials and debris left from the incident. This can be done by either rinsing the wound in water or using tweezers.
2. Clean the inside of the wound by scrubbing the cut directly with gauze or a clean cloth soaked in clean fresh water.
3. Press on the wound to stop the bleeding.
4. If bleeding persists or if the edges of the wound are jagged or gaping, the cut needs stitches.

If a toxin is envenomed in the wound because of the presence of membrane tissue covering the spine, then localized pain and swelling will result. The spines can break inside the wound and are visible on X-ray or by using ultrasound. Removing the parts of the broken spine will prevent infection and foreign body granulation (Thomas and Scot 1997).

Thomas and Scott (1997) reported on incidences of cut caused by acanthurid spine in the Caribbean. In addition, Haddad (2004) reported on several incidences where victims were received a severe cut as result of a contact with species of the family Acanthuridae in Brazil. Sorenson et al. (2013) and Krzyżak and Korzeniewski (2020) suggested that acanthurid species may potentially be dangerous to scuba divers because of their extremely sharp spines on each side of their tail, which can cause a severe injury to anyone who tries to touch or hold the fish in a careless manner. If it is scared and cannot retreat, the fish can attack a human using its scalpel-like spines. The fish can inflict deep lacerated wounds by moving its tail.

3.3 The Species of Squirrelfishes

Family: Holocentridae.
Myripristis berndti (Jordan and Evermann 1903) (Fig. 3.9).
Common name: Blotcheye soldierfish.
Etymology: *Myripristis:* Greek, myros, -ou = male of morey eel + Greek, pristis = saw (Romero 2002).

Diagnostic Features
Body is rhomboidal in shape, with straight dorsal and ventral profiles. The lower jaw is prominently projecting when mouth is closed. The lower half to three fourths of the inner pectoral axil is with small scales. Centers of scales are silvery pink to pale yellowish and edges red. The black part of the opercular membrane is extending below the opercular spine. The naked part of pectoral axil is black. The outer half of the spinous dorsal fin is yellow to orange yellow. The lower half of the remaining fins is red. The lead-

Fig. 3.9 *Myripristis berndti* Jordan & Evermann, 1903. (Courtesy of Paul Asman)

ing edges are white, sometimes with a black submarginal streak.

Habitat, Biology, and Ecology
The blotcheye soldierfish is a marine species that lives in association with reefs at a depth range of 3–159 m (Allen and Erdmann 2012) but usually lives within 3–15 m (Randall and Heemstra 1986). It reaches a maximum total length of 300 mm, with a common total length of 220 mm (Sommer et al. 1996).

Individuals of this species inhabit in caves and hide under ledges of subtidal reef flats, channels, and margins to outer reef slopes (Lieske and Myers 1994; Mundy 2005). They are also found in loose aggregations (Lieske and Myers 1994). They feed mainly on plankton such as crab larvae. They are usually caught at night (Schneider and Krupp 1995).

Distribution
Individuals of this species are distributed in the Indo-Pacific and Eastern Pacific. The species is reported from East Africa and south to Natal (but not from the Red Sea) and east to the Clipperton, Cocos, and Galapagos islands, north to the Ryukyu Islands, and south to the Great Barrier Reef, Norfolk Island, and Lord Howe Island. It is also found all throughout Micronesia (Myers 1991) and common throughout Oceania, being clearly absent only from Easter Island.

Economic Value
This species has good commercial value across its geographical distribution and is marketed fresh.

Conservation Status
Williams and Greenfield (2016a) evaluated *M. berndti* as a Least Concern in the Red List of the IUCN based on the following criteria. This species is widely distributed through its range. It is common in several large marine-protected areas. Although there is a commercial demand for this species as human consumption, this is unlikely to be a major threat to the species.

Pristilepis oligolepis (Whitley 1941).
Common name: Spinyface soldier.
Etymology: *Pristilepis:* Greek, pristis = saw + greek, lepis = scale (Romero 2002).

Diagnostic Features
The body is reasonably deep and compressed, with moderately short snout and a large eye. The mouth is large, with a lower jaw slightly projecting and maxilla extending to or slightly beyond a vertical at the hind edge of the eye. The teeth are in villiform bands in jaws. The external bones of the head are rugose. The free margins of suborbital, postorbital, and opercular bones and lower edge of second-to-fourth branchiostegal rays have numerous spinules. The corner of the preopercle of adults

is with no enlarged spines. Scales are coarsely ctenoid, the ctenii arising from strong ridges on posterior surface of scales. Scales are present dorsally on the head and extending forward slightly anterior to a vertical at the upper end of the preopercle. The preopercle is with four diagonal rows of scales. The base of the opercle is with a single row of scales. The spinous portion of the dorsal fin is without scales. The anal fin and soft portion of dorsal fin are with a basal scaly sheath. The scales do not adhere to the fins. The caudal and pectoral fins have small scales basally. The pelvic fins are with a mid-ventral posterior projection from base of two moderately large scales. Pectoral fins are slightly longer than pelvic fins. The origin of pelvic fins is below the base of pectoral fins. Its color in alcohol is light brown, with a small whitish spot on each scale sometimes. The opercular and fin membranes are pale. Color in life is bright red with a small white spot on each scale, therefore making longitudinal rows.

Habitat, Biology, and Ecology

This holocentrid species is a marine species living in association with coral reefs at a depth range of 14–220 m (Randall and Heemstra 1986). Individuals of this species reach a maximum total length of 300 mm (Burgess et al. 1990). This species usually lives in a benthopelagic environment in or near crevices and caves at 9–348 m (Mundy 2005) and has a nocturnal habit (Paxton et al. 1989).

Distribution

Individuals of this species are distributed in the Indian Ocean region, Reunion, and Western Australia. They have also been reported from the Pacific Ocean, coasts of Japan, and Lord Howe, Hawaiian, Kermadec (Francis 1993), and Easter islands.

Economic Value

This species has good commercial value across its geographical distribution and is marketed fresh.

Conservation Status

Williams and Greenfield (2016b) evaluated *P. oligolepis* as a Least Concern in the Red List of the IUCN based on the criteria that the species is not facing no known major threats.

3.3.1 Injuries Caused by Squirrelfishes

Squirrelfishes cause wounds of various kinds. These wounds are not poisonous, but they can be severely harmful especially with careless handling during fish catch. A secondary infection usually happens from such wound that needs prompt medical action.

3.4 Species of Groupers, the Red Snapper, Marlin, Makaria, Sailfish, and Swordfish

Family: Serranidae.

Acanthistius cinctus (Günther 1859) (Fig. 3.10).

Common name: Yellowbanded perch.

Etymology: *Acanthistius:* Greek, akantha = thorn + Greek, istion = sail (Romero 2002).

Diagnostic Features

The body is elongated and high in the region just behind the head. The head has a low profile and an elongated rostrum. The eyes are near the dorsal profile of the head. The operculum has strong spines directed posteriorly. The spinous part of the dorsal fin is lower than the soft ray part of this fin. The body is yellow in color, with six bands, and a black stripe radiating from the eye. Except for the dorsal fin, the remaining fins are gray in color, with black, white, and dotted edges.

Habitat, Biology, and Ecology

The yellowbanded perch is a marine species that lives in association with reefs at depth range of more than 30 m. It is a non-migratory species (Lee Marsh 1986). This species is a benthic and occurs inshore, near reefs (Paxton et al. 1989; Hoese et al. 2006), and also on gravely, sandy bottoms under rocky and coral overhangs (Lee Marsh 1986). Individuals of this species reach a maximum total length of 500 mm (Burgess et al.

Fig. 3.10 *Acanthistius cinctus* (Günther 1859). (Courtesy of Bray, D.J., Fishes of Australia)

1990). They occupy the same area for months or years, so they are probably home-ranging. They occur singly, in pairs, or in small groups and eat fishes (Francis 2012).

Distribution
This serranid species is reported from Australia, Lord Howe Island, and Norfolk Island to Kermadec Island (Hutchins and Kuiter 1982; Paulin et al. 1989; Heemstra 2010). It is also reported from the Great Barrier and Middleton Reefs, as well as off northern New Zealand (William et al. 2016a, b).

Economic Value
As with the members of the family Serranidae, it has good commercial value.

Conservation Status
William et al. (2016a) evaluated *A. cinctus* as a Least Concern for the IUCN Red List as it is known from a relatively small range and occurs in a variety of habitat types and there are no known major threats.

Caesioperca lepidoptera (Forster 1801) (Fig. 3.11).
Common name: Butterfly perch.
Etymology: *Caesioperca:* Latin, caesius, bluish-gray, 1835 + Greek, perke = perch (Romero 2002).

Diagnostic Features
The body is oval, with a small head, and light pink in color. The eyes are big. The dorsal fin is high. The caudal fin is lunate. The operculum has no spines. There is a large black, central blotch located below the lateral line below the end of the spinous part of the dorsal fin. The body, head, and the dorsal fin are covered with irregular, distributed small to medium black spots. Iridescent blue markings are on the head and mainly behind the eyes. The edges of the dorsal, pelvic, and anal fins are bluish. The pectoral fin is dark pink in color.

Habitats, Biology, and Ecology
The butterfly perch is a marine-inhabiting demersal with a habitat depth range of 10–140 m and reaches a maximum total length of 300 mm (May and Maxwell 1986). Individuals of this species may be found in small schools over rocky reefs (May and Maxwell 1986). In deeper water, they form large aggregates and usually occur in areas that are exposed to continual strong tidal water movement (Russell 1983). This species is planktivorous, feeding on small crustaceans and scallops. It reaches maturity at the end of its first year, at a total length of 100, and reported to live for up to 8 years (Francis 2012).

Distribution
Individuals of this species are found distributed in the Eastern Indian Ocean and Southwest

Fig. 3.11 *Caesioperca lepidoptera* (Forster 1801). (Courtesy of Dubosc, Jeff)

Pacific in southern Australia and New Zealand (Paulin et al. 1989).

Economic Value
As with the members of the family Serranidae, it has good commercial value.

Conservation Status
This species has not yet been evaluated for the IUCN Red List.

Caprodon longimanus (Günther 1859) (Fig. 3.12).
Common name: Pink maomao.
Etymology: *Caprodon:* Latin, capra = goat + Greek, odous = teeth (Romero 2002).

Diagnostic Features
The body is fusiform, with a large head. The mouth is terminal, and the eyes are large. The pectoral fins reach the anterior edges of the anal fin. The posterior edge of the operculum has a prominent broad spine. The imaginary line passing through the snout passes also through the lower third of the eye. The body is bright pink in color, with red-orange marks on the face. Fins are with light-blue margins. A rare color variety is present of this species, with black markings on rear part of the dorsal fin and the upper back of the body and the presence of lines of yellow spots on face. In this variety, all fins have yellow color, except for the dorsal fin, and all are with blue margins, except for the pectoral, which is in a yellow color.

Habitat, Biology, and Ecology
The pink maomao is a marine species that lives in association with reefs at a depth from surface and down to 400 m (Mundy 2005), but it is usually found at a depth range of between 15 and 60 m (Ayling and Cox 1982). Individuals of this species reach a maximum total length of 550 m (McMillan et al. 2011), with a common total length of 300 mm (Ayling and Cox 1982). This species inhabits inshore areas near reefs (Paxton et al. 1989). Individuals of this species feed on plankton and small nekton (Ayling and Cox 1982).

Distribution
The pink maomao is reported from off southwestern Australia (Rottnest Island), southeastern Australia, Lord Howe Island, Norfolk Ridge, New Zealand, the Kermadec Islands, Easter Island, the Nazca Ridge, Desventuradas Island, Juan Fernandez Island, and Sala y Gomez Islands (Paxton et al. 1989; Anderson and Baldwin 2000; Dyer and Westneat 2010). It is also recorded from the Hawaiian Islands (Mundy 2005).

Fig. 3.12 *Caprodon longimanus* (Günther 1859). (Courtesy of Florian Schneider)

Economic Value

As with the members of the family Serranidae, it has good commercial value.

Conservation

William et al. (2016b) evaluated the pink maomao for the IUNCN Red List as Least Concern because it is a wide-ranging anti-tropical species across the entire Pacific. In addition, it is abundant in parts of its range. No major threat to its population is present.

Fig. 3.13 *Hypoplectrodes huntii* (Hector 1875). (Courtesy of Ian Skipworth)

Hypoplectrodes huntii (Hector 1875) (Fig. 3.13).
Common name: Redbanded perch.
Etymology: *Hypoplectrodes:* Greek, hypo = under + Greek, plektron = sting, spur + Greek, oides = similar (Romero 2002).

Diagnostic Features

The body is deep at the anterior edge of the dorsal fin. The head has a pointed snout. The eyes are large and located near the dorsal profile of the head. The operculum has a prominent broad spine. The mouth is terminal, with lower jaw longer than the upper.

Its body has seven oranges to brown stripes reaching the ventral side of the body. The head is orange to brown and often with greenish trace on top. All fins are red or orange.

Habitat, Biology, and Ecology

The redbanded perch is a marine species that lives in association with reefs at a depth from surface to 100 m (Paulin and Roberts 1992). Individuals of this species reach a maximum total length of 200 mm (Paulin and Roberts 1992). This species is solitary and lives in deep holes and rocks ledges. It feeds on small fishes, crabs, and benthic invertebrates (Russell 1983). Males grow larger than females (Francis 2012).

Distribution
Individuals of this species are distributed in the Southwest Pacific, and they are endemic to New Zealand (Francis 2012).

Economic Value
As with the members of the family Serranidae, it has good commercial value.

Conservation
This species has not been evaluated for the Red List of the IUCN.

Family: Epinephelidae.
Epinephelus daemelii (Günther 1876) (Fig. 3.14).
Common name: Saddletail grouper.
Etymology: *Epinephelus:* Greek, epinephelos = cloudy (Romero 2002).

Diagnostic Features
Its body is oval in shape and deep at the vertical line passing through the anterior edge of the dorsal fin and through the base of the pectoral fin. Eyes are well above the imaginary line passing through the snout. Its body color varies from white to brown-black. The body has five oblique brown-black bands. The caudal peduncle has a darker saddle. The body and head are covered with numerous white spots. The pelvic, anal, soft part of the dorsal fin, and caudal fins have black and white edges.

Habitats, Biology, and Ecology
The saddletail grouper is a marine species that lives in association, with reefs at depth down to 50 m (Heemstra and Randall 1993). It reaches a maximum total length of 2000 mm and a maximum-recorded weight of 68 Kg (Randall and Heemstra 1991). Individuals of this species live on rocky substrata. They are aggressive territorial species and may occupy a particular cave for life. Small individuals feed on crabs and small fishes. Sexual transition reported to occur at 100–110 cm (Francis 1988). Divers often attract large individuals (Lieske and Myers 1994).

Distribution
Individuals of this species are distributed in the Southwest Pacific. They are reported from the coasts of Australia, Lord Howe Island, Norfolk Island, Kermadec Islands, and New Zealand (North Island and Poor Knights Island) (Francis 1988). The Australian range extends from southern Queensland to Kangaroo Island off South Australia; they are also reported from the Bass Strait.

Fig. 3.14 *Epinephelus daemelii* (Günther 1876). (Courtesy of John Randall, Wikipedia)

Economic Value

This species has high commercial value along its geographical distribution.

Conservation

Pollard and Sadovy (2018) evaluated this species as Near Threatened for the Red List of the IUCN based on the following justifications. This species has a moderately small distribution range, encompassing mainly only from New South Wales in Australia, northern New Zealand, and the Kermadec Islands. Its meat has a high commercial value, and its territorial and intrusive nature leads to having been captured easily by recreational (spear and line) and commercial line fishers. Moreover, intrinsic biological features make it to be highly vulnerable to speedy population failure. With the heavy fishing activities along its geographical range, its populations, and most probably fish size, dropped severely during the greater part of its distribution since early in the twentieth century (Pollard and Sadovy 2018).

Epinephelus lanceolatus (Bloch 1790) (Figs. 3.15 and 3.16).

Common name: Giant grouper.

Etymology: *Epinephelus:* Greek, epinephelos = cloudy (Romero 2002).

Diagnostic Features

The giant grouper is well-known by the following characteristics: The body is robust. The preopercle is finely serrate, with a rounded corner. The upper edge of the operculum is convex. The mid-lateral part of the lower jaw has 2–3 rows of teeth increasing to 15–16 rows in specimen of 177 cm SL. The canine teeth at the front of the jaws are small or absent. The dorsal fin's third to eleventh spines are sub-equal and shorter than the longest soft rays. The pelvic fins are short. The caudal fin is rounded. The lateral line scales 54–62, while the anterior scales with branched tubules (except small juveniles). The color of the small juveniles (less than 15 cm) is yellow, with three irregular black areas, the first from spinous dorsal fin to belly and chest and extending onto the head, the second from base of the soft dorsal fin to the anal fin, and the last at the base of the caudal fin. The color of the subadult individuals (25–60 cm) is with irregular white or yellow spots on black areas and fins with black spots. Adults (90–165 cm) are dark brown in color, with faint mottling. Adult fins have numerous small black spots. Large adults (180–250 cm) are dark brown in color, with darker fins (Heemstra and Randall 1999; Craig et al. 2011; Allen and Erdmann 2012).

Habitat, Biology, and Ecology

This grouper is a marine species living in association with reefs and sometimes enter a brackish water at a depth range of 1–200 m (Bacchet et al. 2006), but usually it lives down to 50 meters (Myers 1999). Individuals of this species reach a maximum total length of 2700 mm (Fischer et al. 1990), with a common total length of 1900 mm (Bouhlel 1988) and a maximum published weight of 400 kg (Foale 1998).

Individuals of this species are familiar in shallow waters. They live in caves or wrecks and in estuaries. The juveniles are enigmatic in reefs and rarely seen (Kuiter and Tonozuka 2001). This

Fig. 3.15 *Epinephelus lanceolatus* (Bloch 1790). (Courtesy of Moazzam Khan, Pakistan)

Fig. 3.16 Specimen of *Epinephelus lanceolatus* (Bloch 1790) just collected. (Courtesy of Moazzam Khan, Pakistan)

species feeds on spiny lobsters; fishes, including small sharks and batoids; and juvenile sea turtles and crustaceans.

Distribution
This species of grouper distributed in the Indo-Pacific region from the Red Sea to Algoa Bay, South Africa, eastward to the Hawaiian and Pitcairn islands, north to southern Japan, and south to Australia and New Zealand. It is also reported from the Persian Gulf (Heemstra and Randall 1999).

Economic Value
The meat of this species has high commercial value as it is used for human consumption.

Conservation
Individuals of this species are nearly wiped out due to heavy fishing activities along its geographical distribution (Lieske and Myers 1994). Fennessy et al. (2018) evaluated this species as Data Deficient for the Red List of the IUCN based on the following justifications: It is a large-size species that is widely distributed and rare in nature. It has high commercial value for its meat. Due to intrinsic biological features, containing forming spawning aggregations, this species is highly vulnerable to declines under heavy fishing pressure.

Epinephelus rivulatus (Valenciennes 1830) (Fig. 3.17).
Common name: Halfmoon grouper.
Etymology: *Epinephelus:* Greek, epinephelos = cloudy (Romero 2002).

Diagnostic Features
This species of grouper has the following set of characteristics: head is red-brown, and body is mottled pale brown. There are 5–6 irregular dark-brown bars on its body. Caudal fin is dark brown, with pale speckling. The scale on its body is ctenoid, except a cycloid scale on the anterior-dorsal side of the body above the lateral line and on the nape, thorax, and abdomen. The caudal fin is round in shape. The head is flat, with a slightly convex interorbital area. The dorsal head profile

Fig. 3.17 *Epinephelus rivulatus* (Valenciennes 1830). (Courtesy of Stef, South Africa)

is convex. The preopercle is angular in shape, with enlarged serrae at angle. The straight upper edge of the operculum is almost horizontal. The posterior and anterior nostrils are subequal in size. The maxilla reaches to or past vertical at the rear edge of the eye. There are 2–3 rows of teeth on the mid-lateral part of the lower jaw. The digestive system is equipped with a pyloric caecum.

Habitats, Biology, and Ecology

The halfmoon grouper is a marine species living in association with reefs at a depth range of 1–150 m (Lieske and Myers 1994). It reaches a maximum total length of 450 mm (Allen and Erdmann 2012), with a maximum weight recorded of 1.4 Kg (Heemstra and Randall 1993). Individuals of this species live in rocky substrate, algal flats, and seagrass beds areas. They feed on fishes and crustaceans (Sommer et al. 1996). This species forms spawning aggregations (Allsop and West 2003).

Distribution

Individuals of this species are distributed in the Indo-West Pacific region from East Africa to the Western Pacific and southward into temperate waters of South Africa, Australia, and New Zealand. It is not known to be present in the Red Sea or Persian Gulf.

Economic Value

As with other groupers species, its meat has high commercial value.

Conservation

Fennessy et al. (2018) evaluated this species of grouper as Least Concern to the Red List of the IUCN based on the following justifications: This species is distributed. It inhabits rocky and coral reefs and can be common. It is considered among the important commercial species in certain areas along its geographical distribution. There is no major threat on a global level currently.

Family: Lutjanidae.
Etelis carbunculus Cuvier, 1828 (Fig. 3.18).
Common name: Deep-water red snapper.
Etymology: *Etelis:* Greek, etelis, -idos = a fish, perhaps the fish Sparus aurata (Romero 2002).

Diagnostic Features

The body is moderately elongated. The head is small, and the eye is large. The interorbital space is flat. The lower jaw is slightly project-

Fig. 3.18 *Etelis carbunculus* Cuvier, 1828. (Courtesy of Jean-Lou Justine)

ing. A single row of conical teeth is in jaws; it usually has one or two pairs of enlarged canines anteriorly. The maxilla is scaled, extending to below middle of the eye. The vomerine tooth patch is narrowly V-shaped. The number of gill rakers on the lower limb of the first arch (including rudiments) is 11–14, dorsal fin with 10 spines and 11 soft rays, anal fin with 3 spines and 8 soft rays, and pectoral fins with 15–17 rays. Dorsal and anal fin bases are scaleless. The caudal fin is forked. Scale rows on-back parallel have a line. Its color goes from mainly pink to red and then becoming white on lower sides and belly.

Habitat, Biology, and Ecology

The deep-water red snapper is a marine species living in benthopelagic habitats at a depth range of 90–400 m (Anderson and Allen 2001), but usually, it is reported from a depth range of 200–350 m (Fry et al. 2006). Individuals of this species reach maturity at a total length of 610 mm. The maximum total length reported for this species is 1270 mm (Smith and Kostlan 1991), with a common total length reaching to 650 mm (Bouhlel 1988). The maximum reported age reached by individuals of this species is 32 years (Andrews et al. 2011).

Adults of this species usually inhabit rocky bottoms (Sommer et al. 1996). This species feeds on fishes and larger invertebrates such as squids, shrimps, and crabs and also partakes planktonic organisms, including pelagic urochordates.

Distribution

This species is reported from the Indo-Pacific region. It is recorded from the East African coasts to the Hawaiian Islands, north to southern Japan, and south to Australia, as well as in Three Kings Island, New Zealand (Francis et al. 1999). This name has been wrongly used for *Etelis coruscans* by some previous authors.

Economic Value

This species has a high commercial value as it is used as food for humans in most localities across its geographical distribution, where it is marketed fresh or frozen (Allen 1985).

Conservation

Russell et al. (2016a) evaluated this species as Least Concern to the Red List of the IUCN based on the following justifications. This species is widely distributed. Individuals of this species are abundant in some localities of its distribution. A decline in the population of this species in certain localities, but not a major population decline, is reported.

Etelis coruscans Valenciennes, 1862 (Fig. 3.19).
Common name: Deepwater longtail red snapper.
Etymology: *Etelis:* Greek, etelis, -idos = a fish, perhaps the fish Sparus aurata (Romero 2002).

Fig. 3.19 *Etelis coruscans* Valenciennes, 1862. (Courtesy of Noel Rault)

Diagnostic Features

The body is slender and elongated. The head is small, and the eye is large. The snout is short (shorter than eye diameter). The interorbital space is flat. The lower jaw is slightly projecting. The maxilla is covered with small scales. Both jaws are with bands of small teeth, those in the outer series are better developed and more widely spaced, and a few anterior teeth in the upper jaw are canine-like. The vomerine tooth patch is narrowly V-shaped. The number of gill rakers on the lower limb of the first arch (including rudiments) is 15–18, dorsal fin with 10 spines and 11 soft rays, anal fin with 3 spines and 8 soft rays, and pectoral fins with 15 or 16 rays. Dorsal and anal fin bases are scaleless. The caudal fin is deeply forked. The upper caudal lobe becomes greatly elongated with increased growth. Scale rows on the back are parallel with lateral line. Its color is deep pink to red at the back and on the upper sides, pink on the lower sides and belly, and pink to red on fins.

Habitat, Biology, and Ecology

The deepwater longtail red snapper is a marine species living in association with coral reefs at a depth range of 45–400 m, but individuals of this species are usually found at 90–400 m (Anderson and Allen 2001). This species reaches maturity at a total length of 663 mm and a maximum total length of 1200 mm (Anderson 1986), with a common total length of 500 mm (Anderson and Allen 2001).

Adults of this species usually inhabit rocky bottoms (Ref. 30,573) of the continental shelf and continental slope (Sommer et al. 1996). This species feeds on small fishes, squids, and crustaceans (Sommer et al. 1996).

Distribution

Individuals of this species live in the Indo-Pacific region. They are reported to be present along East African coasts and distributed eastward to the Hawaiian Islands, north to the Sea of Japan, south to Australia (Queensland and New South Wales), and Lord Howe Island, as well as off the Kermadec Islands, northeast of New Zealand (Anderson and Allen 2001). This species is also reported to be present on Marquesas (Russell et al. 2016a).

Economic Value

This species is considered an important food fish in some areas. It is usually marketed as fresh and frozen, and it is highly valued for the quality of its flesh (Frimodt 1995).

Conservation

Russell et al. (2016e) evaluated this species as Least Concern to the Red List of the IUCN based on the following justifications: This species is widely distributed in its geographical region. It is rather slow-growing and late to mature, taking an estimated 5–6 years to reach maturity. It met localized population depletions due to heavy fishing.

Lutjanus kasmira (Forsskål 1775) (Fig. 3.20).
Common name: Bluestripe snapper.
Etymology: *Lutjanus:* Malay, ikan lutjan, name of a fish (Romero 2002).

Diagnostic Features
The body is reasonably deep. The dorsal profile of its head is sharply sloped. The preorbital width is usually greater than the eye diameter but sometimes less in small specimens. The preopercular notch and knob are well developed. The vomerine tooth patch is crescentic, without a medial posterior extension. The tongue is smooth, without teeth. The number of gill rakers on the lower limb of the first arch (including rudiments) is 13 or 14, with a total number of rakers on the first arch to 20–22; dorsal fin with 10 spines and 14 or 15 soft rays; and anal fin with 3 spines and 7 or 8 soft rays. The posterior profile of dorsal and anal fins is somewhat angular. Pectoral fins have 15 or 16 rays. The caudal fin is slightly emarginate. Scale rows at the back are rising obliquely above the lateral line. Its color is bright yellow at its back and sides and white on lower sides and underside of the head. There are four bright-blue stripes on its side. Fins are yellow.

Habitat, Biology, and Ecology
The bluestripe snapper is a marine species living in association with reefs at a depth range of 3–265 m (Randall 1995), but individuals of this species are usually found at a depth range of 30–150 m (Myers 1991). This species reaches maturity at a total length of 205 mm, with maximum total length reaching 400 mm (Bykov 1983) and a common total length of 250 mm (Allen 1985). Adults of this species live in the coral reef area and are found occurring in both shallow lagoons and on outer reef slopes. They are regularly found in large groups in the vicinity of coral formations, caves, or wrecks during the day. Juveniles of this species live in seagrass bed areas around patch reefs (Lieske and Myers 1994). This species feeds on fishes, shrimps, crabs, stomatopods, cephalopods, and planktonic crustaceans (Bagnis et al. 1972).

Distribution
Individuals of this species are distributed in the Indo-Pacific region. They are reported to be present in the Red Sea and East African coasts to the Marquesas and the Line islands, north to the southern Japan, and south to Australia, as well as in the Southeast Atlantic at the East London, South Africa (Anderson 1986).

Fig. 3.20 *Lutjanus kasmira* (Forsskål 1775). (Courtesy of Richard Field)

Economic Value

This species is considered among high-commercial-value fish species along its geographical distribution line.

Conservation

Russell et al. (2016b) evaluated this species as Least Concern to the Red List of the IUCN based on the following justifications: It is a widely distributed species. It is abundant in some localities along its distributional range. The longevity of this species is at least 10 years, while the age of maturity is 3 years. No documented declines in this species have been reported; in fact, it is increasing in some areas of its range.

Paracaesio xanthura (Bleeker 1869) (Fig. 3.21).
Common name: Yellowtail blue snapper.
Etymology: *Paracaesio:* Greek, para = the side of + Latin, caesium = bluish gray (Romero 2002).

Diagnostic Features

Body moderately deep and fusiform; eye large, about 3.5 times in head length; snout short, about equal to or shorter than the eye diameter; interorbital space convex; upper and lower jaws about equal; both jaws with an outer series of well-developed canine teeth and an inner band of small villiform teeth; maxilla with or without scales; gill rakers on lower limb of first arch (including rudiments) 18– 20. Dorsal fin with 10 spines and 10 (rarely 11) soft rays; anal fin with 3 spines and 8 (rarely 9) soft rays; dorsal and anal fins scaleless; pectoral fins long, reaching level of anus, with 16 rays; caudal fin deeply forked; scales small, about 70–72 in lateral line; scale rows on back parallel to lateral line. Color: overall blue, sometimes whitish on belly and lower part of head; a bright yellow area at the back extending from the forehead to the base of the caudal fin; dorsal and caudal fins yellow; other fins whitish or translucent.

Fig. 3.21 *Paracaesio xanthura* (Bleeker 1869). (Courtesy of Izuzuki, Japan)

Habitat, Biology, and Ecology

The yellowtail blue snapper is a marine species that lives in association with coral reefs at a depth range of 5–250 m (Fricke et al. 2011). Individuals of this species reach a maximum total length of 500 mm (Anderson 1986). Adults of this species are usually found in rocky bottom, occasionally establishing large schools. This species feeds mainly on zooplanktons (Sommer et al. 1996).

Distribution

The yellowtail blue snapper is distributed in the Indo-Pacific region. It reportedly present from the East African coast to the Austral Islands, as well as from north to southern Japan and south to southeastern Australia. *Paracaesio xanthura*, a complex of variable populations, may create a single or two (or more) species.

Economic Value

This species has high commercial value along its geographical distribution and is marketed fresh or frozen.

Conservation

Russell et al. (2016c) evaluated *Paracaesio xanthura* as a Least Concern in the Red List of the IUCN for the following reasons. It is widely distributed in the Indo-West Pacific. Handlines and longlines are the only two fishing methods used to catch this species. Now, there is no conservation measure to protect this species. Nevertheless, it is found in marine-protected areas throughout its range. It is usually found in low fishing-pressure areas, where populations endure stability.

Fig. 3.22 *Pristipomoides filamentosus* (Valenciennes 1830). (Courtesy of Hamid Badar Osmany, Pakistan)

Pristipomoides filamentosus (Valenciennes 1830) (Fig. 3.22).

Common name: Crimson jobfish.

Etymology: *Pristipomoides:* Greek, pristis = saw + Greek, poma, -atos = cover, operculum + Greek, oides = similar to (Romero 2002).

Diagnostic Features

The body is elongated and robust. The interorbital space is flat. The teeth are in bands with an outer row of canines. The tongue is free of teeth. The last soft rays of the dorsal and anal fins are extended into short filaments. The pectoral fin is as long as the head and slightly falcate. The caudal fin is forked. There are 60 to 65 on small scales on lateral line. The base of dorsal and anal fins is free of scales. The ground color of the body is variable; it ranges from lavender with a blue tinge to reddish purple. There are small blue spots on top of the head. The dorsal fin has two yellowish longitudinal lines. The margin of the soft dorsal and caudal fins is red.

Habitat, Biology, and Ecology

The crimson jobfish is a marine species living in benthopelagic habitats at a depth range of 40–400 m (Opic et al. 1994). It reaches a maximum total length of 1000 mm (Anderson 1986) and a maximum weight of 8.2 kg (Honebrink 1990). The maximum age reported for this species is 44 years (Andrews et al. 2012). Adults of this species are usually found on rocky bottoms (Sommer et al. 1996). During nighttime, individuals of this species move to the upper part of their habitat for feeding (Opic et al. 1994). Crimson jobfish feeds on small fishes, shrimps, crabs, amphipods, ascidians, and salps.

Distribution

This species is distributed in the Indo-Pacific region. It was reported from the East African coast to Hawaii and Tahiti, north to southern Japan, south to eastern Australia, and Lord Howe Island.

Economic Values

This species has good commercial value. It is a good-quality fish and marketed fresh along its geographical distribution line (Allen 1985), where it is caught mainly with bottom longlines and deep handlines (Anderson and Allen 2001).

Conservation

Russell et al. (2016d) evaluated this species as Least Concern for the Red List of the IUCN based on the following reasons: It is widespread in the Indo-Pacific. This species displays great genetic connectivity through its range with the exclusion of Hawaii, which shows some genetic isolation. The maximum age of this species is debated, which may be as high as 40 years old. This is a chief fisheries species that is trapped using bottom longlines and deep handlines. There is a management of different degrees for this species along its range as it usually lives in marine-protected areas.

Family: Istiophoridae
Istiompax indica (Cuvier 1832) (Fig. 3.23).
Common name: Black marlin.

Diagnostic Features
The body is elongated and moderately compressed. The upper jaw is protruding, forming a robust but not very long beak. Two dorsal fins are present. The height of the first dorsal fin is less than the greatest body depth and becomes shorter posteriorly. The pectoral fins are falcate and rigid, with 19–20 rays. The body is densely covered with small, embedded scales with one or two sharp points. The back of the body is dark blue, with a silvery-white belly. The membrane of the first dorsal fin is blue-black, without spots. Flanks have no spots. The body sometimes has light-blue vertical stripes. The first dorsal fin is blackish to dark blue, while other fins are dark brown with tinges of dark blue in some specimens.

Habitat, Biology, and Ecology
Black marlin is a marine species that lives in a pelagic-oceanic habitat, with an oceanodromous habit (Riede 2004). It is found at a depth range of 0–915 m (Nakamura 1985). This species reaches a maximum forked length of 4650 mm (IGFA 2001) and a maximum reported weight of 750 kg (Cornic 1987). Individuals of this species are usually found in surface water above the thermocline, frequently near the shore close to landmasses, islands, and coral reefs. This species feeds mainly on fishes, squids, cuttlefishes, octopods, large decapod crustaceans, and mostly small tunas when abundant (Radtke 1966).

Distribution
This species is distributed in the Indo-Pacific region. It is reportedly found in tropical and subtropical waters and occasionally in temperate waters. Some are in the Atlantic Ocean by way of the Cape of Good Hope, but the presence of Atlantic breeding stocks is doubtful.

Fig. 3.23 *Istiompax indica* (Cuvier 1832). (Courtesy of Michael Janke, Kenya)

Economic Values

This species is considered among the fish species with high commercial value as the flesh is of good quality. It is usually marketed as refrigerated or frozen and prepared as sashimi in Japan (Nakamura 1985).

Conservation

Collette et al. (2011a) evaluated this species as Data Deficient in the Red List of the IUCN based on the following reasons. It is widespread in the Indo-Pacific. No stock assessments are present for this species. Only very inadequate catch and effort data is recorded. As it is usually caught with other Marlin species, it is probable that its population has weakened as well. Its influence on commercial fishing and sports fishing is unknown.

Istiophorus platypterus (Shaw 1792) (Fig. 3.24).
Common name: Indo-Pacific sailfish.
Etymology: *Istiophorus:* Greek, istios = sail + Greek, pherein = to carry (Romero 2002).

Diagnostic Features

The Indo-Pacific sailfish has the following characteristics: The body is slender, elongated, and fairly compressed with a high, sail-like first dorsal fin. The upper jaw protrudes into a very long beak, slender and round in cross-section. Jaws and palatines are with villiform teeth, with the absence of gill rakers. Two dorsal fins are present, the first very large (42–49 rays) and the second small (6–7 rays). Pectoral fins have 18–20 rays. Pelvic fins have one spine and two soft rays, which are very long and narrow, depressible into a groove, and are fused together. The caudal peduncle has double keels on each side. The body is covered with small, embedded scales. Back is dark with about 20 bluish vertical bars. The belly is pale silver. The membrane of the first dorsal fin is blue-black with numerous dark spots. Bases of the first and second anal fins are frequently tinged with silvery-white color. The remaining fins are blackish brown or dark blue.

Habitat, Biology, and Ecology

Istiophorus platypterus is a marine species that lives in the pelagic-oceanic habitat and has an oceanodromous habit (Riede 2004). It is usually found at a depth range of 0–200 m (FAO-FIGIS 2001). It reaches a maximum forked length of 3480 mm (IGFA 2001) and a maximum reported weight of 100.2 Kg (IGFA 2001). This species can live a maximum of 13 years (Prince et al. 1986). This species forms schools of different sizes (Riede 2004). It undertakes spawning migrations in the Pacific. It feeds mainly on fishes, crustaceans, and cephalopods. It is usually caught with longlines and set nets and sometimes by trolling and harpooning from boats (Nakamura 1985).

Fig. 3.24 *Istiophorus platypterus* (Shaw 1792). (Courtesy of Moazzam Khan, Pakistan)

Distribution

This species is distributed in the Indo-Pacific region. It is reported from tropical and temperate waters, approximately 45°–50°N and 40°–35°S in the western Pacific, 35°N and 35°S in the eastern Pacific, 45°S in the western Indian Ocean, and 35°S in the eastern Indian Ocean. It is usually found in the Mediterranean Sea from the Red Sea via the Suez Canal.

Economic Value

The meat of this species is utilized fresh, smoked, and frozen. It is also used for sashimi and sushi and is eaten broiled and baked (Frimodt 1995).

Conservation

Collette et al. (2011d) evaluated this species as Least Concern for the Red List of the IUCN based on the following reasons: It is a widespread and common species in the Atlantic and Indo-Pacific. It is mainly aimed at sport and artisanal fisheries and is also caught as bycatch in longline and purse seines. In some localities like Central America, Iran, and India, the population of this species has shown a decline. No stock assessment is available for this species.

Kajikia audax (Philippi 1887) (Fig. 3.25).
Common name: Striped marlin.

Diagnostic Features

The body is elongated and compressed. The upper jaw protrudes into a robust and medium-sized beak. Two dorsal fins are present, the height of the first is greater than the greatest depth, short anteriorly, taller in the middle, and then becoming shorter posteriorly. Pectoral fins are falcate and flexible, having 18–22 rays. The body is densely covered by small, embedded scales, and its back is dark blue in color. The belly is silvery. The membrane of the first dorsal fin is blue black without dark spots. There are 20 bluish stripes on the flanks. The first dorsal fin is dark blue. The other fins are dark brown, sometimes with a tinge of dark blue. Anal fin bases have a tinge of silvery white.

Fig. 3.25 *Kajikia audax* (Philippi 1887). (Courtesy of Jackie Mora, Wikipedia)

Habitat, Biology, and Ecology
The striped marlin is a marine species inhabiting the pelagic-oceanic region. It has an oceanodromous habit (Riede 2004). It lives at a depth range of 0–200 m (Nakamura 1985). It reaches a maximum total length of 4200 mm (Allen and Steene 1988) and a maximum reported weight of 440 Kg (Allen and Steene 1988). This species is the most dominant and widely distributed of all billfishes. Its abundance increases with distance from the continental shelf (Kailola et al. 1993). Individuals of this species are regularly solitary but form small schools by size during the spawning season (Frimodt 1995). This species feeds on fishes, crustaceans, and squids.

Distribution
This species is distributed in the Indo-Pacific region. It is reported from tropical, subtropical, and temperate waters. The distribution in the Pacific Ocean is unique among billfishes and tunas (Squire and Suzuki 1990). It is also reported from the Indian Ocean at eastern Africa coasts and in the western Arabian Sea, the Bay of Bengal, and off northwestern Australia (Pillai and Ueyanagi 1977).

Economic Value
This species is considered among important commercial fish species for its high-quality flesh. It is usually used for sashimi and sushi, and it is marketcd mostly frozen, sometimes fresh (Nakamura 1985).

Conservation
Collette et al. (2011c) evaluated this species as Nearly Threatened in the Red List of the IUCN based on the following reasons: It is widespread in the Indian and Pacific oceans. This species is commercially important as food fish throughout its geographical distribution line. There is a report on the decline of the populations of this species globally. It is considered not well managed, and stock assessment is not available.

Makaira nigricans (Lacepède 1802) (Fig. 3.26). Common name: Atlantic blue marlin.

Fig. 3.26 *Makaira nigricans* Lacepède, 1802. (Courtesy of Antonio Amaral)

Etymology: *Makaira:* Greek, makaira, -as = happy, rich (Romero 2002).

Diagnostic Features
The color of the body is blue-black from above and silvery-white from below. There are 15 rows of pale-cobalt-colored stripes on the side of the body. The first dorsal fin is plain blackish or sometimes dark blue. Other fins are brown and black with a tinge of dark blue in some specimens. Anal fin bases are with a tinge of silvery-white color. The lateral line is a network of interconnecting canals (Smith 1997).

Habitat, Biology, and Ecology
The blue marlin is a marine species that inhabits the pelagic-oceanic region. It has an oceanodromous habit (Riede 2004). It lives at a depth range of 0–1000 m (Nakamura 1985). It reaches a maximum total length of 5000 mm cm (Randall 1995) and a maximum reported weight of 636.0 kg (IGFA 2001). This species is sensitive to water color as it affects its occurrence, at least in the northern Gulf of Mexico, where the fish show a preference for blue water. It is infrequently met in schools and frequently found as dispersed single individuals. It feeds chiefly on fish but also eats octopods and squids. It feeds during daytime. Individuals of this species reach maturity at 800 mm for males and 500 mm for females (Bianchi et al. 1999). Usually, females are larger than males (Nakamura 1986). This species may reach to the age of 20 years (Andrews et al. 2018). Premature ova are opaque in color, white to yellow, and 0.3–0.5 mm in diameter.

Distribution
This species is distributed in the Atlantic Ocean and reported from tropical and temperate waters. It is a highly migratory species.

Economic Values
The flesh of this species is considered among high-quality fish meat and usually marketed fresh or frozen (Nakamura 1985).

Conservation
Collette et al. (2011e) evaluated this species as Vulnerable in the Red List of the IUCN based on the following reasons: A decline in the populations of this species. It is not well managed. Fishing efforts in the Pacific and Indian Oceans are increasing. There is a vital requisite for an efficient stock assessment of the Pacific Ocean, and a reevaluation of this species may be necessary when newer information becomes accessible.

Tetrapturus angustirostris (Tanaka 1915) (Fig. 3.27).
Common name: Shortbill spearfish.
Etymology: *Tetrapturus:* Greek, tetra = four + Greek, pteron, fin, wing (Romero 2002).

Diagnostic Features
The color of the body is dark blue from above, blue splattered with brown on the sides, and sil-

Fig. 3.27 *Tetrapturus angustirostris* Tanaka, 1915. (Courtesy of N. Holtzha, Wikipedia)

very white from below. The first dorsal fin plain is dark blue, and the other fins are brown or dark brown. The anal fin bases are with tinges of silvery white.

Habitat, Biology, and Ecology
This is a marine species inhabiting the pelagic-oceanic region. It has an oceanodromous habit (Riede 2004) at a depth range of 0–1830 m (Fricke et al. 2011). It reaches a maximum total length of 2300 mm (Nakamura 1995) and a maximum reported weight of 52.0 kg (Nakamura 1985). Individuals of this species usually live above the thermocline (Nakamura 1995). Fishes, cephalopods, and crustaceans are the main food items of this species. Spawning occurs chiefly through the winter months, particularly in warm offshore currents with a surface temperature of about 25 °C.

Distribution
The shortbill spearfish is distributed in the Indian and Pacific Oceans. It lives in tropical and temperate waters, with some wandering individuals found in the Atlantic Ocean. This species is thought to have its spawning grounds and principal populations only in the Pacific and Indian Oceans. It is a highly migratory species.

Economic Value
The flesh of this species is usually marketed frozen and prepared as fish cakes and sashimi in Japan (Nakamura 1995) and also as sausages.

Conservation
Collette et al. (2011b) evaluated this species as Data Deficient in the Red List of The IUCN based on the following reasons: It is a rare and widespread species, found in bycatch of longline fisheries along its geographical distribution line. There are also no stock assessment for this species and a lack of data on its life history. Its population is reported to be declining as it is found in the same areas where the blue marlin and striped marine are present.

Family: Xiphiidae.
Xiphias gladius Linnaeus, 1758 (Fig. 3.28).
Common name: Swordfish.
Etymology: *Xiphias:* Greek, xiphos = sword (Romero 2002).

Diagnostic Features
The body is blackish-brown fading to light brown at the abdomen. The first dorsal fin has blackish-brown membrane. Other fins are brown or blackish brown. This species has a long, flat, swordlike bill and no pelvic fins.

Habitat, Biology, and Ecology
The swordfish is a marine species that lives in the pelagic-oceanic region. It has an oceanodromous habit (Riede 2004). It is found at a depth range of 0–2878 m (Collette et al. 2011f). It reaches a maximum fork length of 4550 mm (IGFA 2001) and a maximum reported weight of 650 Kg (Nakamura 1986). Individuals of this

Fig. 3.28 *Xiphias gladius* Linnaeus, 1758. (Courtesy of Hamid Badar Osmany, Pakistan)

species are usually oceanic, but they are also found in coastal waters and normally live above the thermocline (Collette 1995a, b). The larvae of this species normally live at temperatures above 24 °C (Nakamura 1995). They migrate toward temperate or cold waters in the summer and back to warm waters in the winter. It feeds on fishes such as Atlantic mackerel, barracudinas, silver hake, redfish, herring, and lanternfishes (Scott and Scott 1988) and crustaceans and squids (Collette 1995a, b). This species uses its sword to kill its prey (Collette 1995a, b). The spawning usually takes place in the Atlantic Ocean during spring in the southern Sargasso Sea. Individuals of this species migrate to cooler waters to feed (Nakamura 1995). The body growth in females is faster than that in males. It is difficult to assess the age of this species due to the small size of the otolith and the absence of the scales, but rings for ageing have been effectively counted on cross sections of the fin rays (Muus and Nielsen 1999).

Distribution
This species is distributed in the Atlantic, Indian, and Pacific Oceans. It is reported from tropical and temperate and sometimes cold waters, including the Mediterranean Sea, the Sea of Marmara, the Black Sea, and the Sea of Azov. It is a highly migratory species.

Economic Value
This species is considered a high-quality food for its flesh. It is usually marketed fresh or frozen and can be made into sashimi, teriyaki, or fillets (Collette 1995a, b).

Conservation
Collette et al. (2011f) evaluated this species as Least Concern in the Red List of the IUCN based on the following reasons. This species has a worldwide distribution and shown only 28% decline over three generation lengths (20 years). The Mediterranean stock of this species lacks assessment.

3.4.1 Injuries Caused by Groupers, the Red Snapper, Marlin, Makaria, Sailfish, and Swordfish

Haddad (2021) mentioned that the species of grouper and red snapper can cause severe bites and wounds due to the presence of strong spines, sharp teeth, and sharp blades on the operculum for grouper species. As to the swordfish species like marlin, these species have a traumatogenic beak used to capture prey and for defense purposes.

Gooi et al. (2007) reported on one case of swordfish attack that led to death by penetrating head injury. They reported that it was an unusual case of death that resulted from intracranial penetrating injury.

Haddad and de Figueiredo (2009) reported a case of a bather injured by a swordfish in shallow water off the coast of southern Brazil. The victim suffered multiple injuries from the fish's bill, including a puncture wound of his left knee and the tip of the bill remaining as a foreign body. The victim made a full recovery after field management and conservative care.

Georgiadou et al. (2010) reported on a unique case of abdomen penetration by swordfish. The other similar cases resulted in one thoracic trauma, two to head trauma, and one to knee trauma. Among the victims, three were professional or amateur fishermen, whereas in the last reported case, the victim was a bather. Georgiadou et al. (2010) reported that their case is the only case where organic debris of animal's origin remained in the spinal canal after penetrating trauma. They also concluded that much has been written about the management of penetrating abdominal and spinal cord trauma, and controversy remains about its optimal management. Moreover, there is little experience in the management of patients with such spinal injuries, since such cases are extremely rare. In this report, they focused on the patient's treatment about abdominal and spinal trauma.

Sriram et al. (2017) reported on a case of swordfish attacks on humans. The authors described the first known case of a penetrating spinal injury from a swordfish, in which the patient presented with a small stab wound and hemiparesis. The presentation of a fisherman with hemiparesis and a harmless-looking stab wound must alert clinicians to the possibility of penetrating swordfish injuries to the spine.

In 2022, a Maryland woman was stabbed in the groin by a sailfish that jumped out of the water while her two companions were fishing off the coast of Florida, police said (Washington Post 2022). Katherine Perkins, 73, of Arnold, MD, was airlifted to a hospital after the boat returned to shore in Stuart, FL, according to an incident report from Martin County Sheriff's Office. Perkins was on the boat with Louis Toth, 75, and Dominic Bellezza, 77, when Toth hooked the sailfish, which the two men estimated to weigh 100 pounds, the report said. As Toth tried to reel it in, it "began to charge at the boat" and leaped out of the water, stabbing Perkins in the groin as she stood next to the boat's center console. According to police, Toth and Bellezza "immediately put pressure on the wound" and called for help as they returned to shore. Perkins told the responding officer that "the incident occurred so fast, she did not have time to react," police said. Regarded as the fastest fish in the ocean by some scientists, sailfish can swim up to 68 mph and grow to 11 feet long.

Front Line (2022) brought some news about a fisherman (40 years) was attacked by the giant fish with its spear-like snout in the Bay of Bengal off Visakhapatnam coast when he, along with four others, was fishing in the deep sea 60 nautical miles from the shore. The group netted a huge fish, and since it was heavy, they were unable to lift the net. The victim ventured into the waters to check. Suddenly, the fish popped out and pierced Joanna's stomach with its sharp snout. The other fishermen tried to rescue him, but they were in vain. According to the fishermen accompanying Johanna, he was attacked by "Kommu Konam," as the Black Marlin Fish is called locally. Joanna's body was brought to the shore by the other fishermen. The body was shifted to a government hospital at Anakapalle for autopsy. A police officer said while there have been incidents of giant fish attacking fishermen in the past, this was the first incident where a fisherman lost his life in the attack. Marlin fishes are said to be very aggressive and fast swimmers and attack with their sharp upper snouts. Marlin is often spotted off Visakhapatnam beach during winter months.

References

Allen GR. FAO Species Catalogue. Vol. 6. Snappers of the world. An annotated and illustrated catalogue of lutjanid species known to date. FAO Fish. Synop. 125(6): Rome: FAO; 1985. 208 p.

Allen GR, Erdmann MV. Reef fishes of the east indies. Volumes I-III. Tropical reef research. Perth: University of Hawai'i Press; 2012.

Allen GR, Steene RC. Fishes of Christmas Island Indian Ocean, 6798. Christmas Island: Christmas Island Natural History Association; 1988. 197 p.

Allsop DJ, West SA. Constant relative age and size at sex change for sequentially hermaphroditic fish. J Evol Biol. 2003;16(2003):921–9.

Anderson WD Jr. Lutjanidae. (Genus *Lutjanus* by G.R. Allen). In: Smith MM, Heemstra PC, editors. Smiths' sea fishes. Berlin: Springer; 1986. p. 572–9.

Anderson WD Jr, Allen GR. Lutjanidae. Jobfishes. In: Carpenter KE, Niem V, editors. FAO species identification guide for fishery purposes. The living marine resources of the Western Central Pacific, Bony fishes part 3 (Menidae to Pomacentridae), vol. 5. Rome: FAO; 2001. p. 2840–918.

Anderson WD Jr, Baldwin CC. A new species of Anthias (Teleostei: Serranidae: Anthiinae) from the Galápagos Islands, with keys to Anthias and eastern Pacific Anthiinae. Proc Biol Soc Wash. 2000;113(2):369–85.

Andrews AH, Kalish JM, Newman SJ, Johnston JM. Bomb radiocarbon dating of three important reef-fish species using indo-Pacific D14C chronologies. Mar Freshwat Res. 2011;62:1259–69.

Andrews AH, DeMartini EE, Brodziak J, Nichols RS, Humphreys RL. A long-lived life history for a tropical, Deepwater snapper (Pristipomoides filamentosus): bomb radiocarbon and lead-radium dating as extensions of daily increment analyses in otoliths. Can J Fish Aquat Sci. 2012;69:1850–69.

Andrews AH, Humphreys RL Jr, Sampaga JD. Blue marlin (Makaira nigricans) longevity estimates confirmed with bomb radiocarbon dating. Can J Fish Aquat Sci. 2018;75:17–25.

Ayling T, Cox GJ. Collins' guide to sea fishes of New Zealand. Auckland: Collins; 1982. 343 p.

Bacchet P, Zysman T, Lefèvre Y. Guide des poissons de Tahiti et ses îles. Tahiti (Polynésie Francaise): Éditions Au Vent des Îles; 2006. 608 p.
Baensch HA, Debelius H. Meerwasser atlas. 3rd ed. Melle: Mergus Verlag GmbH; 1997. p. 1216.
Bagnis R, Mazellier P, Bennett J, Christian E. Fishes of Polynesia. Tahiti: Les Editions du Pacifique; 1972. 368 p.
Barss PG. Injuries caused by garfish in Papua New Guinea. Br Med J. 1982;284:77–9.
Barton M, Bond CE. Bond's biology of fishes. 3rd ed. Michael Barton, Thomson, Brooks/Cole; 2007. p. 820.
Beebe W. The Arcturus adventure. New York: Putnam's Sons; 1926.
Bendet E, Wolf M, George Levenion G, Faibel M, Kronenberg J. Penetrating cervical injury caused by a needlefish. Ann Otol Rhinol Laryngol. 1995;104:248–50.
Bianchi G, Carpenter KE, Roux J-P, Molloy FJ, Boyer D, Boyer HJ. FAO species identification guide for fishery purposes. Field guide to the living marine resources of Namibia. Rome: FAO; 1999. 265 p.
Bouhlel M. Poissons de Djibouti. Placerville: RDA International; 1988. 416 p.
Breder CM, Rosen DE. Modes of reproduction in fishes. Neptune City: T.F.H. Publications; 1966. 941 p.
Burgess WE, Axelrod HR, Hunziker RE III. Dr. Burgess's atlas of marine aquarium fishes. Neptune City: T.F.H. Publications; 1990.
Bykov VP. Marine fishes: chemical composition and processing properties. New Delhi: Amerind Publishing; 1983. 322 p.
Cervigón F, Cipriani R, Fischer W, Garibaldi L, Hendrickx M, Lemus AJ, Márquez R, Poutiers JM, Robaina G, Rodriguez B. Fichas FAO de identificación de especies para los fines de la pesca. Guía de campo de las especies comerciales marinas y de aquas salobres de la costa septentrional de Sur América. Rome: FAO; 1992. 513 p. Preparado con el financiamento de la Comisión de Comunidades Europeas y de NORAD.
Chapman LB, Cusack P. South Pacific commission deep sea fisheries development project report on second visit to Tuvalu 30 august—7 December 1983. Noumea: South Pacific Commission; 1990.
Choat JH, McIlwain J, Abesamis R, Clements KD, Myers R, Nanola C, Rocha LA, Russell B, Stockwell B. Acanthurus dussumieri. The IUCN Red List of Threatened Species 2012: e.T177981A1510126; 2012a.
Choat JH, McIlwain J, Abesamis R, Clements KD, Myers R, Nanola C, Rocha LA, Russell B, Stockwell B. Acanthurus nigrofuscus. The IUCN Red List of Threatened Species 2012: e.T178019A1523035; 2012b.
Claro R. Características generales de la ictiofauna. In: Claro R, editor. Ecología de los peces marinos de Cuba. Instituto de Oceanología Academia de Ciencias de Cuba and Centro de Investigaciones de Quintana Roo; 1994. p. 55–70.
Clements KD, Choat JH, Abesamis R, McIlwain J, Myers R, Nanola C, Rocha LA, Russell B, Stockwell B. Prionurus maculatus. The IUCN Red List of Threatened Species 2012: e.T178025A1524949; 2012.
Collette BB. Belonidae. Agujones, maraos. In: Fischer W, Krupp F, Schneider W, Sommer C, Carpenter KE, Niem V, editors. Guia FAO para Identification de Especies para lo Fines de la Pesca. Pacifico Centro-Oriental, vol. 3 Vols. Rome: FAO; 1995a. p. 919–26.
Collette BB. Xiphiidae. Peces espada. In: Fischer W, Krupp F, Schneider W, Sommer C, Carpenter KE, Niem V, editors. Guia FAO para Identification de Especies para lo Fines de la Pesca. Pacifico Centro-Oriental, vol. 3 Vols. Rome: FAO; 1995b. p. 1651–2.
Collette BB. Belonidae. Needlefishes. In: Carpenter KE, Niem VH, editors. FAO species identification guide for fishery purposes. The living marine resources of the Western Central Pacific, Bony fishes part 2 (Mugilidae to Carangidae), vol. 4. Rome: FAO; 1999. p. 2151–61.
Collette BB, Parin NV. Belonidae. In: Quero JC, Hureau JC, Karrer C, Post A, Saldanha L, editors. Check-list of the fishes of the eastern tropical Atlantic (CLOFETA), vol. 2. Lisbon/Paris: JNICT/SEI/UNESCO; 1990. p. 592–7.
Collette B, Matsuura K, Nelson J, Dooley J, Fritzsche R, Carpenter K. Platybelone argalus. The IUCN Red List of Threatened Species 2010: e.T154844A115241089; 2010.
Collette B, Acero A, Canales Ramirez C, Carpenter KE, Di Natale A, Fox W, Miyabe N, Montano Cruz R, Nelson R, Schaefer K, Serra R, Sun C, Uozumi Y, Yanez E. Istiompax indica. The IUCN Red List of Threatened Species 2011: e.T170312A6742465. 2011a. https://doi.org/10.2305/IUCN.UK.2011-2.RLTS.T170312A6742465.en. Accessed 18 Oct 2022.
Collette B, Acero A, Canales Ramirez C, Carpenter KE, Di Natale A, Fox W, Miyabe N, Montano Cruz R, Nelson R, Schaefer K, Serra R, Sun C, Uozumi Y, Yanez E. Tetrapturus angustirostris. The IUCN Red List of Threatened Species 2011: e.T170315A6744759. 2011b. https://doi.org/10.2305/IUCN.UK.2011-2.RLTS.T170315A6744759.en. Accessed 20 Oct 2022.
Collette B, Acero A, Boustany A, Canales Ramirez C, Cardenas G, Carpenter KE, Di Natale A, Die D, Fox W, Graves J, Hinton M, Juan Jorda M, Minte Vera C, Miyabe N, Montano Cruz R, Nelson R, Restrepo V, Schaefer K, Schratwieser J, Serra R, Sun C, Uozumi Y, Yanez E. Kajikia audax. The IUCN Red List of Threatened Species 2011b: e.T170309A6738801. 2011c. https://doi.org/10.2305/IUCN.UK.2011-2.RLTS.T170309A6738801.en. Accessed 20 Oct 2022.
Collette B, Acero A, Amorim AF, Boustany A, Canales Ramirez C, Cardenas G, Carpenter KE, de Oliveira Leite Jr. N, Di Natale A, Die D, Fox W, Fredou FL, Graves J, Guzman-Mora A, Viera Hazin FH, Hinton M, Juan Jorda M, Minte Vera C, Miyabe N, Montano Cruz R, Nelson R, Oxenford H, Restrepo V, Salas E, Schaefer K, Schratwieser J, Serra R, Sun C, Teixeira

Lessa RP, Pires Ferreira Travassos PE, Uozumi Y, Yanez E. Istiophorus platypterus. The IUCN Red List of Threatened Species 2011d: e.T170338A6754507. 2011d. https://doi.org/10.2305/IUCN.UK.2011-2.RLTS.T170338A6754507.en. Accessed 19 Oct 2022.

Collette B, Acero A, Amorim AF, Boustany A, Canales Ramirez C, Cardenas G, Carpenter KE, de Oliveira Leite Jr. N, Di Natale A, Die D, Fox W, Fredou FL, Graves J, Guzman-Mora A, Viera Hazin FH, Hinton M, Juan Jorda M, Minte Vera C, Miyabe N, Montano Cruz R, Nelson R, Oxenford H, Restrepo V, Salas E, Schaefer K, Schratwieser J, Serra R, Sun C, Teixeira Lessa RP, Pires Ferreira Travassos PE, Uozumi Y, Yanez E. Makaira nigricans. The IUCN Red List of Threatened Species 2011e: e.T170314A6743776. 2011e. https://doi.org/10.2305/IUCN.UK.2011-2.RLTS.T170314A6743776.en. Accessed 20 Oct 2022.

Collette B, Acero A, Amorim AF, Bizsel K, Boustany A, Canales Ramirez C, Cardenas G, Carpenter KE, de Oliveira Leite Jr. N, Di Natale A, Die D, Fox W, Fredou FL, Graves J, Guzman-Mora A, Viera Hazin FH, Hinton M, Juan Jorda M, Minte Vera C, Miyabe N, Montano Cruz R, Masuti E, Nelson R, Oxenford H, Restrepo V, Salas E, Schaefer K, Schratwieser J, Serra R, Sun C, Teixeira Lessa RP, Pires Ferreira Travassos PE, Uozumi Y, Yanez E. Xiphias gladius (errata version published in 2016). The IUCN Red List of Threatened Species 2011: e.T23148A88828055. 2011f. https://doi.org/10.2305/IUCN.UK.2011-2.RLTS.T23148A9422329.en. Accessed 20 Oct 2022.

Collette B, Polanco Fernandez A, Aiken KA Ablennes hians. The IUCN Red List of Threatened Species 2015: e.T13486514A15603320; 2015.

Cornic A. Poissons de l'Ile Maurice. Stanley Rose Hill: Editions de l'Océan Indien; 1987. 335 p.

Craig MT, Sadovy de Mitcheson YJ, Heemstra PC. Groupers of the world: a field and market guide. North America: CRC Press/Taylor and Francis Group; 2011. 356 p.

Deutsche Press Agenteur. Needlefish stabs diver to death in Vietnam. Deutsche Press Agenteur; 2007. http://www.digitaljournal.com/article. 10 Sep 2007.

Domeier ML, Colin PL. Tropical reef fish spawning and aggregations: defined and reviewed. Bull Mar Sci. 1997;60(3):698–726.

Dyer BS, Westneat MW. Taxonomy and biogeography of the coastal fishes of Juan Fernández Archipelago and Desventuradas Islands. Chile Rev Biol Mar Oceanogr. 2010;45(S1):589–617.

Ebner Y, Golani D, Ophir D, Finkelstein Y. Penetrating injury of the maxilla by needlefish jaws. J Craniomaxillofac Surg. 2009;37(4):235–8.

Eschmeyer WN. Eschmeyer's catalog of fishes online database. 2022. http://researcharchive.calacademy.org//research/ichthyology/catalog/fishcatget.asp?spid=27224. Visited 17 Feb 2022.

FAO-FIGIS. A world overview of species of interest to fisheries. Chapter: Istiophorus platypterus. Retrieved on 05 May 2005, from www.fao.org/figis/servlet/species?fid=3301. 4p. FIGIS Species Fact Sheets. Species Identification and Data Programme-SIDP, FAO-FIGIS; 2001.

Feitoza BM, Rocha LA, Luiz-Júnior OJ, Floeter SR, Gasparini JL. Reef fishes of St. Paul's rocks: new records and notes on biology and zoogeography. J Ichthyol Aquat Biol Aqua. 2003;7(2):61–82.

Fennessy S, Pollard DA, Samoilys M. Epinephelus lanceolatus. The IUCN Red List of Threatened Species 2018: e.T7858A100465809. 2018. https://doi.org/10.2305/IUCN.UK.2018-2.RLTS.T7858A100465809.en

Fischer W, Sousa I, Silva C, de Freitas A, Poutiers JM, Schneider W, Borges TC, Feral JP, Massinga A. Fichas FAO de identificaçao de espécies para actividades de pesca. Guia de campo das espécies comerciais marinhas e de águas salobras de Moçambique. Publicaçao preparada em collaboraçao com o Instituto de Investigaçao Pesquiera de Moçambique, com financiamento do Projecto PNUD/FAO MOZ/86/030 e de NORAD. Rome: FAO; 1990. 424 p.

Fishelson L, Montgomery LW, Myrberg AH Jr. Biology of surgeonfish *Acanthurus nigrofuscus* with emphasis on changeover in diet and annual gonadal cycles. Mar Ecol Prog Ser. 1987;39:37–47.

Foale S. What's in a name? An analysis of the West Nggela (Solomon Islands) fish taxonomy. SPC Traditional Marine Resource Management and Knowledge Information Bulletin No. 9; 1998. p. 3–19.

Francis M. Coastal fishes of New Zealand, a diver's identification guide. Auckland: Heinemann Reed, Division of Octopus Publishing Group (NZ); 1988. 63 p.

Francis MP. Checklist of the coastal fishes of Lord Howe, Norfolk, and Kermadec Islands. Southwest Pac Ocean Pac Sci. 1993;47(2):136–70.

Francis MP. Coastal fishes of New Zealand. Nelson: Craig Potton; 2012. 268pp.

Francis MP, Worthington CJ, Saul P, Clements KD. New and rare tropical and subtropical fishes from northern New Zealand. NZ J Mar Freshwat Res. 1999;33:571–86.

Fricke R, Kulbicki M, Wantiez L. Checklist of the fishes of New Caledonia, and their distribution in the Southwest Pacific Ocean (Pisces). Stuttgarter Beiträge zur Naturkunde A Neue Serie. 2011;4:341–463.

Frimodt C. Multilingual illustrated guide to the world's commercial warmwater fish. Oxford: Fishing News Books; 1995. 215 p.

Froese R, Pauly D. Editors. FishBase. 2022. World Wide Web electronic publication. www.fishbase.org, version (02/2022).

Front Line. https://frontline.thehindu.com/dispatches/marlin-kills-fisherman-off-visakhapatnam-coast/article65220379.ece. 2022.

Fry GC, Brewer DT, Venables WN. Vulnerability of Deepwater demersal fishes to commercial fishing: evidence from a study around a tropical volcanic seamount in Papua New Guinea. Fish Res. 2006;81:126–41.

Gasparini JL, Floeter SR. The shore fishes of Trindade Island, western South Atlantic. J Nat Hist. 2001;35:1639–56.

Georgiadou D, Zografos GN, Vaidakis D, Avlonitis S, Katopodi A, Tzirakis EN, Sioutos P, Drossos C, Lampropoulou P, Papastratis G. Swordfish bill injury involving abdomen and vertebral column: case report and review. BMC Surg. 2010;10(1):1–8.

Gooi BH, Khamizar W, Suhani MN. Swordfish attack—death by penetrating head injury. Asian J Surg. 2007;30(2):158–9.

Haddad VH Jr. Medical emergencies caused by aquatic animals: a biological and clinical guide to trauma and envenomation cases. Springer; 2021.

Haddad V Jr, de Figueiredo JL. Attack upon a bather by a swordfish: a case report. Wilderness Environ Med. 2009;20(4):344–6.

Haddad Junior V. Infecções cutâneas e acidentes por animais traumatizantes e venenosos ocorridos em aquários comerciais e domésticos no Brasil: descrição de 18 casos e revisão do tema. An Bras Dermatol. 2004;79(2):157–67.

Halstead BW. Poisonous and venomous marine animals of the world, vertebrates, vol. 2. Washington, DC: U.S. Government Printing Office; 1967.

Heemstra PC. Taxonomic review of the perciform fish genus *Acanthistius* from the east coast of southern Africa, with description of a new species and designation of a neotype for Serranus sebastoides Castelnau, 1861. Zootaxa. 2010;2352:59–68.

Heemstra PC, Randall JE. FAO Species Catalogue. Vol. 16. Groupers of the world (family Serranidae, subfamily Epinephelinae). An annotated and illustrated catalogue of the grouper, rock cod, hind, coral grouper and lyretail species known to date. FAO Fish. Synop. 125 (16). Rome: FAO; 1993. 382 p.

Heemstra PC, Randall JE. Serranidae. In Carpenter KE, Niem VH, editors. FAO species identification guide for fishery purposes. In The living marine resources of the Western Central Pacific. Vol. 4. Bony fishes part 2 (Mugilidae to Carangidae). Rome: FAO; 1999. p. 2442–547, 2069–790 p.

Hoese DF, Bray DJ, Paxton JR, Allen GR. Fishes. In: Beesley PL, Wells A, editors. Zoological catalogue of Australia. Australian Biological Resources Study and Australia's Commonwealth Scientific and Industrial Research Organisation (CSIRO); 2006. p. 2178.

Honebrink, R. Fishing in Hawaii: a student manual. Honolulu: Education Program, Division of Aquatic Resources; 1990. 79 p.

Hutchins JB, Kuiter RH. A new species of Acanthistius (Pisces: Serranidae) from eastern Australia. Rec West Aust Mus. 1982;10(2):127–31.

IGFA. Database of IGFA angling records until 2001. Fort Lauderdale: IGFA; 2001.

Jordan DS, Evermann BW. Descriptions of new genera and species of fishes from the Hawaiian Islands. Bull U. S. Fish Comm. 1903;22(1902):161–208.

Kailola PJ, Williams MJ, Stewart PC, Reichelt RE, McNee A, Grieve C. Australian fisheries resources. Canberra: Bureau of Resource Sciences; 1993. p. 422.

Kapoor D, Dayal R, Ponniah AG. Fish biodiversity of India. Lucknow: National Bureau of Fish Genetic Resources; 2002.

Kite Magazine. Issue 5, Sept 2012.

Krupp F. Acanthuridae. Sangradores, cirujanos, navajones. In: Fischer W, Krupp F, Schneider W, Sommer C, Carpenter KE, Niem V, editors. Guia FAO para Identification de Especies para lo Fines de la Pesca. Pacifico Centro-Oriental, vol. 3 Vols. Rome: FAO; 1995. p. 839–44.

Krzyżak J, Korzeniewski K. Medicine for divers [in Polish]. Poznań: Publishing House 4Font; 2020. p. 674–707.

Kuiter RH, Tonozuka T. Pictorial guide to Indonesian reef fishes. Part 3. Jawfishes—sunfishes, Opistognathidae—Molidae. Zoonetics. 2001:623–893.

Labbe JL, Bordes JP, Fine X. An unusual surgical emergency (a knee joint wound cause by a needlefish). Arthroscopy. 1995;16:503–5.

Lee Marsh S. Fish fauna observations at the Kermadec Island Group (N.Z.). TANE. 1986;31:145–53.

Lieske E, Myers R. Collins pocket guide. Coral reef fishes. Indo-Pacific & Caribbean including the Red Sea. Haper Collins; 1994. 400 p.

May JL, Maxwell JGH. Trawl fish from temperate waters of Australia. Tasmania: CSIRO Division of Fisheries Research; 1986. 492 p.

McCabe MJ, Hammon WM, Halstead BW, et al. A fatal brain injury caused by a needle fish. Neuroradiology. 1978;15:137–9.

McIlwain J, Choat JH, Abesamis R, Clements KD, Myers R, Nanola C, Rocha LA, Russell B, Stockwell B. Acanthurus triostegus. The IUCN Red List of Threatened Species 2012: e.T177965A1504553; 2012a.

McIlwain J, Choat JH, Abesamis R, Clements KD, Myers R, Nanola C, Rocha LA, Russell B, Stockwell B. Naso unicornis. The IUCN Red List of Threatened Species 2012: e.T177970A1506556; 2012b.

McMillan PJ, Francis MP, Paul LJ, Marriott PJ, Mackay E, Baird S-J, Griggs LH, Sui H, Wei F. New Zealand fishes. Volume 2: A field guide to less common species caught by bottom and midwater fishing. New Zealand Aquatic Environment and Biodiversity Report No. 78; 2011.

Moreno L. Needlefish victim recalls attack. Honolulu Advisor; 2005.

Mundy BC. Checklist of the fishes of the Hawaiian archipelago. Bishop Mus Bull Zool. 2005;6:1–704.

Muus BJ, Nielsen JG. Sea fish. Hedehusene: Scandinavian Fishing Year Book; 1999. 340 p.

Myers RF. Micronesian reef fishes. 2nd Ed. Barrigada: Coral Graphics; 1991. 298 p.

Myers RF. Micronesian reef fishes: a comprehensive guide to the coral reef fishes of Micronesia, 3rd revised and expanded edition. Barrigada: Coral Graphics; 1999. 330 p.

Nakamura I. FAO species catalogue. Vol. 5. Billfishes of the world. An annotated and illustrated catalogue of marlins, sailfishes, spearfishes and swordfishes known to date. FAO Fish. Synop. 125(5). Rome: FAO; 1985. 65p.

Nakamura I. Istiophoridae. In: Whitehead PJP, Bauchot M-L, Hureau J-C, Nielsen J, Tortonese E, editors. Fishes of the North-Eastern Atlantic and the Mediterranean, vol. 2. Paris: UNESCO; 1986. p. 1000–5.

Nakamura I. Istiophoridae. Agujas, marlines, peces-vela. In: Fischer W, Krupp F, Schneider W, Sommer C, Carpenter KE, Niem V, editors. Guia FAO para Identification de Especies para lo Fines de la Pesca. Pacifico Centro-Oriental, vol. 3 Vols. Rome: FAO; 1995. p. 1186–93.

Nawasiwatte BMTP, Bandusena S, Wadanamby S, Gunaratne PS. A deadly strike of a needlefish. Sri Lanka J Neurol. 2014;3:31–2.

Nelson JS. Fishes of the world. New York: Wiley; 1976. 416 p.

Ohtsubo M, Fujita K, Tsunekawa K. Penetrating injury of the orbit by a needlefish. ePlasty OAJ Plast Surg. 2013;13

Opic P, Conand F, Bourret P. Poissons commerciaux du sud-ouest de l'océan Indien. ORSTOM Éditions; 1994. 91p.

Osmany HB, Moazzam M. Naso Unicornis (Forsskål, 1775): record of a new unicornfish from Pakistan; 2021.

Paulin C, Roberts C. The rockpool fishes of New Zealand (Te ika aaria o Aotearoa). Museum of New Zealand (Te Papa Tongarewa); 1992. 177 p.

Paulin C, Stewart A, Roberts C, McMillan P. New Zealand fish: a complete guide. National Museum of New Zealand Miscellaneous Series No. 19; 1989. 279 p.

Paxton JR, Hoese DF, Allen GR, Hanley JE. Pisces. Petromyzontidae to Carangidae. Zoological Catalogue of Australia, Vol. 7. Canberra: Australian Government Publishing Service; 1989. 665 p.

Pillai PP, Ueyanagi S. Distribution and biology of the striped marlin, Tetrapterus [sic] audax (Philippi) taken by the longline fishery in the Indian Ocean. Bull Far Seas Fish Res Lab. 1977;16:9–32.

Pollard DA, Sadovy Y. Epinephelus daemelii. The IUCN Red List of Threatened Species 2018: e.T61337A100465433. 2018. https://doi.org/10.2305/IUCN.UK.2018-2.RLTS.T61337A100465433.en

Prince ED, Lee DW, Wilson CA, Dean JM. Longevity and age validation of tag-recaptured Atlantic sailfish, Istiophorus platypterus, using dorsal spines and otoliths. Fish Bull. 1986;84:493–502.

Radtke LD. Distribution of smelt, juvenile sturgeon, and starry flounder in the Sacramento-San Joaquin Delta with observations on food of sturgeon. In Turner JL, Kelly DW (comp.) Ecological studies of the Sacramento-San Joaquin Delta. Part II Fishes of the Delta. Fisheries Bulletin 136; 1966. p. 115–29.

Randall JE. Report of a caudal-spine wound from the surgeonfish Acanthurus limeatus in the society islands. Wasman J Biol. 1959;17:245–8.

Randall, J.E. Acanthuridae. In Fischer W, Bianchi G, editors. FAO species identification sheets for fishery purposes. Western Indian Ocean (Fishing Area 51). Vol. 1. Rome: FAO. pag. var; 1984.

Randall JE. Acanthuridae. In: Smith MM, Heemstra PC, editors. Smiths' sea fishes. Berlin: Springer; 1986. p. 811–23.

Randall JE. Coastal fishes of Oman. Honolulu: University of Hawaii Press; 1995. 439 p.

Randall JE. Acanthuridae. Surgeonfishes. In: K.E. Carpenter (ed.), The living marine resources of the Western Central Atlantic. Bony fishes part 2 (Opistognathidae to Molidae), sea turtles and marine mammals. Rome: FAO; 2002. pp. 1801–1805.

Randall JE, Heemstra PC. Holocentridae. In: Smith MM, Heemstra PC, editors. Smiths' sea fishes. Berlin: Springer; 1986. p. 415–27.

Randall JE, Heemstra PC. Revision of indo-Pacific groupers (Perciformes: Serranidae: Epinephelinae), with descriptions of five new species. Indo-Pac Fish 1991; 20: 332 p.

Randall JE, Allen GR, Steene RC. Fishes of the great barrier reef and coral sea. Honolulu: University of Hawaii Press; 1990. p. 506.

Riede K. Global register of migratory species—from global to regional scales. Final Report of the R&D-Projekt 808 05 081. Bonn: Federal Agency for Nature Conservation; 2004. 329 p.

Roberts C, Stewart A, Struthers C. The fishes of New Zealand. Auckland: The National Museum of New Zealand Te Papa Tongarewa; 2015. 2000p.

Robins CR, Ray GC. A field guide to Atlantic coast fishes of North America. Boston: Houghton Mifflin; 1986. 354 p.

Romero P. An etymological dictionary of taxonomy. Madrid, unpublished; 2002.

Russell BC. The food and feeding habits of rocky reef fish of North-Eastern New Zealand. NZ J Mar Freshwat Res. 1983;17(2):121–45.

Russell B, Carpenter KE, Smith-Vaniz WF, Lawrence A. Etelis carbunculus. The IUCN Red List of Threatened Species 2016: e.T154999A46634266. 2016a. https://doi.org/10.2305/IUCN.UK.2016-3.RLTS.T154999A46634266.en

Russell B, Lawrence A, Myers R, Carpenter KE, Smith-Vaniz WF. Lutjanus kasmira. The IUCN Red List of Threatened Species 2016: e.T194337A2314753. 2016b. https://doi.org/10.2305/IUCN.UK.2016-3.RLTS.T194337A2314753.en

Russell B, Lawrence A, Myers R, Carpenter KE, Smith-Vaniz WF. Paracaesio xanthura. The IUCN Red List of Threatened Species 2016: e.T194349A2318370. 2016c. https://doi.org/10.2305/IUCN.UK.2016-3.RLTS.T194349A2318370.en

Russell B, Lawrence A, Myers R, Carpenter KE, Smith-Vaniz WF. Pristipomoides filamentosus. The IUCN Red List of Threatened Species 2016: e.T194331A2312944. 2016d. 10.2305/IUCN.UK.2016-3.RLTS.T194331A2312944.en.

Russell B, Smith-Vaniz WF, Lawrence A, Carpenter KE, Myers R, Thaman R. Etelis coruscans. The IUCN Red List of Threatened Species 2016: e.T194382A2327142. 2016e. https://doi.org/10.2305/IUCN.UK.2016-3.RLTS.T194382A2327142.en

Sabiq. 2015. http://sabq.org/EbFfde

Schneider W, Krupp F. Holocentridae. Candiles. In: Fischer W, Krupp F, Schneider W, Sommer C, Carpenter KE, Niem V, editors. Guia FAO para Identification de Especies para lo Fines de la Pesca. Pacifico Centro-Oriental, vol. 3 Vols. Rome: FAO; 1995. p. 1182–5.

Schober UM. Zum agonistischen Verhalten von einigen Acanthuriden. 1984. Ms. Thesis University of Vienna. Downloaded by McGill University Library at 16:14 08 Jan 2022.

Schober UM. Zum Stachelapparat bei Acanthuriden. DATZ. 1986;5:27–30.

Schober UM, Ditrich M. Anatomy and use of the caudal spines in the aggressive behaviour of a surgeonfish (Osteichthyes: Acanthuridae). Mar Behav Physiol. 1992;21(4):277–84.

Scott S. Ocean watch: those needlefish are not totally harmless after all. Honolulu Star-Bulletin. 16 Dec 1996.

Scott WB, Scott MG. Atlantic fishes of Canada. Can Bull Fish Aquat Sci 1988; 219:731 p.

Siddiqueki T, Islam MJ, Habib AHMS, Hasan SJ, Islam ML, Mahmud Y, Habib KH. First record of two marine fish species from the Northern Bay of Bengal, Bangladesh. Bangladesh J Fish. 2021;33:129–36.

Smith CL. National Audubon Society field guide to tropical marine fishes of the Caribbean, the Gulf of Mexico, Florida, the Bahamas, and Bermuda. New York: Alfred A. Knopf; 1997. 720 p.

Smith MM, Heemstra PC, editors. Smith's sea fishes. Berlin: Springer; 1986.

Smith MK, Kostlan E. Estimates of age and growth of ehu Etelis carbunculus in four regions of the Pacific from density of daily increments in otoliths. Fish Bull. 1991;89:461–72.

Sommer C, Schneider W, Poutiers J-M. FAO species identification field guide for fishery purposes. The living marine resources of Somalia. Rome: FAO; 1996. 376 p.

Sorenson L, Santini F, Carnevale G, Alfaro ME. A multi-locus timetree of surgeonfishes (Acanthuridae, Percomorpha), with revised family taxonomy. Mol Phylogenet Evol. 2013;68(1):150–60.

Squire JL Jr, Suzuki Z. Migration trends of striped marlin (Tetrapturus audax) in the Pacific Ocean. In Stroud RH, editor. Planning the future of billfishes. Research and management in the 90's and beyond. Proceedings of the second international billfish symposium, Kailua-Kona, Hawaii, 1–5 August 1988. Part 2. Contributed papers; 1990. p. 67–80.

Sriram PR, Jien TCT, Sellamuthu P. Backstabbing swordfish: a rare cause of traumatic hemiparesis. J Neurosurg Spine. 2017;27(2):158–60.

Thakker MM, Usha KR. Orbital foreign body and ruptured globe from needlefish impalement. Arch Ophthalmol. 2006;124:284.

Thomas C, Scott S. All stings considered. Honolulu: University of Hawaii Press; 1997. p. 96–8.

Titcomb M. Native use of fish in Hawaii. Honolulu: The University Press of Hawaii; 1972. 175 p.

van der Elst R. A guide to the common sea fishes of southern Africa. 3rd Ed. Cape Town: Struik Publishers; 1993. 398 p.

Walsh W, Cotton S, Carman B, Livnat L, Osada K, Barnett C, Tissot B, Stevenson T, Wiggins C, Tarnas D, Bourdon K, Peck S. Report on the Findings and Recommendations of Effectiveness of the West Hawaii Regional Fishery Management Area. Department of Land and Natural Resources State of Hawaii, State of Hawaii; 2010.

Warrell DA. Animals hazardous to humans. In Hunter's tropical medicine and emerging infectious disease; 2013, p. 938–65.

Washington Post. 2022. https://www.washingtonpost.com/science/2022/07/25/sailfish-stabs-woman/

Williams I, Greenfield D. Myripristis berndti. The IUCN Red List of Threatened Species 2016: e.T67869566A115434172. 2016a. https://doi.org/10.2305/IUCN.UK.2016-1.RLTS.T67869566A67871805.en

Williams I, Greenfield D. Pristilepis oligolepis. The IUCN Red List of Threatened Species 2016: e.T67900591A115441358. 2016b. https://doi.org/10.2305/IUCN.UK.2016-1.RLTS.T67900591A67906272.en

Williams JT, Carpenter KE, Lawrence A, Myers R. Acanthistius cinctus. The IUCN Red List of Threatened Species 2016: e.T69584026A69592247; 2016a.

Williams JT, Carpenter KE, Lawrence A, Myers, R. Caprodon longimanus. The IUCN Red List of Threatened Species 2016: e.T69584359A69592277; 2016b.

Zwisler H, Beigel A. Case report: a traumatic facial paralysis caused by a fish. Laryngo-Rhino-Otologie. 1997;76(1):53–4.

4 Electric Fishes

4.1 Electric Rays and Their Electric Organ

In the animal world, electric organs are only present in fish. These organs are mainly intended to produce an electric discharge. The electric organ is established in dissimilar families of both Chondrichthyes and Osteichthyes living in a wide range of environments. Notwithstanding the diversity of the families that share the occurrence of the electric organ, they share a common feature their electric organs are built up from a large number of disc-like cells called electroplaxes or electroplates prepared in a more or less orderly way and have the same direction. Each electroplate is inserted in a jellylike extracellular substance and disconnected from other electroplates by a layer of connective tissue. Blood and sensation are supplied by nerves and blood vessels, where nerves are dispersed on one face of the electroplates, while blood vessels form a network of branches. The electroplates are single cells with several nuclei situated just beneath the two surfaces. The gelatinous masses of the electric organ can simply be marked from the surrounding muscle owing to their transparent cytoplasm. In torpedoes, the electric organs form two large flat kidney-shaped masses on either side of the midline, with the horizontal electroplates packed firmly into hexagonal columns (Keynes 2013).

Going back to the history of the use of electric rays, Pliny (AD 23–AD 79) has given guidelines using torpedoes or torpedo parts which may seem bizarre to us. He tracked the steps of the physician Hippocrates (460–c. 370 BC) and recommended that those with difficulties with their digestive system, adding some of the torpedo's foodstuffs to their diet would be helpful. For splenic problems, he proposed putting on a live torpedo ray in the area so the electric charges would cure the discomfort (Finger and Picolino 2011).

In the Roman period, gout was the chief aching illness among the aristocracies and royals as they used to pamper themselves with rich food. This discomfort is linked to the occurrence of uric acid of which the meat is rich, in addition to less removal of this acid through the kidney because of heavy drinking of wine which contends with uric acid for extraction (Finger and Picolino 2011).

It was established by coincidence that a shock from a live torpedo would numb the pain in the foot. Subsequently, the Greek doctor Scribonius Largus (about AD 47) directed using the torpedo's shock for both foot pain "gout," the hot one *Podagra calda*, characterized by hotness and red swelling, and the cold *Podagra frigida*, where the pain is not related with these signs of inflammation (Keynes 2013). It is recognized that aching sensations can be stopped by peripheral stimuli transmitted by rapidly conducting nerve pathways, which primarily utilize slowly running nerve fibers (Wall and Sweet 1967). It is also acknowledged that releasing endogenous opioids in the brain caused by aching stimuli can diminish pain. Moreover, electrical shock can decrease the

L. A. Jawad, *Dangerous Fishes of New Zealand*, https://doi.org/10.1007/978-3-031-69698-5_4

twinges and pains instigated by chronic diseases containing numerous headaches, sometimes for extended periods (Slavin 2008). Based on these ideas originating from modern medicine, it is known that the Romans made the correct choice of applying live torpedoes to reduce the pain of headaches and chronic gout. Among the early Islamic physicians who wrote on torpedoes and their usage in medicine is Abu Muhammad Abdullah Ibn al-Bitar (or Ibn al-Bitar) (1197–1248) from Malaga, a part of Muslim Spain. He stated in his textbook, *Al-Jami li mufradat al-adwiya wa-l-aghadhiya*, which translates as *Compendium of Simple Drugs and Foods* about the use of a live torpedo in curing headache if the fish were close to the head of the person who suffered from this pain. During the second half of the eighteenth century, electric fish became a significant matter for naturalists, physicists, physicians, and philosophers (Finger and Picolino 2011). At that moment, awareness was focused on the numbness caused by the South American eel, but also some thought was given to the electric ray, the torpedoes.

Bini (1967) has recommended that Europeans have known torpedoes for a long time as disc-shaped rays with moderately small tails. Its name originates from the Latin because it could cause "torpor," meaning sluggishness or sleepiness.

The torporific fishes convey their influences to animals and humans without direct physical contact. Even when there are media through which the electric discharge can travel, such as water, metal rods, and wet nets, such discharges are still strong enough to restrain, numb, and discourage other fishes and humans. The fishermen are experienced people in perceiving the distance effect of the electric organ because they use several fishing equipment to capture fish (Finger and Picolino 2011). In the torpedo where the electric organs are large and robust enough to produce an influential discharge, this discharge is able to stun the prey. Additionally, fish species with weak electric discharge are frequently used for navigation as these fishes have a nocturnal habit and live in turbid water (Nelson 2011).

The electric field can be perceived with the aid of thousands of specialized sense organs that cover the body of the electric fish. Each of these organs comprises a pit filled with several sensory cells. These cells act as small voltmeters and control the voltage across the fish's body. Once a strange object comes closer to the fish, the electric current flow across the skin will change and, in turn, alter the transdermal voltage monitored by the sense organs. The electric fishes can analyze the changes in voltage across the surface of their body to recognize the impending object. They also can identify electric signals given by other electric fishes located in the environment.

Firm electric fish such as torpedoes frequently have thick layers of connective tissue that guard their electric organs against damage by electric charges produced by another electric fish.

The electric shock of a torpedo fish can cause arterial fibrillation. An uncommon case of this kind occurred to a pipeline inspector working for an oil company in Dubai, United Arab Emirates (Taimur and Hussaini 2008). He was shocked by four electric shocks from a torpedo ray fish. His verbal commands became unintelligible after the fourth shock. The video footage shows that the diver was shouting, "electrical fish." The diver became unconscious and was saved by another diver in 8 min. He was mixed up and restless before being transported to a nearby hospital. His pulse was uneven and he was referred to a cardiologist. The rhythm was confirmed as slow arterial fibrillation (87 min^{-1}) with a BP of 140/80 (Taimur and Hussaini 2008). The investigations showed raised white blood cell count.

An electric shock frequently happens when the human body comes into interaction with a source of high voltage that makes it conceivable for the electric current to flow through the body. Typically, high-tension shocks (>1000 V) cause severe internal burns and cardiac arrest (Forrest et al. 1992). Low-tension shocks (under 300 V at 50 Hz) can lead to ventricular fibrillation. The shock produced by these electric fish is low voltage, which likely caused the diver's arrhythmia. But why doesn't every low-voltage electric shock trigger arrhythmia? Besides individual susceptibility, wet skin conditions lower the body's resistance to electrical current. A shock received on wet skin can potentially cause more severe tissue damage (Taimur and Hussaini 2008).

4.2 Species of Electric Fishes

Family: Torpedinidae.
Tetronarce nobiliana (Bonaparte 1835) (Fig. 4.1).
Common name: electric ray.

Diagnostic Features

Tetronarce nobiliana is a large-sized ray with a huge, paddle-shaped caudal fin and with no papillae around spiracles. The disc is broad and subcircular, with a short snout anterior to the eyes. The skin is smooth skin and the tail is thick and short. The color of the dorsal side of the body is shiny black or dark gray with a white underside. Spots on the body are absent. The edges of the disc and pelvic fins are with the same color as the dorsal side of the body. The tail is with irregular dark margins.

Habitat, Biology, and Ecology

This marine species is benthopelagic and has an oceanodromous habit. It is found at a depth range of 2–800 m (Anonymous 2000). It reaches a maximum total length of 1800 mm (Stehmann and Bürkel 1984) and a maximum reported weight of 90.0 kg (Robins and Ray 1986). The juveniles of this species prefer a benthic environment, with soft bottoms or near coral reefs (Michael 1993). The adults are semi-pelagic to pelagic and usually swim in the water column and have been reported to migrate over long distances (Stehmann and Bürkel 1984). Individuals of this species feed chiefly on pelagic and benthic bony fishes and small sharks (Compagno et al. 1989). The females are ovoviviparous (Dulvy and Reynold 1997). The number of litters is up to 60 pups. In the South Pacific region, males reach maturity at ca. 600 mm TL (Last et al. 2016). The gestation period is 1 year (Michael 1993). The size at birth is about 230 mm (Michael 1993). It packs a powerful electric shock of up to 220 volts (Lieske and Myers 1994).

Distribution

This species has a circumglobal distribution type. It is found in the Eastern Atlantic from Scotland (rare in North Sea) to Morocco, the whole of the Mediterranean Sea, but not the Black Sea. It is reported from the Cap Blanc in Mauritania to the Gulf of Guinea, São Tomé Island, and Walvis Bay, Namibia, to Mossel Bay, South Africa. It is

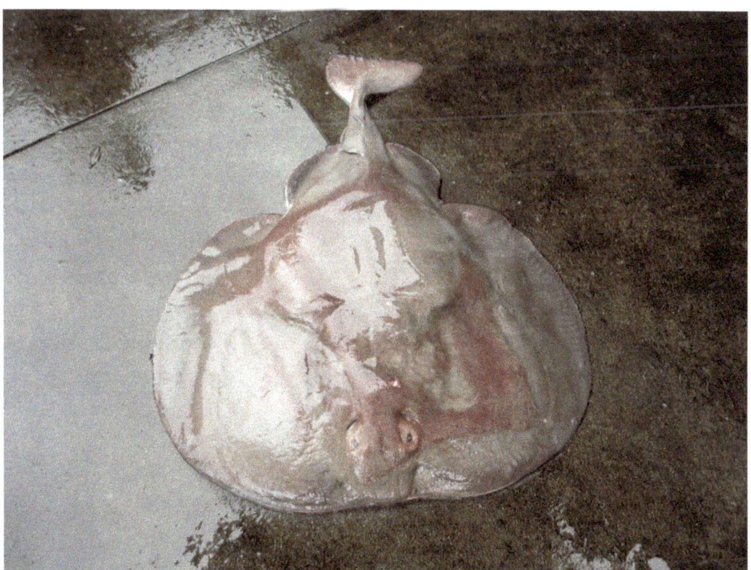

Fig. 4.1 *Tetronarce nobiliana* (Bonaparte 1835). (Courtesy of Clinton Dufy, New Zealand)

also found in the Western Atlantic from Nova Scotia, Canada, south to Brazil. It is recorded from the Eastern Indian Ocean from Australia and from the Southwest Pacific in New Zealand.

Economic Value
This species has no economic value.

Conservation Status
Finucci et al. (2021) evaluated this electric ray as Least Concern in the Red List of the IUCN on the basis of the following reasons. This species is large in size and has a wide distribution line. This species is present in the demersal trawl, longline set net, and trammel net gear. It is a migratory species. There is a lack of information about the biology and ecology of this species. The populations of this species show no indication of decline.

Family: Narkidae.
Typhlonarke aysoni (Hamilton 1902) (Fig. 4.2).
Common name: Blind electric ray.
Etymology: *Typhlonarke:* Greek, *typhlops* = blind + Greek, *narke* = numbness (Romero 2002).

Diagnostic Features
The body is greatly depressed and almost circular. The skin is smooth but lacking dermal denticles. The greatest width of the disc is at least of the last gill slit. The greatest depth of the body is at the pectoral symphysis. The sic is very much thin at its periphery. The caudal peduncle is stout and slightly compressed. The spiracles are irregularly oval, with their long axes directed outwards anteriorly. The margin of the spiracles is produced into a definite rim. The eyes are small and not visible externally. The gill slits are small. The nostrils are moderately closely located. The posterior vales of the nostrils are expanded and curved round forming a nearly complete tube. The mouth is small and slightly curved down deep. The teeth are arranged in a mosaic. The tooth plate is very narrow externally across the median part of the jaws. The dorsal fin is rectangular in shape and originates above the posterior insertion of the pectoral fin. The origin of the caudal fin is just anterior to the rear edge of the dorsal fin. The pectoral fins are completely fused into the body, with a smooth margin. The origin of the pelvic fins is not quite half of the total body length. The anterior portion of each pelvic fin is free and produced into a broad flat appendage. The posterior portion of the pelvic fin is fused with the pectoral as an extension of the disc without any notch between it and the pectoral fin. The upper surface of the body is brown and darkens posteriorly. The undersurface is creamy brown. The underside of the pelvic fin appendage is almost white.

Fig. 4.2 *Typhlonarke aysoni* (Hamilton 1902). (Courtesy of Brit Finucci)

Habitat, Biology, and Ecology
The blind electric ray is a marine species that inhabit a bathydemersal at a depth range of 65–900 m (Cox and Francis 1997). The maximum recorded total length of this species is 450 mm (Last et al. 2016). Individuals of this species usually live on the continental shelf and slope (Cox and Francis 1997). Females produce up to 11 pups a litter. The males can reach maturity at 350–400 mm in total length. The size of the pups at birth is 90–100 mm in total length (Last et al. 2016). The females are ovoviviparous (Cox and Francis 1997).

Distribution
This species is distributed in the Southwest Pacific, and it is endemic to New Zealand.

Economic Value

This species has no economic value.

Conservation

Finucci and Kyne (2018) evaluated this specie as a Least Concern in the Red List of the IUCN on the basis of the following reasons. This species is small in size and inhabits deep-water regions. It is endemic to New Zealand and found in the Southwest Pacific. The population studies of this species are lacking, but the species is deliberated to have a moderately reasonable population size. It is usually found among the deep-water bycatch as an incidental catch. There is no species-specific management for this species.

References

Anonymous. Fish collection database of the Intitut für Seefischerei (ISH). Hohenwestedt: BIOLAB Forschungsinstitut; 2000.

Bini G. Atlante dei pesci delle coste italiane, 1, Leptocardi, Ciclostomi, Selaci. Milano: Mondo Sommerso; 1967.

Compagno LJV, Ebert DA, Smale MJ. Guide to the sharks and rays of southern Africa. London: New Holland; 1989. p. 158.

Cox G, Francis M. Sharks and rays of New Zealand. Canterbury University of Press/University of Canterbury; 1997. 68 p.

Dulvy NK, Reynolds JD. Evolutionary transitions among egg-laying, live-bearing and maternal inputs in sharks and rays. Proc R Soc Lond Ser B Biol Sci. 1997;264:1309–15.

Finger S, Picolino M. The shocking history of electric fishes: from ancient epochs to the birth of modern neurophysiology. New York: Oxford University Press; 2011. 470 p.

Finucci B, Kyne PM. Typhlonarke aysoni. The IUCN Red List of Threatened Species 2018: e.T116863205A116737518. 2018. https://doi.org/10.2305/IUCN.UK.2018-2.RLTS.T116863205A116737518.en. Accessed 29 Oct 2022.

Finucci B, Derrick D, Dossa J, Williams AB. Tetronarce nobiliana. The IUCN Red List of Threatened Species 2021: e.T116861529A116861706. 2021. https://doi.org/10.2305/IUCN.UK.2021-2.RLTS.T116861529A116861706.en. Accessed 28 Oct 2022.

Forrest FC, Saunders PR, McSwinney M, Tooly MA. Cardiac injury and electrocution. J R Soc Med. 1992;85:642–3.

Keynes RD. Electric organs. In: Brown ME, editor. The physiology of fishes. Behaviour, vol. 2. New York: Academic; 2013. 538 p.

Last PR, White WT, de Carvalho MR, Séret B, Stehmann MFW, Naylor GJP. Rays of the world. CSIRO Publishing/Comstock; 2016. i-ix + 1–790.

Lieske E, Myers R. Collins pocket guide. Coral reef fishes. Indo-Pacific & Caribbean including the Red Sea. Haper Collins; 1994. 400 p.

Michael SW. Reef sharks and rays of the world. A guide to their identification, behavior, and ecology. Monterey: Sea Challengers; 1993. 107 p.

Nelson ME. Electric fish. Curr Biol. 2011;21:528–9.

Robins CR, Ray GC. A field guide to Atlantic coast fishes of North America. Boston: Houghton Mifflin; 1986. 354 p.

Slavin KV. Peripheral nerve stimulation for neuropathic pain. Neurotherapeutics. 2008;5:100–6.

Stehmann M, Bürkel DL. Torpedinidae. In: Whitehead PJP, Bauchot M-L, Hureau J-C, Nielsen J, Tortonese E, editors. Fishes of the north-eastern Atlantic and Mediterranean, vol. 1. Paris: UNESCO; 1984. p. 159–62.

Taimur Z, Hussaini S. Arterial fibrillation in a commercial diver. Occup Med. 2008;58:144–6.

Wall PD, Sweet WH. Temporary abolition of pain in man. Science. 1967;155:3758.

Part II

Poisonous

5 Poisonous Fishes

5.1 Ichthyosarcotoxic Fishes

5.1.1 Elasmobranch Fish Poisoning

For decades, a human came to know that eating the meat of certain shark species might cause intoxication (De Sauvages 1758, 1770). To eliminate these intoxicants, but not thoroughly, shark meat should be boiled, and the water should be changed several times (Fabricius 1780). Coutand (1879) was the first to report shark poisoning. These studies include remarks on cases where the victim ingested a shark's liver in New Caledonia, and intoxication from eating shark meat was written for the first time for animals such as dogs (Jensen 1914). The meat of several species of sharks was recognized as poisonous by Macht and Spencer (1941). The clinical characteristics of shark poisoning were summarized by Halstead and Lively (1954) and Bouder et al. (1962).

There are two types of toxins in the meat of sharks: the first type is a natural chemical usually found in the meat of certain species of sharks, and the second type is the chemical that is in the seawater and that gets access into the body of sharks through its food. Both types of toxins are dangerous to human health.

Usually, pollutants enter the sea and oceans, and marine organisms absorb the toxins and heavy metals but cannot excrete them. While the fish or sharks continue to eat, the concentration of the toxins continues to increase accordingly. Such accumulation of toxins is known as toxin biomagnification in the food chain. Since sharks are apex predators sitting at the top of the food web, these biomagnified toxins will be stored in their body and continuously added to during their long life, eventually passing on to human who eats it.

Among the toxins that are usually found in the meat of sharks sold in the market include the following:

A. Methylmercury—This toxin becomes more dangerous to humans when it bounds with organic molecules in the environment. Sharks are found to have the highest level of methylmercury than any other sea creatures, and this is because of the bioaccumulation and the place of sharks at the top of the food web. Moreover, cooking does not remove mercury. Methylmercury can cause a series of health problems in humans, including loss of motor control, corrosion of skin and mucous membranes, blindness, deafness, impotence, sterility, and being linked to Hodgkin's lymphoma, Parkinson's disease, and Alzheimer's disease. Vision Dive (2022) reported that a typical meal of blue shark steak could raise the level of methylmercury to 50 times the legal amount.
B. BMAA—A neurotoxin that has been found in the flesh, cartilage, and organs of sharks. This toxin influences human health. It is related to neurodegenerative diseases, including Alzheimer's and Parkinson's.

L. A. Jawad, *Dangerous Fishes of New Zealand*, https://doi.org/10.1007/978-3-031-69698-5_5

C. Ciguatera—The most common type of seafood poisoning, although many cases are unreported or misdiagnosed. The toxins originate from a type of microorganism which attaches to a species of algae on dead or damaged coral. As coral damage is increasing so are ciguatera outbreaks. This is a typical food poisoning symptom but can also be fatal. In Australia, nearly all ciguatera poisoning has occurred from fish caught in Queensland or the Northern Territory.
D. DDT—This pesticide enters the body of aquatic animals when drained from agricultural lands. This is banned for agricultural use but is still used in certain places worldwide. Sharks are highly exposed to this toxin. It causes dizziness, convulsions, and irritability with long-term exposure leading to lasting cognitive and neurological problems.
E. PCBs—These toxins are very stable and degrade quickly. They are banned from use, but they are still found in the environment and reported from the shark's meat and oil. They can cause liver disease and reduced immune response. Women exposed to these toxins during pregnancy tend to have babies with a variety of problems, and children exposed to these poisonous chemicals have suffered in their development.

A study—given by the shark protection organization Shark project—proved that there are up to 1400 micrograms of methylmercury per kilogram of blue shark steak. These are 420 micrograms of methylmercury in a normal 300 gram shark steak portion, and this is 60 times more than a 70 kg heavy consumer per day may have. The risk value is 0.1 microgram per kilogram of body weight and day. This value was specified by EPA (Environmental Protection Agency) in a toxicologist committee and is considered a new international standard. A consumer might take only 5 grams of blue shark steak or 12.7 grams of smoked rock salmon (smoked dogfish) daily. Each higher dose can have serious consequences.

The Causative Agent of Naturally Present Toxins in Sharks

For shark species that have some chemicals that are considered toxins to humans, these chemicals are usually present in the flesh of the shark and appear to be a kind of liposoluble toxin and are found mainly in the liver and usually named carchatoxin-A and carchatoxin-B (Boisier et al. 1995). The shark's poison is different from that of ciguatera. It does not include the pathognomonic paradoxical sensory disturbance (temperature reversal sensation), the characteristic clinical symptoms of ciguatera. Bouder et al. (1962) suggested that shark poisoning resulted from hypervitaminosis A. The variety and severity of the neurological disturbances occurring in shark poisoning do not support hypervitaminosis A (Halstead 1965).

Symptoms

Shark poisoning is usually mild, and the symptoms include gastroenteritis and diarrhea. Eating shark liver could cause more austere intoxication. The symptoms for such ingestion frequently occurred within 30 min. They include vomiting, diarrhea, abdominal pain, anorexia, headache, prostration, rapid, weak pulse, malaise, insomnia, cold sweats, oral paresthesias, and burning sensation of the tongue and throat and esophagus. Problems of the intoxication might develop with time, and neurological and other symptoms become clear, causing extreme weakness, trismus, muscular cramps, the sensation of heaviness of the limbs, blepharospasm, dilatation of pupils, or spasmodic contractions of upper eyelids. In severe cases of intoxication, complications in the health of the victim might lead to death (Halstead 1965).

Treatment and Prevention

There is no actual treatment practice to follow in cases of shark poisoning. The first step in preventing such intoxication is to avoid eating shark flesh that originates from large tropical shark species. Another precautionary measure is to cut the meat of the shark into thin slices and hang them up to dry in the air. The juice containing the toxin will drain, leaving the meat poison-free. Later, the meat slices should be sun-dried. Poorly dried meat is more poisonous than fresh meat.

5.1.1.1 Poisonous to Eat: Sharks and Rays

Family: Hexanchidae
Heptranchias perlo (Bonnaterre 1788) (Fig. 5.1)
Common name: sharpnose sevengill shark
Etymology: *Heptranchias*: Greek, *heptra* = with seven arms + Greek, *agchein* = throttle

Diagnostic Features

The head is pointed, with big eyes. There are seven small gill slits. The mouth is narrow. The lower jaw is with five rows of combed-shaped teeth. The body is slender to fusiform. The dorsal fin is small and originates over the inner margins of the pelvic fins. The anal fin is small. The color of the body is brownish gray on top and paler ventrally. Indistinct dark spots are present on the body. The juveniles are with dark-tipped dorsal and caudal fins.

Habitat, Biology, and Ecology

This is a marine species that inhabit bathydemersal environments at a depth range of 0–1000 m (Hennemann 2001). Individuals of this species irregularly gathered near seamounts. It is an active and ravenous predator that feeds on pelagic fishes, squid, and crustaceans. It reaches 1400 mm in total length as a maximum size. Males and females mature at 750–850 mm and 900–1050 mm in total length, respectively. Females are ovoviviparous with 6–20 in a litter. Young are born 250 mm in total length. Individuals can breed year around. The species is aggressive when captured and, even if not retained, is likely to be killed (Paul and Fowler 2003).

Fig. 5.1 *Heptranchias perlo* (Bonnaterre 1788). Courtesy of K.V. Akhilesh, India

Distribution

This species of shark has a circumglobal type of distribution. It is found in all tropical and temperate seas except the northeast Pacific (Compagno and Niem 1998). It is reported from the western and eastern Atlantic. In the Indian Ocean, it is found in southwestern India and westward to South Africa. It is also reported from Japan to China, Indonesia, Australia, and New Zealand (Froese and Pauly 2022). It is also reported from the Gulf of Aden by Ebert et al. (2013) and it is not found in either the Arabian/Persian Gulf or the Arabian Sea.

Economic Value

This species has a commercial value as food for humans and as fishmeal. In Japan, they are kept in captivity (Paul and Fowler 2003).

Conservation

This species has been evaluated as a Near Threatened status in IUCN Red List because of the following reasons. Its wide range distribution. It is uncommon in the areas it is found, and its center of abundance lies in the outer shelf, slope, and oceanic seamounts, which make it vulnerable to fishery activities. Because it lives in deep waters, it is included in deep sea fishery activities, which may cause population declines (Paul and Fowler 2003).

Hexanchus griseus (Bonnaterre 1788) (Fig. 5.2)
Common name: bluntnose sixgill shark
Etymology: Hexanchus: Greek, *exa* = six + Greek, *agcho*, narrow; *griseus*: From the Latin *griseus* meaning gray

Fig. 5.2 *Hexanchus griseus* (Bonnaterre 1788)., Courtesy of Hakan Kabasakal, Turkey

Diagnostic Features

The body is large and fusiform, with a broad head and the presence of sixgill slits. The eyes are small with dark pupils surrounded by white. The mouth is ventral in position, with bladelike, comb-shaped teeth on the lower jaw. The snout is rounded. The dorsal fin is larger than the anal fin. The body is gray above and white below. Light lines are present along the sides. The fins are white edges.

Habitats, Biology, and Ecology

This shark is a marine species living in a demersal environment at depths of 1–2500 m (Riede 2004; Mundy 2005). The bluntnose is a recognized predator with a wide spectrum of food items including sharks, rays, skates, and small and large fish species. They appear very sensitive to light, but this information needs to be investigated further as some individuals were photographed underwater using floodlights and no avoidance of lighted areas was noticed (Cook and Compagno 2005). It is ovoviviparous and females bear many litters from 22 to 108 with size at birth ranging from 650 to 740 mm. Males and females mature at 3150 and 4200 mm in total length, respectively. The young prefer shallow water areas and when they grow move to deeper waters. Adults show vertical movement when they rise to the surface at night for feeding (Cook and Compagno 2005).

Distribution

Individuals of this shark have circumglobal distribution in tropical and temperate waters. They are usually found in both the western and eastern Atlantic and they are also reported from the Mediterranean Sea. In the Indian Sea, it is recorded from South Africa to the northwestern Arabian Sea. The distribution of this species extends eastward to reach Japan, New Zealand, and Hawaii (Compagno and Niem 1998). It has not been recorded from the Arabian/Persian Gulf or the south coasts of Oman. It is only found in the south of the Gulf of Aden (Ebert et al. 2013).

Economic value
This species has a limited economic value as a food source.

Conservation
Finucci et al. (2020a) evaluated this species of shark as Near Threatened in the Red List of the IUCN based on the following reasons. This species has a large size body and inhabit deep water areas. It has a global distribution, but this distribution is patchy. The females put many litters, but an assessed late age-at-maturity decreases its capacity to recover from fishing pressure. It is frequently caught by deep sea artisanal fisheries vessels.

Family: Somniosidae
Centroscymnus owstonii (Garman, 1906) (Fig. 5.3)
Common name: roughskin dogfish

Diagnostic Features
The body is sturdy and weakly tapering back from the pectoral fin region. The snout is reasonably long. The preoral fin length is as long as the distance from the mouth to the first gill slits. The lips are relatively thick and fleshy. The upper labial furrows are very short, and their lengths are much less than the distance between their front ends. The lower teeth are with short, oblique cusps and are high, with narrow roots. The second dorsal fin is largely higher than the first, with small fin spines and with tips protruding from the fins. The first dorsal fin is extending forward as a short and inconspicuous ridge, and its origin is behind the pectoral fin bases. The second dorsal fin base is much longer than the space between it and the upper caudal fin origin, and its free rear tip is just in front of the upper caudal fin origin. The pectoral fins are rather large, with apices falling well in front of the first. The free rear tips of the pelvic fins are below or slightly in front of the second dorsal fin insertion. The lateral trunk denticles are large, with mostly smooth, oval, partly ridged, and cuspidate crowns. The body is dark brown or black in color.

Habitats, Biology, and Ecology
The roughskin dogfish is a marine species that inhabit the bathydemersal area at a depth of 100–1500 m (Cox and Francis 1997). It reaches a maximum total length of 1480 and a maximum reported weight of 21.3 kg (Finucci 2017). Individuals of this species are living on the upper continental slopes, on or near the bottom (Compagno 1984). They usually feed on fish and cephalopods (Last and Stevens 1994). The female is ovoviviparous (Breder and Rosen 1966), with 16–28 young born at 27–30 cm (Cox and Francis 1997). This species is usually included in the trawls and longlines net at depths greater than 400 m (Kiraly et al. 2003).

Distribution
This dogfish is distributed in the Atlantic, Pacific, and Indian Oceans. In the Western Indian Ocean, it reported from off the Seychelles to the Madagascar Ridge and South Africa. In the Eastern Indian Ocean, it is recorded from off Indonesia and southwestern Australia. It is also

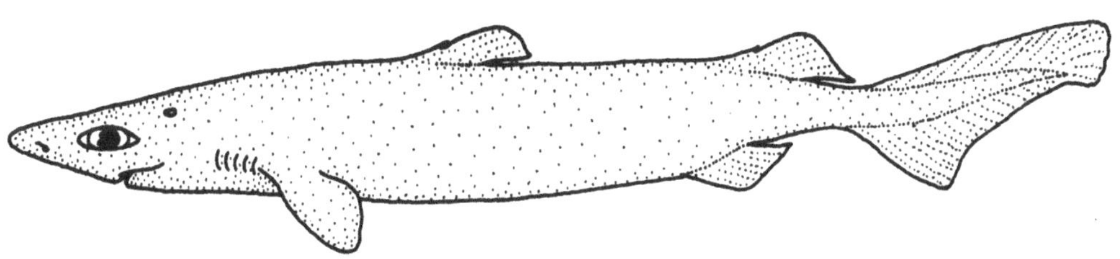

Fig. 5.3 Centroscymnus owstonii Garman, 1906. Courtesy of Tony Ayling

reported from Southwestern Pacific off Australia, New Caledonia, and New Zealand. In Northwestern Pacific, it is found off Japan and from the Hawaiian-Emperor seamount chain. In Southeastern Pacific, it is recorded from the west of Chile. In Southwestern Atlantic, it is usually found off French Guiana, Brazil, and Uruguay. This species is also recorded from Northwestern Atlantic at the Gulf of Mexico, while in the Eastern Atlantic, it is reported from off the Azores in the North to off South Africa in the south.

Economic Value

The flesh is high in mercury (Ref. 6871) and utilized as fishmeal and a source of squalene (liver oil) (Last and Stevens 1994).

Conservation

Finucci and Kyne (2018) evaluated this species as Vulnerable in the Red List of the IUCN based on the following criteria. It is a small-sized species that inhabits deep water. The populations of this species are patchy along its geographical distribution range. The stock assessment of this species is unknown. This species is taken as a bycatch in the benthic trawl and longline (surface and benthic) fisheries across its range. It has been protected in some parts of Australia. Individuals of this species are characterized as being low producers.

Centroselachus crepidater (Barbosa du Bocage and de Brito Capello 1864) (Fig. 5.4)

Common name: longnose velvet dogfish

Etymology: *Centroselachus: centr* [*um*] (L.), prickle or sharp point, referring to spines on both dorsal fins; *selachos* (Gr.), a cartilaginous fish (i.e., shark or ray). (ETYFish 2022); *crepidater: crepida* (L.), low shoe (e.g., slipper); *ater* (L.), black, transliteration of Portuguese vernacular *sapata preta*, black shoe, allusion not explained, possibly referring to superficial resemblance to a black velvet slipper [see also *Deania calceus*, Centrophoridae] (ETYFish 2022)

Diagnostic Features

The body is rather slender and weakly tapering back from the pectoral region. The snout is very long. The preoral fin length is about equal to the distance from the mouth to the pectoral fin origins and much greater than the mouth width. The lips are not thick and fleshy. The upper labial furrows are very long, and their lengths are greater than the distance between their front ends. The lower teeth are with moderately long, semi-oblique cusps and moderately high, broad roots. The dorsal fins are about equal in size and height. The dorsal fin spines are very small but with tips protruding from the fins. The first dorsal fin base expanded forward as a prominent ridge, with its origin over the pectoral fin bases. The second dorsal fin base is longer than the space between it and the upper caudal origin. The free rear tip is nearly reaching the upper caudal origin. The pectoral fins are reasonably large, with apices falling

Fig. 5.4 Centroselachus crepidater (Barbosa du Bocage & de Brito Capello, 1864). Courtesy of Daniel Moore

well in front of the first dorsal spine when laid back. The free rear tips of pelvic fins are extending to about opposite the second dorsal insertion. The lateral trunk denticles are moderately large, with anteriorly smooth but posteriorly ridged, oval, cuspidate crowns. The body is blackish brown in color.

Habitats, Biology, and Ecology

The longnose velvet dogfish is a marine species that inhabits bathydemersal areas at a depth range of 230–1500 m (Cox and Francis 1997). It reaches a maximum total length of 1300 mm (Bass et al. 1986) and a maximum reported age of 54 years (Zidowitz 2005). It is an equitably popular species found on continental and insular slopes (Last and Stevens 1994), on or near the bottom (Compagno et al. 1989). Individuals of this species feed mostly on fish and cephalopods (Last and Stevens 1994). The females are ovoviviparous (Breder and Rosen 1966), with 4–8 young in a litter (Last and Stevens 1994), born at 28–35 cm (Cox and Francis 1997).

Distribution

This species of dogfish is distributed in the Eastern Atlantic. It is reported from Iceland, Faeroe Islands along Atlantic slope to Portugal, Senegal, Madeira, Gabon to Democratic Republic of the Congo, and Namibia. It is also found in the Indian Ocean. It is recorded from Aldabra and the Travancore coast of India. It is also found in Western Pacific. It is recorded in New South Wales, Australia, and New Zealand.

Economic Value

The flesh is high in mercury; it is usually used as fishmeal and a source of squalene (Last and Stevens 1994).

Conservation

Finucci et al. (2020b) evaluated this dogfish as Nearly Threatened in the Red List of the IUCN based on the following reasons. It is a small-sized species that inhabits deep water areas. It is known from a wide range of geographical localities. Although the populations of this species are widely distributed, they are patchy along its distributional range. It is usually present as a bycatch in deep sea trawlers and demersal longlines. Population declines have been stated in part of its range but growing trends in others.

5.1.1.2 Other Cartilaginous Fishes with a Poisonous Flesh

The followings are cartilaginous fish species whose flesh showed to be poisonous and not suitable for human consumption. The description of these species was given in Chap. 2 of this book. Therefore, a list of their names is given here to remind readers of the danger of these species:

1. *Carcharodon carcharias*
2. *Galeocerdo cuvier*
3. *Prionace glauca*
4. *Sphyrna zygaena*

5.2 Ciguatera Fish Poisoning: The Emerging Threat in New Zealand

5.2.1 Introduction

Ciguatera fish poisoning (CFP) is the most customary nonbacterial disease linked with fish consumption (Friedman et al. 2008), affecting between 50,000 and 500,000 people per year (Fleming et al. 1998). Ciguatera is a common ichthyosarcotoxaemia with outstanding clinically vital neurological aspects. The large-sized fishes have eaten dinoflagellates and usually harbor the neurotoxin in their viscera. The ingestion of herbivorous and carnivorous fish that have orally accumulated effective levels of ciguatoxins (CTXs), and possibly maitotoxins (MTXs), can cause CFP in humans (Bagnis et al. 1979; Gillespie 1987; Sims 1987). The latest investigations and studies have shown the global upsurge in the occurrence and intensity of harmful algal events (Glibert et al. 2005; Hallegraeff 2010). Despite being considerably underreported, CFP existence worldwide is increasing, with reports of a 60% increase in CFP in the Pacific Islands over the last decade (Skinner et al. 2011).

The poisonous features of some fish species containing ciguatera have been known since the time of the ancient Greeks in 800 BC (cited in Savtschenko 1886), and ancient Chinese fishermen have struggled with this type of intoxication since the time of the T'ang Dynasty (AD 618–907) mainly by eating the yellowtail amberjack (quoted by Read 1939). Even though ciguatera had such a long history, the first generally recognized reference specifically referring to this fish poisoning was published by Peter Martyr d'Anghiera only in 1457–1526, when the first accounts of fish poisoning in the West Indies were noted.

The Portuguese biologist Don Antonio Parra was the first to use the term ciguatera. It was originally spelled "Siguatera" because Cuban writers of this period substituted "S" for the usual "C" spelling in Spanish words. The name refers to cigua = a person ill with ciguatera, perhaps from the cigua sea snail. The outbreaks of ciguatera around the world were many, but the best known of these took place on His Majesty's ship Resolution in the eighteenth century during the famous world voyage of Captain James Cook at Port Sandwich, Malekula, New Hebrides, on 23 July 1774 (Halstead 1965). The fish was *Lutjanus bohar* and the outbreak was cited in Anderson (1776), Forster (1777), Cook (1777), and Sparrman (1944).

The start of the nineteenth century presented an upsurge in scientific writings about ciguatera poisoning and ciguatoxic fishes (Meyer 1805; Chisholm 1808; Orfila 1817). This era of time saw the publication by Charles Grant in 1801 on the largest ciguatera outbreak in history which took place at Rodrigues Island, east of Mauritius, Indian Ocean, in 1748 (cited by Wheeler 1953). The responsible fish that caused this outbreak of ciguatera was *Serranus lutra = Epinephelus fuscoguttatus*. Due to the augmented occurrence of intoxication resulting from the ingestion of fishes causing ciguatera, this fish intoxication was further investigated and appraised during the second half of the nineteenth century (Meyer-Ahrens 1855; Vinson 1858; De Rochas 1860; D'Arras 1877; Vincent 1883).

At the start of the twentieth century, Becke (1901) recognized that 30–50% of the fishes in the Ralik Lagoon, Marshland Islands, were poisonous to eat. Studies endure being published on the poisonous fishes that cause ciguatera and a classification system for fish poisoning was put forward by Engelsen (1922). The poisonous fishes were divided into groups:

1. Ichthyismus choleriformis: A group of toxins distinguished by causing violent gastroenteritis.
2. Ichthyismus exanthematicus: Those types of poisons characterized by a cutaneous eruption with fever and indisposition, but not very dangerous.
3. Ichthyismus neuroticus: Toxins that take place within a few hours after eating the fish, causing burning and dryness in the throat, violent abdominal pain, dizziness, and visual disturbances and death resulting from respiratory paralysis (Halstead 1965).

Toxicological studies were very scarce during the first half of the twentieth century. Fitch (1952) performed toxicological tests on the musculature and liver of *Tetragonurus cuvieri* from Point Conception, California. Later, Halstead and his associates performed extensive studies on the poisonous fishes of the tropical Pacific and Red Sea areas (Halstead and Lively Jr. 1954; Halstead and Bunker 1954; Halstead and Schall 1955; Halstead and Schall 1956; Halstead 1958; Halstead and Carscallen 1964). Albert Banner, Philip Helfrich, and their associates conducted thorough investigations on ciguatera in the tropical Pacific toward epidemiology, development of a critical bioassay, and isolation of ciguatoxin (Banner and Boroughs 1958; Martin and Banner 1958; Banner et al. 1960; Helfrich and Banner 1963).

With the advancement of science and technology during the twentieth century and the present time, a large number of scientific publications appeared about ciguatera that the present book cannot accommodate in detail (Calvert et al. 1987; Kodama et al. 1989; Cameron et al. 1991;

Benoit et al. 2000; Hamilton et al. 2002a, b; Boydron-Le Garrec et al. 2005; Laurent et al. 2008; Caillaud et al. 2010; Munir et al. 2011; Pearn 2001; Rhodes et al. 2014; Smith et al. 2016; Munday et al. 2017; Smith et al. 2017; Rhodes et al. 2017; Hallegraeff et al. 2021).

Causative Agent

Ciguatoxin is a chemical made through the metabolic events of members of the marine benthic dinoflagellate genus *Gambierdiscus*. Numerous comparable chemicals were revealed which are different in their molecular structure (Caillaud et al. 2010). Dissimilar forms of this toxin were recognized from different geographical locations (e.g., Pacific, Caribbean, and Indian). Some techniques were found to identify and quantify the ciguatoxins in the fish's tissues as the concentrations of these toxins are very low.

It is known that the toxin stretches higher concentrations through the food chain, a hypothesis that suggests that the transport of toxins along the food web is an elucidation for the movement of ciguatoxic fish species. This hypothesis was first proposed by Mills (1956) and was approved when isolation of a toxic marine benthic dinoflagellate, described as *Gambierdiscus toxicus*, was made which designated the source of ciguatoxin (Yasumoto et al. 1977).

The journey of the ciguatoxin starts with the dinoflagellates, then to the herbivorous grazing fish, and then to the carnivorous fish that feed upon grazers (Yasumoto et al. 1979). In the tissue of the fish, this toxin is changed through the metabolic activities of the fish into different configurations of toxins (e.g., Cigutoxin-1B and 51-hydroxy Cigutoxin-3C) from original ciguatoxin structures (e.g., Cigutoxin-3C and Cigutoxin-4B) that are eventually responsible for human intoxication (Caillaud et al. 2010). In such conditions, both herbivorous and carnivorous fish from tropical areas can be toxic. More than 400 fish species have been reported to be potentially ciguateric (Halstead 1978). The concentration of the toxin is less in young individuals as they are subjected to the causative dinoflagellates for a shorter period than adult individuals. Likewise, those species located at a higher level of the food web are more hazardous than those situated at the lower levels (Oshiro et al. 2009). The ciguatera toxin is found to be concentrated in the liver and viscera of the fish but is also found in their muscles (Vernoux et al. 1985).

It is imperative at this stage to give a clue about the biology of the organism that produces ciguatoxin. *Gambierdiscus toxicus* has been deliberated as the chief cause accountable to produce ciguatoxin. It is a marine dinoflagellate found usually in tropical and subtropical waters as an epiphyte on macroalgae in coral reefs, in mangrove systems, and on artificial surfaces or sand (Faust 1995; Villareal et al. 2007). In addition to this species, five more species were described: *G. belizeanus* (Faust 1995), *G. yasumotoi* (Holmes 1998), *G. polynesiensis*, *G. pacificus*, and *G. australis* (Chinain et al. 1999a, b). The factors impelling the existence of toxic *Gambierdiscus* blooms endure the goal of researchers in this field (Chinain et al. 1999a, b). Although the exact nature of reasons inducing the toxicity and abundance of *Gambierdiscus* spp. continues to be unclear, light intensity, salinity, water temperature, nutrients, growth stage, and the presence of bacteria have been shown to influence the growth and toxicity of this algal species (Sakami et al. 1999; Lartigue et al. 2009; Llewellyn 2009; Doucette et al. 1998).

The chemical configuration of ciguatoxin studied by Caillaud et al. (2010) has shown that it is a lipophilic polyether isolated from fish and *Gambierdiscus* spp. cell extracts (wild samples and cultures). It is also odorless, colorless, and lacking heteroatoms other than oxygen and bears few conjugated bonds. The different strains of ciguatoxins have different distinct structures (e.g., Pacific, P-CTX; Caribbean, C-CTX; Indian Ocean, I-CTX; Legrand et al. 1992; Murata et al. 1993; Vernoux and Lewis 1997; Hamilton et al. 2002a, b; Pottier et al. 2002a, b; Satake et al. 1993; Murata et al. 1990).

Within human cells, ciguatoxin aims the membrane excitability ensuing in the release of neurotransmitters (Molgó et al. 1990), axonal and Schwann oedema (Benoit et al. 1996; Mattei et al. 1999), increase of intracellular calcium

(Molgó et al. 1993), and blockage of voltage potassium channels (Hidalgo et al. 2002). The neurological signs the victim regularly has are linked to the fluctuations in the sodium ion contraction in the cell, and the chronic fatigue condition that may last for weeks or months (Racciatti et al. 2001) is associated with high nitric oxide (NO) production (Pall 2001).

Ciguatera fish poisoning (CFP) is the most stated fish-related poisoning in humans and is a substantial health problem in the Pacific region (Parsons et al. 2012), while, with a growing global need for seafood and growing pathways to market, CFP is progressively becoming an international issue (Friedman et al. 2017). Throughout the last few years, there has been an upsurge in records of New Zealand tourists returning from the Pacific Islands and northern Australia with diseases connected to the eating of reef fish, and many of these sicknesses have been identified as CFP (Schep et al. 2010; Rhodes and Smith 2019; Rhodes et al. 2017; Sparrow 2018). In New Zealand, between 2006 and 2014, there were 54 hospitalizations diagnosed as CFP (Rhodes et al. 2020). Onc cluster of cases was credited to the eating of imported moray eel (family Muraenidae), which was confirmed as containing Pacific-CTX-1B (P-CTX) (Armstrong et al. 2016). No complaints resulting from the eating of locally caught fish have been officially documented for New Zealand or Rangitāhua/Kermadec Islands.

As the genera *Gambierdiscus* and *Fukuyoa* have both now been identified in New Zealand waters (Rhodes and Smith 2019; STU 2022), it is only a matter of time before potential CTX producers proliferate in New Zealand's northern subtropical waters (Northland), increasing the risk of CFP occurring locally (Rhodes et al. 2020).

Treatment

Among the procedures of treatments is intravenous mannitol infusion, which is the most used therapy for ciguatoxin poisoning (Schnorf et al. 2002; Bagnis et al. 1992). This injection is given at 0.5–1.0 g/kg body weight over a 30–45 min period and suggested to be provided within 48–72 h of ingestion of toxic fish (Blythe et al. 2001; Palafox et al. 1988), although the outcome has been perceived even up to several weeks after intoxication (Blythe et al. 2001).

Mannitol infusion can be interceded by the osmotic reduction of neural oedema (Pearn 2001) and can get rid of the free radicals that might be generated by ciguatoxin molecules and reduce the action of ciguatoxin on sodium/potassium channels in the cells (Birinyi-Strachan et al. 2005). Support of any depressed vital functions is the other option for treatment. This type of treatment is used with acute cases of ciguatoxin poisoning (Lewis and King 1996), where supportive therapies may be necessary for controlling fluid and electrolyte balance (Lewis 2000) and are necessary for patients in shock.

As to traditional medicine, there are over 60 herbs that were reported as remedies in ciguatoxin poisoning in the Western Pacific (Bourdy et al. 1992) and in New Caledonia. The extract of the leaves of *Argusia argentea* have been reported as medicine to cure ciguatoxin (Benoit et al. 2000).

Patients of ciguatoxin poisoning showed relapse of symptoms with ingestion of alcohol (Lange et al. 1992) or any kind of fish (Ruff and Lewis 1994). Drinking coffee (Fleming et al. 1997) and eating nuts (Lewis 2001) and chicken (Gillespie et al. 1986) were also associated with symptom recurrence or augmentation. Such food and drinks need to be avoided for a period of 3–6 months after intoxication.

Prevention

Among the difficulties of preventing ciguatoxication is the flesh of the ciguatoxic fishes is odorless and tasteless, and toxic fish cannot be identified by their appearance or behavior. The ciguatoxication incidences are not attributable to inadequate food handling, storage, preparation, or procurement methods for the contaminated fish. The toxin is heat-stable, and therefore, cooking, boiling, freezing, baking, or frying does not eliminate or destroy the toxin from the fish tissue (Bagnis 1993). It is very rare to have gastrointestinal symptoms transmitted person-to-person, but there are reports of transmission of GI symptoms from an acutely ill mother across the placenta to the fetus/newborn (Fleming et al. 1997) and to a nursing infant (Blythe and de Sylva 1990) and of transient genital paresthesias in sexual partners of those with acute CFP (Lange et al. 1989).

The first and important step to prevent ciguatoxication is by the individual's avoidance of fish that have a greater likelihood of ciguatoxicity. Local people are aware of certain fish species to be ciguatoxic and therefore they avoid eating them. In addition, they have simple home tests to detect toxic fishes. As the greater illness severity is associated with eating the fish viscera and larger portions (Lewis and King 1996), people are therefore advised to avoid eating the viscera of reef fish. It is also recommended to avoid eating fish in excess of 3 kg (Ting and Brown 2001). To be on the safe side, eating small portions (i.e., <50 g) of different fish is better than eating larger portions of any individual fish that might be associated with ciguatoxin (Lewis 2000). In the field, there is no reliable method of detecting ciguatoxicity in fish based on their appearance or behavior.

The Species of Gambierdiscus and Fukuyoa: The Causative Factors of CFP in New Zealand

The results of Rhodes and Smith (2019) and Rhodes et al. (2017) have confirmed the presence of *Gambierdiscus* species in both the northern and southern groups of Rangitāhua/Kermadec Islands, a New Zealand territory 1000 km northeast of New Zealand. Rhodes et al. (2020) found that *Gambierdiscus* cells have been isolated from samples they collected from macroalgae in the northern, largest, and volcanically active Raoul Island and from nearby North Meyer Island. The cells of this species of algae were also obtained from Macauley Island (100 km SSW of Raoul Island and the second largest island) (Trnski and Schlumpf 2011). In this study, several species of the genus *Gambierdiscus* were isolated comprising *G. honu* (Rhodes, Smith, & S. A. Murray) (first described from Macauley Island; Rhodes et al. (2017)), *G. australes* (Chinain & M. A. Faust), and *G. polynesiensis*.

The other species, *Fukuyoa*, causing ciguatera was retrieved from samples collected from subtropical northern New Zealand waters, and such a result is a confirmation of the establishment of this species in northern New Zealand (Rhodes et al. 2020). In the results of Rhodes et al. (2014a, b), a repeated presence of *Fukuyoa* was perceived in specimens obtained from the Bay of Islands, at Te Uenga Bay and Motuarohia Island. The first appearance of *Fukuyoa* in northern New Zealand goes back to February 2013, when it was stated as *Gambierdiscus* cf. *yasumotoi* M. J. Holmes. The genus, which is further rounded in shape than the allied lenticular genus *Gambierdiscus*, has since been known scientifically as *Fukuyoa* (Gómez et al. 2015).

Identification was inveterate by phylogenetic examinations of DNA sequence data. At the present time, only one record of *Gambierdiscus* cell, using light microscopy, has been known from New Zealand. This was from a sample gathered from the floating macroalga *Sargassum* Agardh (Phaeophyta) collected in Northland waters in the vicinity of Doubtless Bay (STU 2022). The species was stated in 1996 as *G. toxicus* Adachi & Fukuyo, but identifications made by microscope examination alone prior to 2009 are indefinitely owing to the comparable morphologies and morphological plasticity of some species (Parsons et al. 2012; Rhodes and Smith 2019; Litaker et al. 2009). Molecular indicators now permit fast and vigorous decisions (Rhodes et al. 2020).

The Events of Ciguatera Fish Poisoning in New Zealand Because of Imported Tropical Reef Fish

Cases of ciguatera fish poisoning (CFP) recorded in New Zealand until the publication of the report of Murray et al. (2020) resulted from the ingesting polluted reef fish (including eels) by tourists who have come back to New Zealand after eating reef fish in the Pacific Islands and then become sick, or from New Zealanders eating introduced reef fish they have brought back into New Zealand themselves or have purchased from a commercial vendor (Murray et al. 2020). Several cases of CFP, related to the eating of moray eel brought back from Samoa, were reported in Wellington, New Zealand, in 2016 (Armstrong et l. 2016). An additional wave of cases was clinically detected, built on symptomology, in 2019 following the eating of moray eel brought back from the Kingdom of Tonga and home smoked (Waqanivavalagi et al. 2019).

Recently, Murray et al. (2020, 2021) reported on CFP case that includes five people. This incident happened after eating Fiji Kawakawa (camouflage grouper; *Epinephelus polyphekadion*). The fish was bought in New Zealand but imported from Fiji. The meal remains were analyzed for ciguatoxins, and the causative agents of ciguatera poisoning were detected in the food. From 10 March until 21 May 2020, Fiji Kawakawa (camouflage grouper; *Epinephelus polyphekadion*) trade in from Fiji was sold by Krazy Mart in Christchurch, New Zealand. At the commencement of May 2020, five adults from two families became sick after eating some of the imported fish. Two were females in their forties and fifties and three were males aged between 19 and 58 years old. The eldest male, who had eaten four times the amount of the other diners, was introduced to the hospital within 48 hours of eating the fish meal. The signs were typical of CP (inversion of hot and cold, paresthesia, and dysesthesia), and the patients were clinically identified as suffering from CP.

In their results, Murray et al. (2020, 2021) gave a report written by the Canterbury District Health Board showing the quantifiable levels of P-CTX-1B. The following is what the report has said: "The fried fish sample contained 0.29 μg/kg and the curried fish sample contained 0.21 μg/kg P-CTX-1B. The curried fish sample contained less fish, although it had a coconut base. As CTXs are lipophilic, the whole sample was tested. Two additional CTX fish metabolites, 52-epi-54-deoxyCTX-1B (P-CTX-2) and 54-deoxyCTX-1B (P-CTX-3), were also easily detected in both samples. However, as calibrated reference material was not available, they were not able to be quantified. Associated fish from the same shipment as the recalled product were also tested but were negative for the fish metabolite CTXs."

Murray et al. (2020, 2021) suggested that it is important to focus on the increasing sea surface temperatures and changes in the prevailing currents due to climate change. This is instigating a development of the subtropical/temperate latitudes, which in turn is affecting the habitable range of *Gambierdiscus* and results in its expansion. This is incidence as CP in Australia has generally been along the coast of Queensland; however, in recent years, fish caught in New South Wales waters have caused CP events (MPI 2019). The coastal waters around New Zealand are also warming up and therefore the hazard of locally contracted cases of CP is increasing (IAMAT 2022).

Murray et al. (2020, 2021) reached to the conclusion that the case of fish poisoning they stated from Christchurch in May 2020 and diagnosed as CP was instigated by fish that had been introduced from Fiji, which was shown to be polluted with CTXs. The jeopardy of distress from CP after ingestion reef fish from the Pacific region is genuine and must be deliberated in any case where there is illness following intake of such a meal. Analysis of meal leftovers will help confirm a CP finding, but only the fish eaten is of interest, as CTX concentrations vary broadly between individual fish, even from the same catch.

Rolton et al. (2022) suggested that harmful algal blooms have wide-ranging environmental influences, comprising on aquatic species of social and commercial status. In New Zealand, planned development of the aquaculture industry could be harmfully affected by the presence of harmful algal blooms. They also recommended that climate change and anthropogenic impacts may alter harmful algal bloom population structure and dynamics, as well as the physiological reactions of fish and shellfish, possibly further compromising aquatic species. Fluctuations in the phytoplankton may also lead to the propagation of harmful algal blooms novel to NZ. The ciguatera fish poisoning dinoflagellate genus *Gambierdiscus* has been recorded in New Zealand coastal waters and is possible to thrive in warming waters (Rhodes et al. 2020).

5.2.2 Possible Ciguatoxic Fish Species

There are many fish species that belong to many families in a wide phylogenetic spectrum that are involved in ciguatera poisoning around the world. Now the ciguatera fish poisoning dinoflagellate genus *Gambierdiscus* is reported from the Kermadecs about 1000 km North Island. With changes in climate, which induce flourishing of this algae, then it is a matter of time, and it will be in the waters around New Zealand.

In New Zealand, there are several fish species that considered a potential to carry the ciguatera toxin, and once the toxin spread in these species, the ciguatera poisoning will be epidemic. Below is a list of the fish families with their species that contemplate possible to carry the ciguatera toxin, and it is too early to give their morphological description until they confirm a carrier to the ciguatera toxin in the future.

1. Family: Engraulidae
   A. *Engraulis australis* (White, 1790), Australian anchovy
2. Family: Clupeidae
   A. *Sardinops sagax* (Jenyns, 1842), South American Pilchard
   B. *Sprattus antipodum* (Hector, 1872), New Zealand blueback sprat
   C. *Sprattus muelleri* (Klunzinger, 1879), New Zealand sprat
3. Family: Congridae
   A. *Ariosoma howensis* (McCulloch & Waite, 1916), Lord Howe conger
   B. *Bassanago bulbiceps* Whitley, 1948, swollen-headed conger eel
   C. *Bassanago hirsutus* (Castle, 1960), deep sea conger
   D. *Conger verreauxi* (Kaup, 1856), conger
   E. *Gnathophis habenatus* (Richardson, 1848), little conger eel
   F. *Gnathophis longicauda* (Ramsay & Ogilby, 1888), little conger
   G. *Gnathophis umbrellabius* (Whitley, 1948), umbrella conger
   H. *Gorgasia japonica* (Abe, Miki & Asai), 1977
   I. *Scalanago lateralis* (Whitley, 1935), ladder eel
4. Family: Muraenidae
   A. *Anarchias supremus* (McCosker & Stewart), 2006
   B. *Enchelycore ramosa* (Griffin, 1926), mosaic moray
   C. *Gymnothorax eurostus* (Abbott, 1860), Abbott's moray eel
   D. *Gymnothorax nubilus* (Richardson, 1848), gray moray
   E. *Gymnothorax obesus* (Whitley, 1932), Griffin's moray
   F. *Gymnothorax pictus* (Ahl, 1789), paint-spotted moray
   G. *Gymnothorax porphyreus* (Guichenot, 1848), lowfin moray
   H. *Gymnothorax prasinus* (Richardson, 1848), yellow moray
   I. *Gymnothorax prionodon* (Ogilby, 1895), Australian mottled moray
   J. *Gymnothorax ypsilon* (Hatooka & Randall), 1992, Y-bar moray
5. Family: Ophichthidae
   A. *Apterichtus australis* (McCosker & Randall, 2005), snake eel
   B. *Apterichtus flavicaudus* (Snyder, 1904)
   C. *Myrichthys maculosus* (Cuvier, 1816), tiger snake eel
   D. *Ophisurus serpens* (Linnaeus, 1758), serpent eel
   E. *Quassiremus polyclitellum* (Castle, 1996)
   F. *Scolecenchelys australis* (Macleay, 1881), shortfinned worm eel
   G. *Scolecenchelys breviceps* (Günther, 1876), short-headed worm eel
   H. *Scolecenchelys castlei* (McCosker, 2006), deepwater big-eyed worm eel
6. Family: Belonidae
   A. *Ablennes hians* (Valenciennes, 1846), flat needlefish
   B. *Platybelone argalus* (Lesueur, 1821), keeltail needlefish

7. Family: Hemiramphidae
   A. *Euleptorhamphus viridis* (van Hasselt, 1823), ribbon halfbeak
   B. *Hyporhamphus ihi* (Phillipps, 1932)
   C. *Oxyporhamphus micropterus* (Valenciennes, 1847), bigwing halfbeak
8. Family: Syngnathidae
   A. *Cosmocampus howensis* (Whitley, 1948), Lord Howe pipefish
   B. *Leptonotus elevatus* (Hutton, 1872), pipefish
   C. *Leptonotus norae* (Waite, 1910), longsnout pipefish
   D. *Lissocampus filum* (Günther, 1870), shortsnout pipefish
   E. *Solegnathus spinosissimus* (Günther, 1870), spiny pipehorse
   F. *Stigmatopora argus* (Richardson, 1840), spotted pipefish
   G. *Stigmatopora macropterygia* (Duméril, 1870), long-snouted pipefish
   H. *Stigmatopora nigra* (Kaup, 1856), wide-bodied pipefish
9. Family: Holocentridae
   A. *Myripristis berndti* (Jordan & Evermann, 1903), blotcheye soldierfish
   B. *Pristilepis oligolepis* (Whitley, 1941), spinyface soldier
10. Family: Acanthuridae
   A. *Acanthurus dussumieri* (Valenciennes, 1835), eyestripe surgeonfish
   B. *Acanthurus nigrofuscus* (Forsskål, 1775), brown surgeonfish
   C. *Acanthurus triostegus* (Linnaeus, 1758), convict surgeonfish
   D. *Naso unicornis* (Forsskål, 1775), bluespine unicornfish
   E. *Prionurus maculatus* (Ogilby, 1887), yellowspotted sawtail
11. Family: Apogonidae
   A. *Ostorhinchus fukuii* (Hayashi, 1990)
   B. *Pristiapogon kallopterus* (Bleeker, 1856), iridescent cardinalfish
12. Family: Blenniidae
   A. *Cirripectes alboapicalis* (Ogilby, 1899), blackblotch blenny
   B. *Cirripectes castaneus* (Valenciennes, 1836), chestnut eyelash-blenny
   C. *Entomacrodus caudofasciatus* (Regan, 1909), tail-barred rockskipper
   D. *Omobranchus anolius* (Valenciennes, 1836)
   E. *Parablennius laticlavius* (Griffin, 1926)
   F. *Plagiotremus tapeinosoma* (Bleeker, 1857), piano fangblenny
13. Family: Carangidae
   A. *Decapterus koheru* (Hector, 1875), koheru
   B. *Decapterus muroadsi* (Temminck & Schlegel, 1844), amberstripe scad
   C. *Elagatis bipinnulata* (Quoy & Gaimard, 1825), rainbow runner
   D. *Naucrates ductor* (Linnaeus, 1758), pilotfish
   E. *Seriola dumerili* (Risso, 1810), greater amberjack
   F. *Seriola hippos* Günther (1876), samson fish
   G. *Seriola lalandi* (Valenciennes, 1833), yellowtail amberjack
   H. *Seriola rivoliana* (Valenciennes, 1833), longfin yellowtail
   I. *Trachurus declivis* (Jenyns, 1841), greenback horse mackerel
   J. *Trachurus murphyi* (Nichols 1920), Chilean jack mackerel
   K. *Trachurus novaezelandiae* (Richardson, 1843), yellowtail horse mackerel
14. Family: Cirrhitidae
   A. *Cirrhitus pinnulatus* (Forster, 1801), stocky hawkfish
   B. *Cyprinocirrhites polyactis* (Bleeker, 1874), swallowtail hawkfish
   C. *Notocirrhitus splendens* (Ogilby, 1889), splendid hawkfish
15. Family: Coryphaenidae
   A. *Coryphaena hippurus* (Linnaeus, 1758), common dolphinfish
16. Family: Gempylidae
   A. *Epinnula magistralis* (Poey, 1854), domine
   B. *Gempylus serpens* (Cuvier, 1829), snake mackerel
   C. *Lepidocybium flavobrunneum* (Smith, 1843), escolar

    D. *Nealotus tripes* (Johnson, 1865), black snake mackerel
    E. *Nesiarchus nasutus* (Johnson, 1862), black gemfish
    F. *Paradiplospinus gracilis* (Brauer, 1906), slender escolar
    G. *Rexea antefurcata* (Parin, 1989), long-finned escolar
    H. *Ruvettus pretiosus* (Cocco, 1833), oilfish
    I. *Thyrsites atun* (Euphrasen, 1791), snoek
    J. *Thyrsitoides marleyi* (Fowler, 1929), black snoek
17. Family: Istiophoridae
    A. *Istiompax indica* (Cuvier, 1832), black marlin
    B. *Istiophorus platypterus* (Shaw, 1792), Indo-Pacific sailfish
    C. *Kajikia audax* (Philippi, 1887), striped marlin
    D. *Makaira nigricans* (Lacepède, 1802), blue marlin
    E. *Tetrapturus angustirostris* (Tanaka, 1915), shortbill spearfish
18. Family: Labridae. This family contains 30 species reported to be present in New Zealand waters.
19. Family: Lutjanidae
    A. *Etelis carbunculus* (Cuvier, 1828), deep water red snapper
    B. *Etelis coruscans* (Valenciennes, 1862), deep water longtail red snapper
    C. *Lutjanus kasmira* (Forsskål, 1775), common bluestripe snapper
    D. *Paracaesio xanthura* (Bleeker, 1869), yellowtail blue snapper
    E. *Pristipomoides filamentosus* (Valenciennes, 1830), crimson jobfish
20. Family: Mugilidae
    A. *Aldrichetta forsteri* (Valenciennes, 1836), yellow-eye mullet
    B. *Mugil cephalus* (Linnaeus, 1758), flathead gray mullet
21. Family: Mullidae
    A. *Mulloidichthys vanicolensis* (Valenciennes, 1831), yellowfin goatfish
    B. *Parupeneus pleurostigma* (Bennett, 1831), sidespot goatfish
    C. *Parupeneus spilurus* (Bleeker, 1854), black-spot goatfish
    D. *Upeneus francisi* (Randall & Guézé, 1992), Francis' goatfish
22. Family: Priacanthidae
    A. *Priacanthus macracanthus* (Cuvier, 1829), red bigeye
23. Family: Scombridae. This family contains 13 species in the New Zealand waters.
24. Family: Serranidae. This family covers 15 species in the New Zealand waters.
25. Family: Sparidae
    A. *Acanthopagrus australis* (Günther, 1859), yellowfin bream
    B. *Acanthopagrus pacificus* (Iwatsuki, Kume & Yoshino, 2010), Pacific seabream
    C. *Pagrus auratus* (Forster, 1801), silver seabream
26. Family: Sphyraenidae
    A. *Sphyraena qenie* (Klunzinger, 1870), blackfin barracuda
27. Family: Bothidae
    A. *Arnoglossus scapha* (Forster, 1801)
    B. *Asterorhombus filifer* (Hensley & Randall, 2003), longlure flounder
    C. *Bothus mancus* (Broussonet, 1782), flowery flounder
    D. *Chascanopsetta lugubris* (Alcock, 1894), pelican flounder
    E. *Engyprosopon raoulense* (Amaoka & Mihara, 1995)
    F. *Lophonectes mongonuiensis* (Regan, 1914)
28. Family: Balistidae
    A. *Rhinecanthus rectangulus* (Bloch & Schneider, 1801), wedge-tail triggerfish
29. Family: Monacanthidae
    A. *Aluterus monoceros* (Linnaeus, 1758), unicorn leatherjacket filefish
    B. *Aluterus scriptus* (Osbeck, 1765), scribbled leatherjacket filefish
    C. *Meuschenia scaber* (Forster, 1801), velvet leatherjacket
    D. *Nelusetta ayraud* (Quoy & Gaimard, 1824), Chinaman-leatherjacket
    E. *Thamnaconus analis* (Waite, 1904)

F. *Thamnaconus hypargyreus* (Cope, 1871), lesser-spotted leatherjacket
30. Family: Antennariidae
A. *Antennarius striatus* (Shaw, 1794), striated frogfish
B. *Fowlerichthys scriptissimus* (Jordan, 1902)
C. *Histrio histrio* (Linnaeus, 1758), sargassum fish

5.3 Clupeotoxic Fishes

Introduction
Among the types of ichthyosarcotoxism is clupeotoxin. This kind of toxication is caused by clupeoid fishes such as sardines and herrings (Clupeidae) or anchovies (Engraulidae), tarpons (Elopidae), and bonefishes (Albulidae). It is widespread in tropical and subtropical areas of the world but rare. Some fatal cases were reported from different parts of the world because of consuming sardines (McCosker and Randall 2005). Clupeotoxin is a sporadic and an unpredictable public health problem in the tropical Atlantic Ocean, Caribbean Sea, and the tropical Pacific Ocean (Halstead 1967). Background: The history of clupeotoxication may go back to the second half of the eighteenth century when Desportes (1770) reported on the poisoning caused by eating small sardine species in the Dominican Republic. On the other hand, Oldendorp (1777) suggested that the sprat was the most poisonous fish in the Virgin Islands. Death from eating this type of fish happened within 1 h of consuming the meat. Mariner (1820) reported for the first time on cases of clupeotoxism in the tropical Pacific. He wrote about a fish locally known as *ooloo caoo* which is like sprat and caused intoxication during the month of July. Severe cases of intoxication from eating this fish and several fatalities were reported, but the history of the other aspects of type of intoxication such as toxicology, pharmacology, and chemistry are lacking information.

Causing Agent
The food web of the fish seems to be the origin of clupeotoxin. Melton et al. (1984) suggested that sardines, herrings, and anchovies feed on planktonic organisms; thus, the clupeotoxin they produce cannot be of benthic origin. Halstead (1967) reported that D'Arras (1877) was the first to report that sardines become poisonous because they feed on a "green monad." These monads cause conjunctivitis, coryza, and erythema in persons meeting them.

Clepeotoxin differs from ciguatoxin in having a nonbenthic origin and occurs during the warm months of the year in contrast to ciguatoxin, which may occur at any season (McCosker and Randall 2005).

Symptoms
Halstead (1967) gave the following symptoms of clupeotoxism: the first symptoms that the patient feels is the metallic taste that occurs upon ingestion of the fish. Nausea, dryness of the mouth, vomiting, malaise, abdominal pain, and diarrhea will follow. The further step in symptoms will be gastrointestinal upset accompanied with feeble pulse, tachycardia, chills, cold skin, and a drop in blood pressure, and a variety of neurological disturbances such as dilated pupils, violent headaches, numbness, tingling, hypersalivation, muscular cramps, respiratory distress, progressive muscular paralysis, convulsions, coma, and death. Within 15 min of ingestion of the fish, death could occur. Those victims who survived have shown pruritus and various types of skin eruptions including squamation and ulceration.

Treatment
The treatment is symptomatic and should follow the same regimens as those of ciguatoxin.

Prevention
The main thing to remember is to avoid eating clupeiform fishes during the warm season because the poison of these fishes cannot be detected by their appearance. Clupeotoxin cannot be deactivated by heat, thus cooking procedures, salting, and drying will not prevent intoxication.

5.3.1 Clupeotoxic Fish Species

Family: Clupeidae
Sardinops sagax (Jenyns, 1842) (Fig. 5.5)

Fig. 5.5 Sardinops sagax (Jenyns, 1842). Courtesy of Peter J. Bryant

Common name: South American pilchard
Etymology: *Sardinops:* Latin and Greek, *sarda* = sardine; name related to the island of Sardinia + Greek, *ops* = appearance (Romero 2002); *sagax:* From the Latin word *sagax* which means of quick perception, acute, or alert (Hart 1973)

Diagnostic Features
The body is elongated and has the ventral part of operculum with obvious cut bony striae radiating downward, which distinguish this species from all other clupeids. The abdomen is rounded with scutes at its ventral side. The body is blue green at the back, while the flanks are white. In New Zealand, the species appears to grow larger (21.3 cm standard length; cf. 19.7 cm), has slightly larger eggs, and has a higher mean number of vertebrae (50.52; cf. 49 to 50.08 in various samples) (Baker 1972).

Habitat, Biology, and Ecology
The South American pilchard is a marine species that inhabits the pelagic-neritic zone (Riede 2004). It lives at a depth ranging from surface and down to 200 m (Whitehead 1985). Individuals of this species reach a maximum standard length of 395 mm (Whitehead and Rodriguez-Sánchez 1995), with a common standard length of 200 mm (Whitehead 1985). The maximum reported weight of this species is 486.00 g (Hart 1973) and a maximum recorded age is 25 years (Whitehead 1985).

Individuals of this species form bigger school (Whitehead 1985). Phytoplankton and crustaceans are their main food (Anonymous 2001a). Females are oviparous, with pelagic eggs and pelagic larvae (Matarese et al. 1989). This species will possibly live up to 25 years (Matarese et al. 1989).

Distribution
Individuals of this species are distributed in the Indo-Pacific region from southern Africa to eastern Pacific (Parrish et al. 1989). There are three possible lineages of this species: southern Africa (*ocellatus*) and Australia (*neopilchardus*), Chile (*sagax*) and California (*caeruleus*), and Japan (*melanostictus*) (Grant et al. 1998).

Economic Value
This species is usually marketed fresh, frozen, or canned. Utilized mainly for fish meal, but also eaten fried and broiled (Frimodt 1995).

Conservation
Gaughan et al. (2018) evaluated this species as Least Concern in the Red List of the IUCN based on the following reasons. This species has a wide geographical distribution, and its populations are isolated. It has an economic value in most of its geographical distribution range. There is a decline in the spawning over the past 10 years.

Sprattus antipodum (Hector, 1872) (Fig. 5.6)
Common name: New Zealand blueback sprat
Etymology: *Sprattus:* Old German, *sprotte* (1611) = a small fish, *Clupea* sp. (Romero 2002)

Diagnostic Features
The body is slender. The gill cover is without bony radiating stria. The tooth-plate on the tongue is broad and it is almost oval. The lateral teeth are strong. The scales are with fine radiating or horizontal striae on the exposed posterior part. The color of the back of the body is back steel blue when fresh. The upper jaw is rounded and not notched on the frontal view. Pterotic bulla is absent.

Habitat, Biology, and Ecology
The New Zealand blueback sprat is a marine species that lives in the pelagic-neritic region at a depth range between the surface and 72 m depth (Stevenson 2004). This species reaches a maximum standard length of 120 mm, with a common standard length of 90 mm (Whitehead 1985). Individuals of this species are known to form schools in coastal waters. The biology of this species has not been fully investigated. Individuals of this species are usually found in the bottom or midwater during summertime and near the surface during winter (Armitage et al. 1994).

Distribution
Individuals of this species are found in New Zealand coasts (east coast of the North Island, Cook Straight, apparently all coasts of the South Island south to Stewart Island and Foveaux Strait).

Economic Value
This species has no commercial value and is used as a bait.

Conservation
Gaughan et al. (2017a) evaluated this species as Least Concern in the Red List of the IUCN based on the following reasons. This species is widespread throughout coastal New Zealand. This species has no managed fishery but provides in small quantities to mixed landings with *S. muelleri*. Additionally, this species presents in areas with sufficient fishery guidelines, and up-to-date catches are insignificant and below the total allowable commercial catch (TACC) due to a lack of commercial interest.

Sprattus muelleri (Klunzinger, 1879) (Fig. 5.7)
Common name: New Zealand sprat
Etymology: *Sprattus:* Old German, *sprotte* (1611) = a small fish, *Clupea* sp. ((Romero 2002)

Fig. 5.6 Sprattus antipodum (Hector, 1872). Courtesy of Wikipedia

Fig. 5.7 Sprattus muelleri (Klunzinger, 1879). Courtesy of Christopher Caine

Diagnostic Features
The body is relatively deep. The gill cover is without bony radiating striae. The tooth-plate on the tongue is narrow and pointed at each end. The lateral teeth are moderate in size. The scales are without fine striae on the exposed posterior part. The color of the back of the body is green/gray when fresh. The pterotic bulla is absent.

Habitat, Biology, and Ecology
The New Zealand sprat is a marine species that inhabits the pelagic-neritic zone at a depth ranging between 28 and 110 m (Stevenson 2004). This species reaches a maximum standard length of 130 mm, with a common standard length of 100 mm (Whitehead 1985). Individuals of this species are known to form schools in coastal waters and have an obviously long spawning season (July to January) around the South Island, but not a fractional spawner. Eggs are present in midsummer near the Clutha River mouth.

Distribution
Individuals of this species are distributed in the south Pacific at the coasts of New Zealand.

Economic Value
This species has no commercial value and is used as a bait.

Conservation
Gaughan et al. (2017b) evaluated this species as Least Concern in the Red List of the IUCN based on the following reasons. This species is widely distributed in the coastal waters of New Zealand. It has no directed fishery but contributes in small quantities to mixed landings with *S. antipodem*. Additionally, this species is found in regions with sufficient fishery guidelines, and present-day catches are small and below the total allowable commercial catch (TACC) due to a lack of commercial interest.

5.4 Gempylotoxic Fishes

Gempylotoxication is an instance of intoxication happening when fish species of the family Gempylidae are consumed. The incident is also known as keriorrhea (in Greek: "flow of wax"), which is a gastrointestinal disorder defined as an oily orange rectal discharge that occurs after eating of fish with high content of non-saponifiable lipids (wax esters), which are nondigestible by humans (Caro et al. 2011; Givney 2002; Ling et al. 2009; Nichols et al. 2001; Yohannes et al. 2002). The condition has been characteristically linked with the ingesting escolar (*Lepidocybium flavobrunneum*), rudderfish (*Centrolophus niger*), or oilfish (*Ruvettus pretiosus*) (Givney 2002; Ling et al. 2009; Yohannes et al. 2002; Shadbolt et al. 2002).

Gempylotoxication, also known as keriorrhea, particularly describes the pathological symptom of involuntary passage or leakage of oil, or wax esters, through the rectum. This term was coined by Berman et al. (1981) based on the Greek words *keri* and *diarroia*, which means "wax" and "to flow through," respectively. The oil discharged appears orange or brownish green in color, while little water is lost. This ailment is variously described as "oily diarrhea," "oily orange diarrhea," or "orange oily leakage" by the mass media and bloggers on the Internet.

5.5 Background

The work of Lowe (1841) is thought to be the initial published source about the laxative causes of castor-oil fish. He inscribed, "The flesh of the gempylid species is said to be extremely rich, and the bones, it is affirmed, abound in an oil or marrow, which, when they are sucked carelessly, produces speedy diarrhea." Poey (1966) also noted about this type of fishes and mentioned that the bones of the head are soft and spongy filled with oil. An explanatory review on the history of oilfish was written by Gudger (1925). The review includes evidence about the medical properties of the oil and the biology of the fish producing this oil. The pharmacology of the oil of the castor-oil fish is discussed by Macht and Barba-Gose (1931a, b). Despite the cleansing effect of the oil of the castor-oil fish, the gempylid species are reported to be a favorite food in the Gilbert Islands (Cooper 1964).

Causative Agent

The laxative influence gained from ingesting castor-oil fishes is owing to the existence of wax esters in large concentrations in the body of these fishes. Wax esters are esters of long-chain fatty acids with long-chain fatty alcohols. Liquid wax esters are made up of fatty acids and fatty alcohols of between 10 and 30 carbon atoms (Anderson 2010). These esters are found in the skin and muscles of certain deep sea fish species to help in buoyancy control and to provide a source of stored energy (Ling et al. 2008a). Castor-oil fishes are members of the family Gempylidae. They are generally found in deep waters of tropical and subtropical seas. They retain a high level of body lipids, which comprise approximately 20% of their wet weight (Barling and Foong 2015). The known name given to the liquid wax esters in this group of fishes is "gempylotoxin" because of their laxative and purgative effects (Lum 2012; Maralit et al. 2013). In restaurants and fish markets, the castor-oil fishes are known for their various names such as butterfish, codfish, and even white tuna (Hwang ct al. 2012). The origin of this wax is a deep-living zooplankton (Lee et al. 2006) and the bacteria *Acinetobacter* spp. (Rontani 2010). The consumed liquid ester wax found in the flesh of the castor-oil fish will cause keriorrhea within 1–36 h accompanied with a frequent "call to stool" due to the lubricant effect of the indigestible wax esters and other linked lipids such as possibly undigested triacylglycerols that have accumulated in the rectum. The action of the pancreatic lipases will be very slow leading to passing the liquid ester unmodified into the large intestine, which in turn will lead to keriorrhea and other gastrointestinal (GI) symptoms (Ling et al. 2009). The method of the digestion of triacylglycerols in the human body starts in the oral cavity, where the lingual lipase starts its effect, in the stomach, the gastric lipase, and is completed by pancreatic lipase in the jejunum. Pancreatic colipase and bile salts are needed to activate the pancreatic lipase. The resulting compounds are partial hydrolysis of triacylglycerols, free fatty acids, and glycerol (Barling and Foong 2015).

Likewise, the gempylid fish have high levels of histidine in their muscles (Feldman et al. 2005; Kan et al. 2000; Leask et al. 2004). If they are refrigerated incorrectly, bacteria can multiply and transform histidine into histamine, also known as scombrotoxin (FDA 2001), which can lead to cardiovascular, gastrointestinal, and neurological disorders (FDA 2001; Feldman et al. 2005; Leask et al. 2004).

Symptoms

The symptoms of intoxication by gempylotoxin are an involuntary anal discharge of orange or brownish-green liquid associated with gastrointestinal symptoms involving nausea, vomiting, and stomach cramps. No treatment is required and gempylotoxism is not a serious matter; people should be aware of the laxative effect of the castor-oil fish.

Epidemics of keriorrhea have been reported in many continents, including Africa, America, Asia, Australia, and Europe (Ling et al. 2009). However, little is reported in the scientific or medical literature. Consequently, the factual number of affected people over the years is largely undervalued as the Internet is floating with many more reports or communication about personal experiences of having oily diarrhea after consumption of fish. Generally, these fishes came into the spotlight due to the large outbreak that affected a considerable number of people; then, scattered incidents are generally neglected. Australia has recorded numerous keriorrhea outbreaks, permitting additional tracking into the etiology and symptoms in patients. In South Australia, between 1997 and 1999, there were nine cases of gastrointestinal complaints after rudderfish eating. In two instances that took place in 1999, patients complained of diarrhea, frequently oily and orange colored, within hours of consumption. By way of protein fingerprinting, the implicated fish was identified as escolar (Givney 2002). Two epidemics of diarrhea in 1999 and 2001 were recounted to be linked with butterfish consumption in Victoria, Australia. The patients were shown to have diarrhea or yellow oily diarrhea.

The fish causing the outbreaks in 1999 was identified as either escolar or rudderfish and the one in 2001 as escolar (Gregory 2002). Another outbreak of gastrointestinal illness happened during participants of a conference lunch in New South Wales, Australia, in October 2001. Analysis of the oil in the fish samples helps to show a high percentage of wax ester (96–97%) and a close similarity to the oil structure in escolar. A characteristic symptom recorded by many ill persons was the presence of oily diarrhea. Examiners of the outbreak performed a telephone interview of the group of conference participants utilizing a standard questionnaire. Out of 44 attendees, 20 (46%) became ill following the conference. The median incubation period was 2.5 h (range 1–90 h). The most common symptoms recorded were diarrhea (80% including 38% reporting oily diarrhea), abdominal cramps (50%), nausea (45%), headache (35%), and vomiting (25%). None of the food or beverages consumed was considerably linked with the illness. Nevertheless, all individuals who have eaten the fish became sick, but not those who did not (four persons). Among those who consumed the fish, the following possible risk agents did not have a substantial link with the illness: body mass index (BMI), age, health status, and the number of fish consumed (Yohannes et al. 2002).

The Chemistry of Keriorrhea

The members of the family Gempylidae have substantial lipid in their body, reaching up to approximately 20% of their wet weight (Cox and Reid 1932; Gudger 1925; Mori et al. 1966; Nichols et al. 2001; Ukishima et al. 1987). The main elements promoting to more than 90% of the total oil content are indigestible wax esters (Alexander et al. 2004; Ruiz-Gutierrez et al. 1997; Yohannes et al. 2002). Additional lipid groups (hydrocarbon, triacylglycerol, sterol, and polar lipid) are found in small or trivial amounts. Furthermore, the gempylid species have high levels of histidine (8–11 mg/g) in their muscles (Feldman et al. 2005; Kan et al. 2000; Leask et al. 2004).

Effect of Cooking on Oil Content and Composition

Ruello & Associates Pty. Ltd. (Ruello 2004) performed an investigation for the Australian Government Department of Health and Ageing. A comparative study was made on baking and grilling gempylid fish. Far more water than oil is lost during cooking as the total oil content on a wet-weight basis rises. Furthermore, the cooking technique has little influence on the oil composition; wax esters remain as the major oil class in the samples tested. The awareness for the reduction of the volume or potency of the oil/wax ester in these fish species by heavy grilling or other normal cooking method is unfounded (Ling et al. 2009). Battering and frying are unlikely to raise the risk of keriorrhea. There was no proof to support the common perception that wax esters can be "grilled out" of the fish or otherwise considerably decreased by normal cooking methods. The oil will be concentrated by cooking as water is expelled from the flesh. Freezing the (raw) fillet for as long as 9 months also failed to markedly lessen the capability of gempylid species to induce keriorrhea (Ruello 2004).

The Usage of Gempylid Fish Species Around the World

The gempylid fish species have been diversely used as food. Oilfish has been by tradition used by Polynesians and Melanesians as a purgative medicine (Cox and Reid 1932; Gudger 1925). In the Comoros Islands, these species are targeted as a food supply and caught by local people frequently (Helfman et al. 1999; Stobbs and Bruton 1991). In Bermuda, they are used as food and are recommended as an outstanding food (Gudger and Mowbray 1927). In the Canary Islands and other marine regions in Spain, these fish are also taken as a traditional medicine in drinking broth made from the bones to relieve constipation (Raisfeld and Patronite 2006; Ruiz-Gutierrez et al. 1997). In Japan, the gempylid fish are dried and consumed annually and around 10 tons of these fish were dried before the sale of these fish is banned (Mori et al. 1966). Even though defined by health authorities as a purgative in Taiwan, they are openly sold and served as sashimi and fish steaks under the name You-yu (literally oilfish). The roe of these fish is pressed and dried in Tungkang (Dunggang), Taiwan, and acknowledged as one of the three treasures of that little port as a new delicacy (Li 2007). On the other hand, these fish received acceptance from different levels of the populations in the South Asian regions. Berman et al. (1981) suggested that these fish had a very pleasant taste with a soft, butter-like texture. Some people consider the fish as a delicacy; for instance, Bykov (1983) mentioned in his book that "The taste qualities of this fish (oilfish) are high. It is an excellent table fish." Raisfeld and Patronite (2006) commended it as a dream fish, like toro (tuna), but cheaper.

The gempylid fish are prohibited in Italy, Japan, and the Republic of Korea. In the rest of the world, they are generally not regarded as appropriate for cooking. Therefore, they are often marketed and branded as other more expensive commercial fish but at lower prices.

There are several ideas for their use, with the hope that their unusually high levels of wax esters could be exploited for the good of mankind. Wax esters originated from orange roughy and oreo dories have been involved in various industrial processes such as cleaning and degreasing (Nichols et al. 2001). Wax esters enriched in o3 LC-PUFA can be absorbed by rats, and wax esters are less prone to oxidation and can be better formulated than liquid o3 derivatives (Gorreta et al. 2002). Consequently, wax esters supplemented in o3 can be a food addition as there is increasing sign that a diet high in o3 LC-PUFA may help prevent coronary heart diseases (Lacono and Dougherty 1993), and high level of wax esters in gempylid fish can act as a possible source for this use. Wax esters from some gempylid fish were administered by deacidification, decolorization, hydrogenation, and distillation to obtain a semisolid wax at room temperature; the processed wax was confirmed safe and useful as a base for medicine and cosmetics (Ukishima et al. 1987). An additional value-added process involved removal of most of the lipid from the meat of gempylid fish by repeated alkaline washing to make a gel, which was tested and found to be a better raw material than the commercial bigeye snapper used currently for surimi (ground fish meat) production (Pattaravivat et al. 2008).

International Worries

Australia, Canada, the USA, and most of member states of the European Union have declared special guidelines toward trading and utilization of the gempylid fish. Italy and Japan have forbidden their import and sale (Alexander et al. 2004) before 2007. Nevertheless, epidemics of keriorrhea are still reported occasionally across continents. Lately, an epidemic of over 600 cases of keriorrhea happened in Hong Kong at the end of 2006 (Chong 2007; Chung 2007a; Connolly et al. 2007; Jacquet and Pauly 2008; Ling et al. 2008a, b). The packs of oilfish cutlets were mislabeled as codfish (Chong 2007; Ling et al. 2008a, b). The other gempylid species, on the other hand, was found to be a sushi or sashimi under the name of snow-fish or white tuna (Chung 2007b; Ling et al. 2008a, b; Mok 2007). In February 2007, similar fish cutlets were found in Chinatown in Toronto, Canada, and resulted in a keriorrhea mini epidemic there (Ling et al. 2009). Three months later, the gempylid fish was found offered for sale as white tuna sushi in South Korea, and several cases of illness were reported, leading to the South Korean government prohibiting the use of gempylid fish for human food and banning any import of the fish in August 2007 (Stenhouse 2007). The public constantly showed great worry over this dilemma; for example, the oilfish scandal in 2006 awoke the Hong Kong public to this food safety issue, and there were strong demands for regulating and identifying these two fish (Goh 2007).

Species of the Family Gempylidae

Family: Gempylidae
Gempylus serpens Cuvier, 1829 (Fig. 5.8)
Common name: snake mackerel
Etymology: *Gempylus*: Greek, *gempylos* = a kind of fish (Romero 2002)

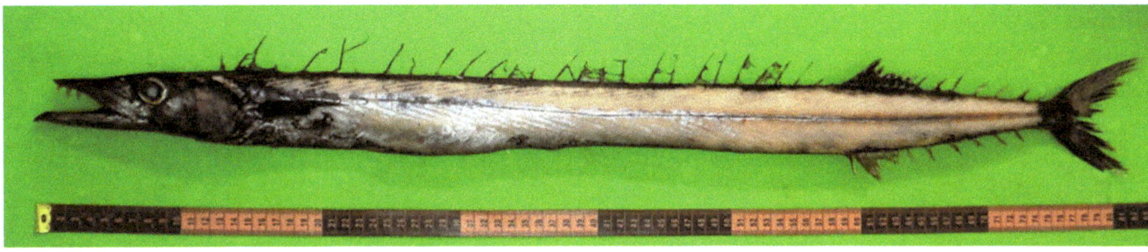

Fig. 5.8 Gempylus serpens Cuvier, 1829. Courtesy of Hamid Badar Osmany, Pakistan.

Diagnostic Features

The body is very much elongated and compressed. The lower jaw reaches to the anterior edge of the upper. The tips of both jaws are with a dermal process. Large fangs in the upper jaw are present, while fangs in the lower jaw are absent. The base of the first dorsal fin is long. The second dorsal fin is equipped with a very small spine. A free and compressed spine is present in the anal fin. The pelvic fins are reduced to one spine and 3–4 soft rays. Two lateral lines on the side of the fish body are present originating both from below the first spine of the first dorsal fin. The upper lateral line is running corresponding to the dorsal contour of the fish body until the spine of the first dorsal fin, while the lower shown is plunging posteriorly to the tip of the pectoral fin and then runs midlaterally. The body is naked with no scales except the posterior part of the body. The body coloration is uniformly dark brown. All fins are dark brown, with darker margins.

Habitat, Biology, and Ecology

The snake mackerel is a marine species that inhabits the pelagic-oceanic region (Reide 2004) at depth range of 0–600 m (Cervigón 1994), but it is usually found at range of 0–200 m (McMillan et al. 2011). Individuals of this species reach a maximum standard length of 1000 mm (Ref. 6181), with a common standard length of 600 mm (Ref. 6181). This species usually lives in solitary (Nakamura and Parin 1993). The grown-up individuals move to the surface at night, while larvae and juveniles remain near the surface during the day (Nakamura and Parin 1993). Fishes, cephalopods, and crustaceans are the main food items of this species (Nakamura 1995). The males mature at 430 mm SL, while the females reach maturity at 500 mm standard length (Bianchi et al. 1999). The eggs and larvae are pelagic (Parin 1986).

Distribution

This species is distributed worldwide and prefers tropical and subtropical seas. The adults are also frequently found in temperate waters.

Economic Value

The meat of the snake mackerel is usually sold frozen, as sausages or fish cake (Nakamura 1995), and eaten cooked in any way and also dried (Titcomb 1972).

Conservation

Collette et al. (2015) evaluated snake mackerel as Least Concern for the Red List of the IUCN based on the following reasons. This species is widely distributed and lacking a major threat.

Lepidocybium flavobrunneum (Smith, 1843) (Fig. 5.9)

Common name: Escolar

Etymology: *Lepidocybium:* Greek, *lepis* = scale + Greek, *kybion*, -ou = tunna (Romero 2002)

Diagnostic Features

The first dorsal fin is very low. The caudal peduncle with a pronounced keel is bordered with two small ancillary keels, one above and one below. A single, sinuous lateral line is present. The body is semi fusiform and somewhat compressed. The lower jaw marginally reaches anterior to the upper jaw. The tip of both jaws is without dermal processes. Two pairs of fangs are present anteriorly in the upper jaw. A vomer and palatines are present, each with uniserial small teeth. The anal

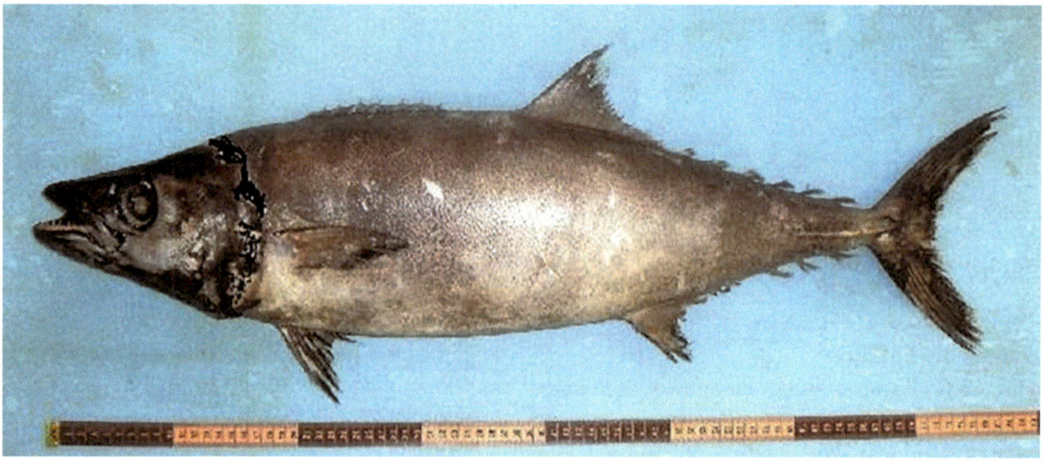

Fig. 5.9 Lepidocybium flavobrunneum (Smith, 1843). Courtesy of Hamid Badar Osmany, Pakistan

fin is with one or two spines and 12–14 soft rays. The pectoral fins have 15–17 soft rays. The pelvic fins are well developed, with one spine and five soft rays. The caudal fin is wide but rather small. Scales are quite small, each surrounded by a network of tubules bearing pores. The body with a uniform dark brown color becomes almost black with age.

Habitat, Biology, and Ecology

The escolar is a marine species that inhabit benthopelagic regions (Riede 2004) at a depth range of 200–1100 m (Shcherbachev 1987). Individuals of this species reach a maximum standard length of 2000 mm (Nakamura and Parin 1993), with a common length of 1500 mm standard length (Nakamura and Parin 1993). This species reaches a maximum reported weight of 45.0 kg (Nakamura 1984). It is found primarily over the continental slope, down to 200 m and above (Ref. 6181). Individuals of this species has an upward migration at night (Nakamura and Parin 1993). They usually feed on squid, crustaceans, and a wide variety of fishes (Nakamura and Parin 1993).

Distribution

Escolar is distributed in the tropical and temperate seas of the world, but perhaps not found in the northern Indian Ocean. It is reported from the Northwest Atlantic in the Canadian waters (Scott and Scott 1988), in the Eastern Atlantic: known from Guinea to Lobutu, Angola (Ref. 10816).

Economic Value

High percentage of oil in the flesh causes keriorrhea (Nakamura and Parin 2001). Sometimes is caught by tuna long-liners (Smith 1997). The meat of this species is usually sold frozen and as fish cakes in Japan (Nakamura 1995).

Conservation

Smith-Vaniz et al. (2015) evaluated this species as Least Concern for the Red List of the IUCN based on the following reasons. The species showed wide geographical distribution. Specimens of this species are abundant in museums around the world. It is among the usual species taken by pelagic longline fisheries. It is not particularly aimed and presently there no major threats are reported.

Family: Gempylidae
Ruvettus pretiosus (Cocco 1833) (Fig. 5.10)
Common name: Castor-oil fish

Diagnostic Features

A spinous bony tubercles are present between scales. The skin is very hard and rough. The mid-ventral keel is located on the ventral side of the

Fig. 5.10 *Ruvettus pretiosus* (Cocco 1833). Courtesy of Bañón Díaz, Rafael, Spain

body. The color of the body is usually brown with black tips of pectoral and pelvic fins. The second dorsal and anal fins are with white margins in young.

Distribution

The castor-oil fish is distributed in the temperate regions, with a circumtropical pattern. There are no reports of this species from the Arabian-Persian Gulf and from the Sea of Oman. It is reported from the southern coasts of the Arabian Peninsula (Manilo and Bogorodsky 2003).

Habitat, Biology, and Ecology

The castor-oil fish is a marine species that inhabits a benthopelagic environment and living at a depth range 100–800 m (Nakamura 1995). Individuals of this species are usually solitary or in pairs. They feed on crustaceans and squids (Nakamura and Parin 1993). Not much is available on the biology and behavior of this species.

Economic Value

The meat of this species is used as a food source and marketed fresh and as fish cakes in Japan (Nakamura 1995) and processed into fishmeal (Cervigón et al. 1992).

Conservation

This species is not evaluated for the Red List of the IUCN.

Other Gempylid Fish Species Present in the New Zealand Waters

There are eight more gempylid fish species other than those described living in the New Zealand waters, which might also cause keriorrhea. Therefore, a great caution is required to identify them before buying them for food. The eight species belonged to seven genera, and these are the following:

1. Domine, *Epinnula magistralis*.
2. False barracouta, *Nealotus tripes*.
3. Black barracouta, *Nesiarchus nasutus*.
4. False frostfish, *Paradiplospinus gracilis*.
5. Longfin gemfish, *Rexea antefurcata*.
6. Gemfish, *Rexea solandri*.
7. Barracouta, *Thyrsites atun*.
8. Black snoek, *Thyrsitoides marleyi*.

5.6 Scomberotoxic Fishes

It is known that scomerotoxine is usually produced by members of the family Scombridae. This toxin can initiate an instance of intoxication called "scombroid poisoning." It is also recognized as histamine fish poisoning. It is a kind of food toxic condition associated with signs and treatment like those related to seafood allergies (Hungerford 2010). Mismanaged fish samples will lead to scombroid poisoning because of their intake. Histamine [2-(1H-imidazol-4-yl) ethanamine] and other decomposition outcomes are generated in time-temperature ill-treated raw fish by bacterial, enzymatic conversion of free histidine (Rawles et al. 1996). The scombroid fish species have in common a high level of free histidine in their muscle tissues (Suyama and Yoshizawa 1973; Perez-Martin et al. 1988; Ruiz-Capillas and Moral 2004). It has been known that non-scombroid fish species of a wide range of fish groups are also included in initiating scombrotoxism (Taylor 1986; Hwang et al. 1997) as their flesh is rich in histidine (Lukton and Olcott 1958; Taylor 1986).

5.7 Background

The initial writing about the scombroid giving rise to poisoning is that of Thomas (1799). With the publication of the work by Burrows (1815), there were a substantial number of reports and publications about occurrences of scombroid poisoning in Europe and the West Indies. Maracacci (1891) was the first to use toxic analysis via the blood of certain scombroids. He determined that the blood of tuna can cause convulsion and paralysis in dogs when given either intravenously or intraperitoneally. Markov (1943) explained the process of forming the poison in the muscles of scombroid fishes. He indicated that selected fermentation processes by bacteria on harmless chemicals turn them into toxic mixtures. Igarashi (1939) recommended that histidine in large amounts will form if the meat of *Scomber japonicus* is left at temperatures of 24–25 °C because of autolysis. Possibly Geiger (1944) was the first to indicate a technique for determination of the freshness of fish using the measurement of histamine level in the muscles.

In the past years, there was a dispute about the exact nature of scomberotoxin. It started when Pergola (1937) anticipated that the poison was a result of special toxigenic bacterium called 'ichthyovenim.' Some scientists are still believing that the toxic substance was histamine which present in decayed scombroid fishes (Legroux et al. 1947; Pergola 1956). The occurrence of histamine by itself will have no poisonous effect (Kawataba 1962), but there are related toxic substances present that boost the effect of histamine (Kawataba et al. 1955).

Causative Agent

Biogenic amines are established mainly through the decarboxylation of specific free amino acids by exogenous decarboxylase bacteria freed by the microbial populations linked with seafood (Rawles et al. 1996). *Morganella* (*Proteus*) *morganii*, *Klebsiella pneumoniae*, and *Hafnia alvei* have been known to be associated with scombroid poisoning (Taylor and Speckard 1983).

In the human body, histamine has an important role as a messenger molecule and hence it is not a natural toxin. It is ample in its distribution and released from mast cells, enterochromaffin-like cells, and neurons. It has many necessary roles ranging from control of gastric acid secretion to neurotransmission in the central nervous system (Katzung 2007; Maintz and Novak 2007). Other significant tasks of histamine incorporate haematopoiesis, wound healing, and day-night rhythm (Kusche et al. 1980; Raithel et al. 1998). The response to histamine implies the immune system and, more particularly, allergic responses, which for some people are most familiar and dramatic (White 1990). Such tasks will help understand the nature, mechanism, and treatment of scombroid poisoning (Hungerford 2010).

In the case of taking in a large amount of histamine through eating fish meat with a high concentration of this substance, a potentiation of histamine toxicity by other compounds present in toxic fish will occur. This compound reaction has been suggested by a number of investigators (Bjeldanes et al. 1978; Paik and Bjeldanes 1979; Taylor and Lieber 1979; Lyons et al. 1983; Taylor 1986; Stratton and Taylor 1991) and requires the presence of dietary histamine. The protective binding of histamine to intestinal mucin is disrupted by potentiators.

Symptoms

After eating the poisoned scombroid fish, the symptoms begin from 10 min to 1 h (Ansdell 2015). There are disparities in the signs and can comprise peppery or metallic taste, oral numbness, headache, dizziness, palpitations, rapid and weak pulse (low blood pressure), difficulties in swallowing, and thirst (Arnold and Brown 1978; Kim 1979; Gilbert et al. 1980; Taylor 1986). Other indications are allergy-like symptoms such as hives, rash, flushing, and facial swelling (Kim 1979; Taylor et al. 1989). Some signs related to the central nervous system such as anxiety are less commonly seen (Russell and Maretić 1986; Sabroe and Kobza Black 1998; Specht 1998). There are some usual medical signs that the victim shows such as nausea, vomiting, abdominal cramps, and diarrhea (Gilbert et al. 1980). Recovery is usually complete within 24 h but could last for days (Taylor 1986). Worrying cardiac and respiratory complications observed seldom occur (Russell and Maretić 1986; Taylor et al. 1989; Ascione et al. 1997; Kounis 2013; Anastasius and Yiannikas 2015). Dermatological effects might occur because of severe poisoning (Tanew 2009).

Treatment

The treatment for anaphylactic shock is needed in some instances (Sanchez-Guerrero et al. 1997; Otani et al. 2004). People may show differences toward the sensitivity to scombroid poisoning (Motil and Scrimshaw 1979). The attack rate is regarded as the key attribute and, in some instances, the initial diagnosis of histamine poisoning versus seafood allergy. Normally, individuals eating poisoned scombroid fish react to the toxin, but only a small percentage of illnesses are expected if the observed signs are initiated by an allergy (Taylor et al. 1989). The administration of antihistamines is the only treatment of scombroid poisoning (Lerke et al. 1978; Blakesley 1983; Guss 1998).

Prevention

Due to scombrotoxin because of unsuitable handling/storage of fish, there are efficient testing techniques to detect toxic fish, and prevention and control of epidemics are possible. Contamination with histidine-decarboxylating bacteria can happen directly after the fish are caught on the fishing vessel, in the processing plant, in the distribution of the fish, and with the consumer, such as at home or in a restaurant (Taylor 1986). The main step to the prevention of the spread of scombrotoxin is appropriate cooling of the fish directly after they have been caught (Lehane and Olley 2000; Ritchie and Mackie 1979). Numerous countries have proposed limits for the maximum-permitted levels of histamine in fish. The amount of histamine produced is a function of the fish type, the part of the fish sampled, temperature, and the types of bacteria found in the fish (Rawles et al. 1996). Normal fish have less than 100 ppm of histamine (1 mg/100 g of flesh) (Stommel 2007). Although the toxic dose and symptoms of HFP are variable (Taylor 1986; Taylor et al. 1989; Motil and Scrimshaw 1979), illness commonly happens at levels of 1000 ppm (100 mg/100 g of flesh), but lower levels (20 mg/100 g of flesh) can also cause illness in some individuals (Morrow et al. 1991).

5.7.1 Fishes of the Family Scombridae

In New Zealand, the family Scombridae comprises 13 species belonging to eight genera. Below is a biological data for some of these species.

Family: Scombridae
Acanthocybium solandri (Cuvier, 1832) (Fig. 5.11)
Common name: wahoo
Etymology: *Acanthocybium:* Greek, *akantha* = thorn + Greek, *kybion*. -ou = a tunna (Romero 2002).

Diagnostic Features

The body is narrow and elongated, with a pointed snout about as long as the rest of the head. The mouth is large, with strong, triangular, compressed, and finely serrate teeth. The posterior part of the maxilla is completely concealed under the preorbital bone. The gill rakers are absent. The interpelvic process is small and bifid. The body is covered with small scales. The color of the back of the body is iridescent bluish green,

Fig. 5.11 Acanthocybium solandri (Cuvier, 1832). Courtesy of Moazzam Khan, Pakistan

while the sides are silvery with 24–30 cobalt blue vertical bars which extend to below the lateral line.

Acanthocybium solandri is distinguished from the morphologically similar narrow-barred Spanish mackerel (*Scomberomorus commerson* (Lacepède, 1800)) by a longer first dorsal fin (XXIII–XXVII versus XV–XVIII spines), lateral line curving abruptly downward below middle of the first dorsal fin (versus the lateral line distinctly sloping ventrally below the second dorsal fin in *S. commerson*), and a preorbital length (i.e., distance from the tip of the snout to the anterior edge of the eye) approximately the same length as the rest of the head (i.e., distance from the anterior edge of the eye to the posterior edge of the opercle), whereas the snout is shorter than the rest of the head in *S. commerson* (Fischer and Bianchi 1984; Collette and Graves 2019).

Distribution

The wahoo is distributed in the Atlantic, Indian, and Pacific Oceans. It is reported from the tropical and subtropical waters, including the Caribbean and Mediterranean Seas.

Habitat, Biology, and Ecology

Acanthocybium solandri is a marine species that inhabit the pelagic-oceanic region (Riede 2004) at a depth range 0–20 m (Bacchet et al. 2006), but it is usually found at range of 0–12 m (Humann and Deloach 1993). Individuals of this species reach maturity at total length of 850–1050 mm and reach a maximum recorded total length of 2500 mm (Sommer et al. 1996), with a common total length of 1700 mm (Collette 1995). The maximum recorded weight is 83.0 kg (Collette and Nauen 1983). Individuals of this species are oceanic and epipelagic and live frequently in solitary or forming small loose aggregations rather than compact schools (Collette 1986). Their main food items are fish and squid. The eggs and larvae are pelagic (Collette 1986). This species is an essential sport fish in some areas (Collette 1995).

Economic Value

Wahoo has a commercial value. It is marketed fresh, salted, or spice-cured slices of meat and also frozen (Frimodt 1995). Its flesh is of very good quality (Collette 2001).

Conservation

Collette et al. (2011a) assessed *Acanthocybium solandri* as Least Concern for the Red List of the IUCN based on the following reasons. Wahoo has a wide geographical distribution. Recent studies showed an increased landing over the past 20 years. There are no signs of influence on its population through fishing services.

Allothunnus fallai (Serventy, 1948) (Fig. 5.12)
Common name: slender tuna
Etymology: *Allothunnus:* Greek, *allos* = other + Greek, *thynnos* = tunna (Romero 2002)

Diagnostic Features

The body is tough, elongated, and rounded. The teeth are very small and conical; there are 40–55 teeth on each side of the upper and lower jaws. There are 70–80 gill rakers on first arch. The dorsal fins are close to each other, the first with 15–18 spines and the second with 12 or 13 rays.

Fig. 5.12 Allothunnus fallai Serventy, 1948. Courtesy of Robbie N. Cada

There are six or seven finlets behind the dorsal fins. The anal fin has 13 or 14 rays, followed by six or seven finlets. The pectoral fins have 24–26 rays. The interpelvic process is small and bifid. The body is naked ventrally behind the long anterior corselet. The dorsal half of the body to the lateral line is covered with small scales. The caudal peduncle is slender with a well-developed lateral keel between the two smaller keels on each side.

Distribution

The slender tuna is distributed in the Southern Ocean, with a circumglobal distribution.

Habitat, Biology, and Ecology

Allothunnus fallai is a marine species that inhabits the pelagic-oceanic region (Riede 2004) at a depth range of 0–20 m (Fricke et al. 2011). The maximum fork length recorded is 1050 mm (Collette 1986), with a common total length of 860 mm (Collette 2001). The maximum reported weight for this species is 13.7 kg (Collette 1986). Individuals of this species form a seldom schooling which feeds mainly on krill (euphausiids) and on squids and small fishes. The juveniles live between 20 and 35°S at surface temperatures ranging from 19 to 24 °C.

Economic Value

The flesh of this species is paler than that of most true tunas and is very oily, but the cooked meat has fine eaten qualities. It is mostly marketed fresh.

Conservation

Collette et al. (2011e) assessed the slender tuna as Least Concern in the Red List of the IUCN based on the following reasons. The slender tuna has a wide geographical distribution and is abundant in the Southern Ocean. Individuals of this species are present in the bycatch of the long line fisheries. Nothing is known about the influence of fishery activities on its populations. Also, there is no threat against this species.

Auxis thazard (Lacepède 1800) (Fig. 5.13)
Common name: frigate tuna
Etymology: *Auxis*: Greek, *auxis* = a variety of tunna (Romero 2002)

Diagnostic Features

The body is large, rounded, and elongated. The teeth are small conical and fall in single series. The pectoral fins are short. There is a large, pointed flap between pelvic fins. There are no scales on the body except in the corselet. There is a strong central keel on each side of the base of the caudal fin. The body is bluish in color on the back and deep purple to black on the head. There are 15 obliques to horizontal dark lateral lines. The ventral side is white. The pectoral and pelvic fins are purple, with black inner side (Froese and Pauly 2016).

Fig. 5.13 *Auxis thazard* (Lacepède 1800). Courtesy of Courtesy of Hamid Badar Osmany, Pakistan

Fig. 5.14 Katsuwonus pelamis (Linnaeus, 1758). Courtesy of Hamid Badar Osmany, Pakistan

Distribution
This species has a worldwide distribution in the Atlantic, Indian, and Pacific Oceans (Froese and Pauly 2016).

Habitat, Biology, and Ecology
The frigate tuna is a marine species living in the pelagic-neritic region (Riede 2004) at a depth of more than 50 m (Collette 1995). The maximum recorded fork length of this species is 650 mm (Cayré et al. 1993), with a common total length of 600 mm (de Figueiredo et al. 2002). The maximum reported weight is 1.7 kg (IGFA 2001) and the maximum recorded age is 5 years (Cayré et al. 1993). Individuals of this species feed on small fish, squids, planktonic crustaceans (megalops), and stomatopod larvae (Fischer et al. 1990). Because of their large number, they are believed to be a significant component of the food web, mainly as search for other species of commercial interest. Larger fish species feed on this species, including other tunas (Frimodt 1995).

Economic Value
The meat of this species is marketed fresh and frozen (Collette 1995) and also utilized dried or salted, smoked, and canned (Frimodt 1995).

Conservation
Collette et al. (2011b) assessed this species for the Red List of the IUCN based on the following reasons. The frigate tuna has a wide geographical distribution. It is essential in artisanal fisheries and is caught as bycatch in commercial fisheries, but landings are often mixed with *Auxis rochei*.

Katsuwonus pelamis (Linnaeus, 1758) (Fig. 5.14)
Common name: skipjack tuna

Diagnostic Feature
Th body is fusiform, elongate, and rounded. The teeth are small and conical, in a single series. The gill rakers are numerous, 53–63 on the first gill arch. There are two dorsal fins separated by a small interspace (not larger than the eye), the first with 14–16 spines and the second followed by

7–9 finlets. The pectoral fins are short, with 26 or 27 rays. The interpelvic process is small and bifid. The anal fin is followed by seven or eight finlets. The body is scaleless except for the corselet and lateral line. There is a strong keel on each side of the caudal fin base between two smaller keels. The swim bladder is absent. The back of the body is dark purplish blue, and the lower sides and belly are silvery, with four to six very conspicuous longitudinal dark bands which in live specimens may appear as discontinuous lines of dark blotches.

Distribution
This species has a cosmopolitan distribution in the tropical and the warm-temperate waters. It is not found in the Black Sea. This species is a highly migratory species.

Habitat, Biology, and Ecology
The skipjack tuna is a marine species that inhabits pelagic-oceanic areas at a depth down to 260 m (Collette 1995). Individuals reach maturity at length ranges of 400–450 mm (McMillan et al. 2011). The maximum recorded fork length is 1100 mm (McMillan et al. 2011), with a common length of 800 mm fork length (Collette and Nauen 1983). The maximum reported weight is 34.5 kg (Collette and Nauen 1983) and the maximum recorded age is 12 years (Collette and Nauen 1983). This species usually lives in offshore waters and larvae are restricted to waters with surface temperatures of 15–30 °C (Kailola et al. 1993). Individuals of this species show a strong affinity to school in surface waters with birds, drifting objects, sharks, and whales and may show a characteristic behavior such as jumping, feeding, and foaming. They feed on fishes, crustaceans, cephalopods, and mollusks; cannibalism is common. Thes spawn during the year in the tropics, and eggs are released in several portions (Ref. 35388). The eggs and larvae are pelagic (Collette 1986). They have been preyed upon by large pelagic fishes (Hart 1973). This species is taken by trolling on light tackle using plugs, spoons, feathers, or strip bait (Collette 2001).

Economic Value
This species has a high commercial value and marketed fresh, frozen, or canned (Collette 1995, 2001) and also dried-salted and smoked (Frimodt 1995).

Conservation
Collette et al. (2021a) assessed this species for the Red List of the IUCN based on the following reasons. Skipjack tuna is heavily exploited but is not considered to be overfished. The species is fast-growing, reasonably short-lived, and highly productive.

Sarda australis (Macleay, 1881) (Fig. 5.15)
Common name: Australian bonito
Etymology: *Sarda:* Latin and Greek, *sarda* = sardine; name related to the island of Sardinia (Romero 2002)

Diagnostic Features
The number of teeth in the upper jaw are 16–26. The number of teeth on the lower jaw teeth are 11–20. The vomerine teeth are sometimes present. The supramaxilla is intermediate. There are

Fig. 5.15 Sarda australis (Macleay, 1881). Courtesy of Wikipedia

19–21 gill rakers on the first gill arch. The first dorsal fin is with 17–19 spines. The number of the dorsal finlets is usually seven. There are 14–17 rays in the anal fin. The anal finlets are usually six. The pectoral fin rays are 25–27, modally 26. Color: the dorsal stripes on the body are closer to being horizontal than in other species of *Sarda* and extending onto the belly in some specimens.

Distribution

The Australian bonito is usually distributed in the Southwest Pacific from the southeastern Australia, Norfolk Island, and New Zealand.

Habitat, Biology, and Ecology

Sarda australis is a marine species that inhabits the pelagic-neritic region (Riede 2004). The maximum reported fork length is 1800 mm (IGFA 2001), with a common fork length of 450 mm (Collette and Nauen 1983). The maximum recorded weight is 9.4 kg (Collette and Nauen 1983). Individuals of this species intended to form schooling by size and maturing from January through April (IGFA 2001).

Economic Value

The meat of this species of tuna is light-colored of delicate flavor and good canning quality.

Conservation

Collette et al. (2011d) assessed this species for the Red List of the IUCN based on the following reasons. This species is confined to southeastern Australia, Norfolk Island, and northern New Zealand. It has less commercial value.

Scomber australasicus (Cuvier, 1832) (Fig. 5.16)
Common name: blue mackerel
Etymology: *Scomber:* Greek, *skombros* = tunny or mackerel, 1623 (Romero 2002)

Diagnostic Features

The body is elongate and rounded. The snout is pointed. The front and hind margins of the eye are covered by an adipose eyelid. The teeth in the upper and lower jaws are small and conical. The teeth are also present on vomer and palatine bones (roof of mouth). The gill rakers are shorter than gill filaments, barely visible through open mouth. There are 25–35 gill rakers on the lower limb of the first gill arch. There are two dorsal fins, which are widely separated; the first with 8–13 spines, while the second dorsal and anal fins are with 12 rays. The anal spine is stiff and strong. There are five dorsal and five anal finlets. The interpelvic process is small and single. The pectoral fin is short, with 18–21 rays. The entire body is covered with rather small scales. The scales behind the head and around the pectoral fins are larger and more obvious than those covering the rest of the body. There are two small keels on each side of the caudal peduncle (at base of caudal fin lobes), but no central keel between them. The swim bladder is either present or absent. The last branchiostegal ray is slightly flattened but not forming a wide plate. The color of the back of the body is steel-blue crossed by faint

Fig. 5.16 Scomber australasicus Cuvier, 1832. Courtesy of Hamid Badar Osmany, Pakistan

wavy lines. The lower sides and belly are silvery yellow unmarked or with numerous dusky, rounded blotches. There are no rows of spots on the back next to the dorsal fin bases as in the genus *Rastrelliger*.

Distribution
This species of tuna is distributed in the Indo-West Pacific region from the Red Sea, Arabian Gulf, from Japan, south to Australia, and New Zealand. It is also reported from the Eastern Pacific area at Hawaii and off Mexico (Socorro Island).

Habitat, Biology, and Ecology
The blue mackerel is a marine species that inhabits the pelagic-neritic region (Riede 2004) at a depth range 87–200 m (May and Maxwell 1986). The maximum reported fork length is 440 mm (IGFA 2001), with a common fork length of 300 mm (Collette and Nauen 1983). The maximum reported weight is 1.4 kg (IGFA 2001). Individuals of this species are found in coastal waters (Collette 1995) and in oceanic waters (May and Maxwell 1986). They form schools by size which may include jack mackerels and Pacific sardines. They feed on plankton, copepods, and other crustaceans, but adults also feed on small fish and squids.

Economic Value
This species can be caught with encircling nets (Collette 1995). The meat is marketed fresh, dried salted, smoked, canned, and frozen (Frimodt 1995).

Conservation
Collette et al. (2011c) assessed this species on the Rd. List of the IUCN based on the following reasons. The predictable spawning stock biomass for at least one stock is increasing in the northwest Pacific, while for the other stock, it is fluctuating, but constantly.

Thunnus alalunga (Bonnaterre, 1788) (Fig. 5.17)
Common name: albacore
Etymology: *Thunnus:* Greek, *thynnos* = tunna (Romero 2002)

Diagnostic Features
The albacore tuna is a large species, with its deepest body part at the posterior part of its body. The gill rakers are 25–31 on the first gill arch. The second dorsal fin is clearly lower than the first dorsal fin. The pectoral fins are very long; they are usually 30% of the fork length or longer in 50 cm which have proportionately smaller pectorals than other tunas, i.e., *T. obesus*. The swim bladder is present, but poorly developed and not obvious in fish smaller than about 50 cm fork length. A presence of a faint lateral iridescent blue band runs along the sides of the body of the fish. The first dorsal fin is deep yellow, the second dorsal and anal fins are light yellow, and the anal finlets are dark. The posterior margin of the caudal fin is white.

Distribution
This species of tuna has a cosmopolitan distribution and found in the tropical and temperate waters of all oceans including the Mediterranean

Fig. 5.17 Thunnus alalunga (Bonnaterre, 1788). Courtesy of Cihant Turkey

Sea but not at the surface between 10°N and 10°S. It is also reported from the Western Pacific at a range extending in a broad band between 40°N and 40°S (Collette 2001). It is frequently mistaken with juvenile *Thunnus obesus* which also have very long pectorals but with rounded tips. The individuals of this species are greatly migratory.

Habitat, Biology, and Ecology

The albacore tuna is a marine species that inhabits the pelagic-oceanic region (Riede 2004). It lives at a depth range of 0–600 m (Collette and Nauen 1983). The maximum fork length recorded for this species 1400 mm (Torres and Pauly 1991), with a common fork length of 1000 mm (Collette 2001). The maximum reported weight is 60.3 kg (IGFA 2001). The maximum recorded age of this species is 9 years (Altman and Dittmer 1962). Individuals of this species are found in waters of 13.5–25.2 °C (Collette and Nauen 1983). They form mixed schools with skipjack tuna (*Katsuwonus pelamis*), yellowfin tuna (*Thunnus albacares*), and bluefin tuna (*T. maccoyii*); schools may be linked with floating objects, including sargassum weeds (Collette and Nauen 1983). This species feeds on fishes, crustaceans, and squids. The eggs and larvae are pelagic (Collette 1986). Its sexual maturity reached at 90 cm (Bianchi et al. 1999).

Economic Value

The meat of this species is highly appreciated and marketed fresh, smoked, deep frozen, or canned and is eaten steamed, broiled, fried, and microwaved (Fromodt 1995).

Conservation

Collette et al. (2021b) assessed this species for the Red List of the IUCN based on the following reasons.

Albacore Tuna is a vital resource for global fisheries. Over the past 20 to 25 years, equivalent to roughly three generations (spanning from 1993/1995 to 2016/2018), there has been an estimated global decline in biomass or spawning stock biomass of around 3–19%, depending on stock size and other indicators. The most significant reduction, up to 40%, occurred in the Mediterranean Sea between 1993 and 2015, raising concern. In the Indian Ocean, where overfishing is a noted issue, declines of 20–40% have been observed over the same time frame. This region accounts for roughly 10% of the global Albacore Tuna population. Despite these declines, the species remains widespread globally with a large population size. It does not meet the thresholds for criteria B, C, or D, and criterion E is not applicable due to the lack of quantitative analyses on future extinction risks. However, with more than 90% of its global range outside areas impacted by overfishing and the species not considered overfished, Albacore Tuna is evaluated under criterion A1. Given the global population's reduction by 3–19% over the past three generations, it is classified as Least Concern.

5.8 Hallucinogenic Fishes

Hallucinogenic or ichthyoallyeinotoxism is a rare form of ichthyosarcotoxism and is indicated by development of central nervous system disturbances, specifically hallucinations and nightmares. Ichthyoallyeinotoxism has been known in many regions around the world (Halstead 1988; Château-Degat 2003). The biotoxication is a result of eating either the head or the flesh of the fish.

5.9 Background

The first information on the ichthyoallyeinotoxism was that of Jordan et al. (1927). They described a case of intoxication that happened to the mayor of Honolulu because of eating the head of a fish. Data on the ichthyoallyeinotoxic fishes were published over the years by Tinker (1944), Nichols and Bartsch (1945), Titcomb and Pukui (1952), Fish and Cobb (1954), and Randall (1958). In most of these cases, the causative fish was *Kyphosus vaigiensis*. On the other hand, Helfrich and Banner (1960) proposed that eating the brain of the members of the family Mullidae can lead to convulsions. They were the first to give the entity the name ichthyoallyeinotoxism and gave it the term "hallucinogenic fish poisoning" or "ichthyosarcephialtilepsis." A clinical summary of the biotoxications was given by Bouder et al. (1962).

Causative Agent

The contributing agents for ichthyoallyeinotoxism are still unidentified, but some authors have implicated the toxic macroalgae (Caulerpaceae family) that are ingested and contaminate the flesh of fish (Helfrich and Banner 1960; Chevaldonne 1990). Because the incidences have been explained after eating of fried, boiled, steamed, or raw fish, these toxins are possibly heat stable (Helfrich and Banner 1960; Raikhlin-Eisenkraft and Bentur 2002). Some instances of intentional consumption of "dreamfish" have been stated (Helfrich 1963; Roughly and Roberts 1960; Cooper 1964).

During the Roman times in the Mediterranean region, consumption of *Sarpa salpa* was reported for recreational reasons (de Haro et al. 1993) and for ceremonial purposes mainly in Polynesian populations (Helfrich 1963; Roughly and Roberts 1960; Cooper 1964). The toxic consequence can be identified from the traditional names given to hallucinogenic species. *Sarpa salpa* is called "the fish that makes dreams" in Arabic; *Siganus spinus* is called "the fish that inebriates" in Mascareignes (southwest coast of Reunion Island; Lebeau 1979; Quod and Turquet 1996); and *Mulloidichthys samoensis* is called "the chief of ghosts" in the Hawaiian Islands of Molokai, Kauai, and Oahu (Helfrich 1963; Jordan et al. 1927; Banner 1973). It should be noted that there are differences in clinical expression between ichthyoallyeinotoxism and ciguatera. The earlier is characterized by central nervous system involvement, whereas the second shows peripheral nervous system engagement. The two diseases are also distinct regarding potential severity and interval. Most patients with ichthyoallyeinotoxism recover within 36 h (Helfrich and Banner 1960; Halstead 1988; Château-Degat 2003), whereas many patients with ciguatera die or present prolonged symptomatology for several months (de Haro et al. 2003).

Symptoms

Symptoms may happen within 2 h after eating fresh fish. These begin with nausea and vomiting, which arise quickly. Later, symptoms endure and may be associated with marked muscle weakness, blurring of vision, and hallucinations such as aggressive and screaming animals (de Haro and Pommier 2006). The typical symptoms include dizziness, loss of equilibrium, lack of motor coordination, hallucinations, and mental depression.

Orsolini et al. (2018) suggested that several hallucinatory and onyroid occurrences,

also called "ichthyoallyeinotoxic or hallucinatory mullet poisoning" or "ichthyoallyeinotoxism," have been reported after ingestion of different species mullets as raw (Helfrich and Banner 1960; Halstead 1988; Château-Degat 2003). The consequences of eating ichthyoallyeinotoxic raw fish may include vivid/terrifying auditory and visual hallucinations, dizziness, loss of equilibrium, lack of motor coordination and mental depression, terror, nightmares, itching, burning of the throat, muscular weakness, and rarely abdominal distress (Helfrich and Banner 1960; Halstead 1988; Château-Degat 2003). Symptomatology may occur within a few minutes to 2 hours after consumption and may last for up to 24 h (De Haro and Pommier 2006, Helfrich and Banner 1960). The first symptoms usually comprise imbalance, loss of coordination, and a generalized malaise, followed by delirium, visual and/or auditory hallucinations (mainly zooptic), depression, and nightmares (Helfrich and Banner 1960; Helfrich 1963; De Haro and Pommier 2006; Froese 2011).

Treatment

Stomach wash should be instantly applied to the victim. The remaining treatments are symptomatic.

Prevention

Care should be taken when a judgment is made to eat reef fish species that are considered as the reason behind the intoxication or ichthyoallyeinotoxism. It is suitable not to eat the head of tropical reef fish. Hallucinogenic fish cannot be detected by their appearance.

5.9.1 Cases of Fish Hallucination

De Haro and Pommier (2006) reported two incidents of hallucination (ichthyoallyeinotoxism) after ingesting the sparid species *Sarpa salpa*. This case happened in the Western Mediterranean.

Case 1

In April 1994, a healthy 40-year-old man started feeling weak and tired within 2 hours after eating freshly caught baked sea bream (*Sarpa salpa*) for dinner during a holiday in Cannes on the French Riviera. Symptoms such as nausea and vomiting started rapidly during the night. In the next day, the symptoms endured and were associated with marked muscle weakness. Then, he began to feel blurring of vision and hallucinations involving aggressive and screaming animals. Physical examination upon arrival at the hospital emergency room demonstrated no notable abnormalities: no fever, no sign of focalization or sensory-motor deficit, and normal hemodynamic status except for sinusal tachycardia linked directly to the mental disturbances. In the hospital, the patient is well again rapidly with complete resolution of symptoms within 36 h post ingestion. He was unable to recall the hallucinatory period.

Case 2

The second incident happened in March 2002. A healthy 90-year-old started having auditory hallucinations 2 hours after eating a sea bream (*Sarpa salpa*) that he had purchased from a professional fisherman in Saint Tropez on the French Riviera. Later, he experienced hallucinations of a particularly terrifying nature (human screams and bird squealing), and he had several bad dreams for the next two nights. Being afraid that these symptoms might indicate the start of a main mental illness, he did not tell his friends or attending physician. The symptoms ended in 3 days after he had eaten the fish. Later, he remembered a warning that he had heard at the fish market concerning the hallucinogenic possibility of sea bream and decided to contact the Poison Control Center in Marseille.

The clinical symptoms of ichthyoallyeinotoxism occurred to the two men within a few minutes to 2 hours after ingestion of toxic fish. Among the symptoms the two victims showed is intoxication with loss of balance and coordination and generalized malaise (Helfrich and Banner 1960; Halstead 1988; Château-Degat 2003). The symptoms started with sore throat and heartburn, and within a few hours, precise signs of poisoning started to show comprising delirium, visual and/or auditory hallucinations (often involving animals), depression, feelings of impending death with reactive tachycardia and hyperventilation, and disturbed behavior. In case the two victims were able to sleep, they experienced a terrifying nightmare (Helfrich and Banner 1960). In addition, gastrointestinal symptoms including nausea, abdominal pain, and diarrhea have also been reported, but are usually low-grade (Helfrich and Banner 1960).

As a treatment, no peculiar method or antidote treatment is available. The two victims were treated with an appropriate management of transient behavioral disturbances (e.g., using benzodiazepine or neuroleptics), where it is significant to prevent self-inflicted or other injuries. Symptomatic treatment for gastrointestinal manifestation can enhance patient comfort. Symptoms generally terminate within 24–36 h, but weakness persisted for several days (Helfrich and Banner 1960; Halstead 1988; Château-Degat 2003).

Although the toxic agents of ichthyoallyeinotoxism are still unknown, some authors have linked toxic macroalgae (Caulerpaceae family) that are ingested and contaminate the flesh of fish (Helfrich and Banner 1960; Chevaldonne 1990). It should be emphasized that tropical fish species that cause ichthyoallyeinotoxism are also involved in incidences of ciguatera poisoning.

5.9.2 Fish Species Causing Hallucination

Family: Mugilidae
Mugil cephalus (Linnaeus 1758) (Fig. 5.18)
Common name: flathead gray mullet
Etymology: Mugil: Latin, *mugil*, -ilis = gray mullet (Romero 2002); *cephalus*: *cephalus* meaning head (Etnier and Starnes 1993)

Diagnostic Features
The body is sturdy, cylindrical in cross-section, and somewhat compressed; the head is broad and flattened (Albaret 2003). The adipose eyelid is well developed (Keith and Allardi 2001; Albaret 2003). The upper lip is thin and without papillae (Albaret 2003). One or two outer rows of small, reasonably close-set, unicuspid teeth and up to six inner rows of smaller, bicuspid teeth are present (Harrison 2008). The lower lip is with an outer row of small unicuspid teeth and sometimes one or more inner rows of smaller, bicuspid teeth (Harrison 2008). The posterior end of the upper jaw is reaching a vertical line from the anterior eye margin. The maxillary pad is not visible below the corner of the mouth when closed. The origin of the first dorsal fin is nearer to the snout tip than to the caudal fin base (Albaret 2003). The anterior parts and the bases of the second dorsal and anal fins are with a moderately dense coverage of scales (Albaret 2003; Harrison 2008). A pectoral axillary process is present (Albaret 2003). There are 13–15 scale rows between the origins of the dorsal and the pelvic fins (Albaret 2003; Harrison 2008).

Fig. 5.18 *Mugil cephalus* (Linnaeus 1758). Courtesy of Hamid Badar Osmany, Pakistan

Distribution
This species of mullet has a cosmopolitan distribution in the coastal waters of the tropical, subtropical, and temperate zones of all seas. It is reported from the Eastern Pacific in waters from California, USA, to Chile (Eschmeyer et al. 1983). It also recorded in the Western Pacific from Japan to Australia (Harrison and Senou 1997). Reports of this species are also available from the Western Indian Ocean from India to South Africa (Smith and Smith 1986). In the Western Atlantic, it is reported from the Nova Scotia, Canada, to Brazil (Robins and Ray 1986) and from Cape Cod to southern Gulf of Mexico (Smith 1997). It is absent in the Bahamas and most of West Indies and Caribbean (Robins and Ray 1986; Humann 1994). In the Eastern Atlantic, it is reported from the Bay of Biscay to South Africa, including the Mediterranean Sea and Black Sea (Thomson 1990).

Habitat, Biology, and Ecology
The flathead gray mullet is a marine species found also in the freshwater and brackish water environments (Riede 2004). It lives at a depth range of 0–120 m (Harrison 1995), but it is usually found at 0–10 m. The adults live in coastal waters (Eschmeyer et al. 1983; Allen et al. 2002; Albaret 2003) and are regularly found in estuaries and rivers (Allen 1991; Thomson 1986; Yamada et al. 1995; Allen et al. 2002; Albaret 2003). Individuals of this species are usually forming schools over sand or mud bottom (Eschmeyer et al. 1983). They feed mainly at daytime on detritus, microalgae, and benthic organisms during the daytime (Blaber 1976; Tung 1981; Cardona 2000).

This species reproduces at sea and at different times of the year depending on the location. The maximum recorded standard length is 1200 mm (Thomson 1990; Albaret 2003; Harrison 2008) and the maximum weight known is 12 kg (Fadeev 2005).

Economic Value
This species has high commercial value. It is marketed fresh, dried, salted, and frozen; roe is

sold fresh or smoked (Harrison 1995); also used in Chinese medicine (Tang 1987).

Conservation
Camara et al. (2019) assessed the conservation status of this species as Least Concern for the Red List of the IUCN based on the following reason. It is widely distributed, with no reported threat.

Family: Kyphosidae
Kyphosus vaigiensis (Quoy & Gaimard, 1825) (Fig. 5.19)
Common name: brassy chub
Etymology: *Kyphosus:* Greek, *kyphos* = bent (Romero 2002)

Diagnostic Features
The brassy chub has a distinctive set of characters such as an elongated and oval-shaped body, the scale rows are horizontal along the body which appear alternately golden and bluish and metallic blue dorsally and white ventrally, the caudal fin and the very low dorsal and anal fins are moderately emarginated, the dorsal fin has 10–11 spines and 13–15 soft rays, the pectoral fin is with 17–20 soft rays, the teeth are incisor in shape, scales on the interorbital region are present, the external side of the first gill arch is with high number of gill rakers on lower limb (16–23), and the total lateral line scale rows 63–80 and 52–63 are with pores. The body has a silvery color, and it is a bluish shine, with golden horizontal lines along the body (around 23–29). The dorsal line above the lateral line is curved along the lateral line. A golden streak is present beneath the eye from the snout to behind the anterior edge of the eye. The fins are gray or slightly darker gray than the body color (Knudsen and Clements 2013).

Fig. 5.19 Kyphosus vaigiensis (Quoy & Gaimard, 1825). Courtesy of Hamid Badar Osmany, Pakistan

Distribution
This species chub has a circumtropical type of distribution. It is reported from the eastern Pacific across the Pacific Ocean to the Indian Ocean, the Red Sea, and the eastern and western Atlantic and the Mediterranean.

Habitat, Biology, and Ecology
Kyphosus vaigiensis is a marine species that inhabits the reef region (Riede 2004). It lives at a depth range of 0–40 m (Bacchet et al. 2006). Individuals of this species reach a maximum total length of 700 mm (Masuda et al. 1984), with a common total length of 500 mm TL (Bouhlel 1988). They usually gather over hard, algal-coated bottoms of exposed surf-swept outer reef flats, lagoons, and seaward reefs (Masuda et al. 1984). The adults usually live near the shore and the coastline, while the juveniles are linked with flotsam and can be encountered in the open ocean close to the surface (Knudsen and Clements 2013). This species usually lives in groups (Allen and Erdmann 2012). Animal source food is the main food item for this species (Masuda et al. 1984).

Economic Value
This species is considered to have good commercial value as it used as food for human (Masuda et al. 1984).

Conservation
Knudsen (2015) evaluated this species as Least Concern for the Red List of the IUCN based on the following reasons. The species has wide geographical variation and it is common along its distribution regions.

5.10 Tetrodotoxic Fishes

Tetrodotoxin (TTX) is considered among the greatest potent natural marine toxins, which performs by carefully obstructing the action potentials of voltage-gated Na + channels along nerves and skeletal and cardiac muscle membranes, without altering the resting membrane potentials. The name of this toxin was given after the Tetraodontidae pufferfish family, from which it was originally isolated and is separated the most lethal toxin found in the marine environment (Jal and Khora 2015; Lago et al. 2015). This toxin is characterized in being water soluble and heat stable, thus not being damaged by heat processing; otherwise, it rather increases its toxic effect, whereas there is no accepted antidote for TTX to date (Bane et al. 2014; Saoudi et al. 2007). Notwithstanding its known poison, or occasionally lethal influences when ingested by humans at high doses (lethal doses range from 1.5 to 2.0 mg TTX, equal to blood level 9 ng/mL), when directed at much lower levels, TTX shows therapeutic characteristics, so far mostly targeted to treating cancer-related, neuropathic, and/or visceral pain (Bucciarelli et al. 2021).

Chemically, tetrodotoxin is a crystalline, weak basic, colorless substance with a molecular formula of C11H17O8N3 (Fig. 5.20). At least 30 structural analogues have been described to date, with varying degrees of toxicity. Rest on structure, these are grouped into three classes: hemilactal, lactone, and 4,9-anhydro types, altogether represented as tetrodotoxins (TTXs). Tetrodotoxins are in various, taxonomically diverse, groups of animals, dwelling in both terrestrial and aquatic (marine, freshwater, and brackish) environments (Reverté et al. 2015; Silva et al. 2012; Biessy et al. 2019a, b). It is thought to be initiated from bacteria belonging to the phylum *Proteobacteria*, including *Pseudomonas*, *Pseudoalteromonas*, and *Vibrio*, though numerous other bacterial phyla (*Actinobacteria*, *Bacteroides*, *Firmicutes*, *Proteobacteria*) are infrequently described as potential TTX sources (Jal and Khora 2015; Magarlamov et al. 2017). The bacteria that produce TTX (i.e., genera *Vibrio*, *Pseudomonas*, *Aeromonas*, *Alteromonas*, *Nocardiopsis*, *Bacillus*, *Shewanella*, *Roseobacter*) have been located in the subcutaneous mucus, ovaries, and the gastrointestinal tract of several aquatic species (Bane et al. 2014; Wu et al. 2005; Noguchi et al. 2006; Knutsen et al. 2017); indication occurs in the literature signifying their link to exact dinoflagellate blooms, such as *Alexandrium tamarense* or *Prorocentrum cordatum* (Kodama et al. 1996; Vlamis et al. 2015; Rodríguez et al. 2017).

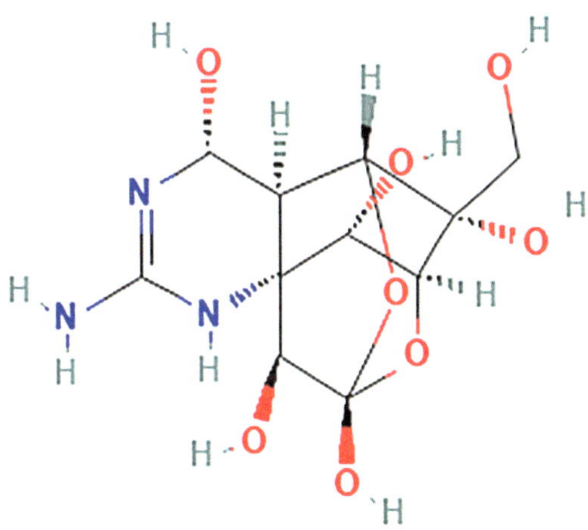

Fig. 5.20 The structure of tetrodotoxin. Courtesy of Wikipedia

This toxin is liable for many—irregularly lethal—human intoxication events and usually associated with pufferfish consumption, particularly in countries of the Far East (particularly Japan), where they establish a delicacy known as "fugu" (Bane et al. 2014). Until the establishment of the twenty-first century, tetrodotoxin was frequently found in tropical waters and was not being known, or yet alleged, as a probable risk in temperate areas, such as the Mediterranean Sea and Europe. In 2003, nevertheless, a recognized TTX vector, the pufferfish species *Lagocephalus sceleratus*, has been gradually reported in eastern Mediterranean coasts, owing to its invasion through the Suez Canal (the so-called Lessepsian migration). In 2007, this species had accomplished to have a sustainable population in these habitats, slowly expanding toward Israel, Lebanon, Turkey, Cyprus, and Greece (Akyol et al. 2005; Bentur et al. 2008; Katikou et al. 2009; Rosenberg and Wrathall 2001; Michailidis 2010) and later in the rest of the Mediterranean (Italy, Croatia, Malta, Libya, Algeria, and Tunisia) and finally reaching Spain by mid-2014 (Katikou and Vlamis 2017; Rambla-Alegre et al. 2017; Ujević et al. 2020).

5.11 Background

From ancient times, puffer poisoning is one of the best-known kinds of ichthyosarcotoxism that has been known to humans owing to its aggressive nature. It was maybe an Egyptian who was first to mention the toxicity of the pufferfish as seen from writings on their tombs (Gaillard 1923). In other regions of the world such as China and Japan, they knew the toxicity of pufferfish since the time of the T'ang dynasty (AD 618–907; Halstead 1967).

Clavijero (1852) wrote the first scientific account about the puffer poisoning in California. He referred to the case of four soldiers who were poisoned by eating the liver of a pufferfish. One of them died in 30 min; the other died a short time later; the third, who only chewed the liver without swallowing it, lost perception until the following day; and the fourth, who had hardly touched it, was sick for several days. The flesh of pufferfish was used by natives in Baja California for poisoning stray dogs (Phisalix 1922). Puffer poisoning ended the famous Captain Cook's voyage (Forster 1777).

There is an extensive study that dealt with the toxicology of the puffer poison by Rémy (1883). He based his experiments on dogs, where he used to feed them puffer gonads and injected crude extracts of puffers into their bodies in order to make observations. With the start of the year 1885, the Japanese literature about puffer poisoning increased very fast (Suehiro 1947). Among the remarkable Japanese chemical studies were those of Tahara (1894, 1896, 1897, 1910). He was the pioneer in isolating the puffer poison. The term "tetrodotoxin" or "fugutoxin" was established to signify the toxic assumption that was found to be the same to what was earlier known as tetrodonic acid. The Brazilians established their work on the toxicology of puffer poisoning in 1903, when Furtado started his research in this field (Halstead 1967). The publications about puffer poisoning have risen significantly with the turn of the nineteenth century led by Japanese researchers (Ishihara 1917; Yano 1938; Suehiro 1947; Fukuda 1951). Advances in research about the pharmacology of tetrodotoxin were attained by Hamada (1960), Ogura (1963), Mosher et al. (1964), Halstead (1964), and Russell (1965).

Origin and Sources of Tetrodotoxin

Recognizing the source of TTX in edible marine organisms and the procedures by which this toxin gathers in the food chain is fundamental for efficient risk evaluation and management. Regrettably, in the present time, the data accessible on revealing these practices continue to be debated. The most popular ideas indicate that TTXs are either generated by symbiotic bacteria (endogenous route) or that they are exogenously gathered through the diet [8]. An in-between description involving the participation of microalgae, mainly dinoflagellates, bearing the symbiotic TTX-produced bacteria has also been recommended; the TTX-bearing microalgae are then eaten by aquatic organisms during their feeding courses, resulting in TTX accumulation in the latter (Rodríguez et al. 2017).

As regards the bacterial symbiosis theory, though the proof is inconsistent, a wide variety of bacterial species and strains has been linked to TTX making. A minimum of 150 bacterial strains have been suggested as TTX makers, with the *Vibrio* genus being the key representative, containing more than 30% of the total TTX-producing strains. For example, *V. alginolyticus* strains found in the aquatic animals' microflora have been linked with TTX making. Strains of the *Bacillus* genus, on the other hand, form about another 15% of the isolated TTX producers, while the genera *Pseudomonas*, *Alteromonas*, *Aeromonas*, *Streptomyces*, and *Roseobacter* cover up to 7% of TTX-producing bacteria, with the lasting genera only represented by maybe a single strain each (Magarlamov et al. 2017). Concurring to other investigators, *Vibrio* is still counted the most common bacterial TTX-making genus, followed by *Bacillus*, *Pseudomonas*, *Actinomyces*, and *Micrococcus* (Turner et al. 2018). Nonetheless, no proof of TTX production in cultured bacterial strains separated from the blue-ringed octopus *Hapalochlaena* sp., the sea slug *Pleurobranchaea maculata*, and the gastropod *Nassarius semiplicatus* was achieved by the method of advanced analytical practices, such as liquid chromatography attached to mass spectrometry (LC-MS), despite the wide variety of bacterial species contained in these species (Chau et al. 2013).

To describe this argument, an overestimate concerning the number of TTX-making bacteria, assigned to analytical constraints, has been suggested (Biessy et al. 2019a, b). It is also feasible, however, that certain TTX-making organisms are not easily culturable under standard laboratory environments or the strains isolated were not able to produce TTX ex situ, despite their genetic ability for TTX biosynthesis (Chau et al. 2013). The latter is more confirmed by the reported presence of a *Bacillus* bacterial strain (*Bacillus* sp. 1839) showing a stable TTX making constant after 5 years of cultivation under laboratory conditions since its isolation (Melnikova et al. 2019), although loss of making with further inoculations is usually detected with many cultured strains, demonstrating potential contamination by the starting material (Biessy et al. 2019a, b).

More results in backing of the link between TTX existence and toxin-making marine bacteria were the first two individuals of the marine nemertean species *Cephalothrix simula* recently discovered in the UK (Turner et al. 2017). *Cephalothrix simula* starts off from the Pacific Ocean and is long known to be linked with high TTX levels. The 16S gene sequencing of the *C. simula* microbiome performed to show its taxonomic responsibility proving the occurrence of numerous bacterial genera linked to TTX making, such as *Alteromonas*, *Vibrio*, and *Pseudomonas*. Research backing a wholly endogenous origin of TTX even back that, in the case of some pufferfish species, where TTX plays a defence mechanism role, extant bacteria are not involved anymore in TTX biosynthesis, despite the initial bacterial origin of their biosynthetic genes (Arakawa et al. 2017). Remarkably, TTX is not found in completely freshwater-living fauna, not returning to aqueous systems for breeding reasons [57], which may indicate the link between TTX making and marine bacteria.

The engagement of microalgal species in the bacterial interaction theory has also been proposed. Substantial TTX quantities were uncovered in cultured cells of the dinoflagellate *A. tamarense*, where a link to endocellular bacteria was proposed (Kodama et al. 1996). Additionally, the presence of an association between the presence of TTX-making bacteria and *Prorocentrum cordatum* (*P. minimum*) dinoflagellates was presumed when TTX was in shellfish harvested from Greek coastal waters during a concurrent *P. cordatum* bloom [14]. This thought was later reinforced by investigations including these specific shellfish specimens and selected reference strains of *P. cordatum* stemming from Ecuador and Johor Strait, between Singapore and Malaysia (Rodríguez et al. 2017).

Up to recent years, the accurate biosynthetic pathways and genes for TTX production are still vague (Katikou 2019). However, it is assumed that TTX biosynthetic pathways of terrestrial and marine animals vary, as most TTX variants are only identified in either terrestrial or marine organisms, with only few exclusions of similarities found in both (Yotsu-Yamashita et al. 2013). Latest investigations by Ueyama et al. (2018) separated seven novel spiro-bicyclic-guanidino compounds (compounds 2–8) from the pufferfish species *Tetraodon biocellatus*; the carbon skeleton and relative configuration of six of them (compounds 2–5 and 7–8) were similar to TTX. These seven compounds were contemplated to be biosynthetic intermediates of TTX for marine environment organisms and, more precisely, expected to be predecessors of 5,6,11-trideoxyTTX, as none of them had been perceived in toxic terrestrial species, such as newts.

As to the occurrence of TTXs in marine bivalves or edible gastropods, the current limited indications available so far show an exogenous resource, with TTX accumulation in bivalves mainly associated to *Gammaproteobacteria*, particularly *Vibrio* and *Pseudomonas* sp. The latest studies (Leão et al. 2018; Turner et al. 2017) have associated TTX presence in shellfish to the immediate presence of *Vibrio* and *Pseudomonas*, although no cultures of any TTX-related bacteria were feasible to attain from the samples analyzed.

The Tetrodotoxin in the Possibly Safe to Eat Aquatic Organisms

The presence of tetrodotoxin in possibly edible aquatic organisms, specifically pufferfish, has been long recognized. The earliest data goes back to the start of the twentieth century, when Tahara and Hirata (Tahara and Hirata 1909) separated a toxin from the ovaries of fish belonging to the Tetraodontidae family, which was named tetrodotoxin (Tahara 1911). Tetrodotoxin was assumed to be present solely in pufferfish, until the mid-1960s, when Mosher et al. (1964) suggested that the eggs of the California newt *Taricha torosa* included TTX (Mosher et al. 1964). Therefore, TTXs have been noticed in a wide selection of aquatic animals, including bivalve mollusks, gastropods, and echinoderms, as well as other mollusks and crustaceans. Some of these aquatic creatures may be hardly found in markets or are not regarded as edible. Nevertheless, intoxications thence are regularly stated due to unintentional consumption, frequently linked to recreational fishing, or harvesting and consumption by local populations, unconscious of potential limitations forced by the legislation or supervising programs.

Tetrodotoxin in Pufferfish

Pufferfish are the furthermost disreputable TTX cause, accountable for innumerable human intoxication incidences, chiefly in regions of the Indian Ocean or the Far East, such as Japan, where they have been historically inspired as a delicacy and have been recognized a unique food culture linked with this fish group (Arakawa et al. 2010; Tamele et al. 2019). TTXs found in pufferfish and linked to human poisoning events worldwide have been greatly assessed by the works of Arakawa et al. (2010), Guardone et al. (2020), Tamele et al. (2019), and Bédry et al. (2021).

The greatest TTX concentrations are in the gonads, liver, and gastrointestinal tract, followed by the muscle and skin. The levels in all incidences, however, are higher than these reflected as safe for human consumption, since either the Japanese safety limit of 10 mouse units (MU)/g, equivalent to 2000 μg TTX/kg pufferfish tissue (1 MU = 0.2 μg TTX), or the stricter provisional concentration of 44 μg TTX/kg shellfish meat indicated by the recent EFSA opinion on TTXs (Katikou 2019). As the least lethal dosage for humans is assessed to be around 10,000 MU, equivalent to a total ingestion of 2 mg TTX for a 50 kg human (Katikou and Vlamis 2017; Noguchi and Arakawa 2008), it is obvious by the TTX concentrations stated that all these pufferfish species are dangerous for human to be considered as a food. In this situation, marketing of pufferfish, explicitly those of the Tetraodontidae, Molidae, Diodontidae, and Canthigasteridae families, is banned by the European or national legislation of the countries affected (Katikou 2019; European Union 2004; Nader et al. 2021; Kosker et al. 2019; European Union 2019). Nonetheless, unintentional pufferfish consumption by people unconscious of the risks involved has continually happened, causing acute or occasionally lethal human intoxication events.

Tetrodotoxin in Bivalve Molluscs

Around the world, the existence of TTXs has been stated so far in 14 bivalve mollusk species harvested from nine different countries. It is first recorded in the digestive glands of Japanese scallops of the species *Mizuhopecten yessoensis* (previously *Patinopecten yessoensis*) with TTX contents calculated at approximately 8000 μg/kg (Kodama et al. 1993). The next oldest world record derives from New Zealand and suggested to shellfish collected during 2001–2003 but assessed much later as frozen archived samples (Boundy et al. 2020). Katikou et al. (2022) reported 11 cases of intoxication in New Zealand from the period from 2001 to 2017. They were isolated from the whole flesh of New Zealand rock oyster, *Saccostrea glomerata*; Pacific oysters, *Magallana gigas* (*Crassostrea gigas*); pipi clams, *Paphies australis*; greenshell mussels, *Perna canaliculus*; and unknown species of cockles and oysters. Biessy et al. (2018) suggested that TTXs are in the siphons, digestive gland, and adductor muscles of the pipi clams *Paphies australis*. Their results showed that the siphons contained the highest concentrations of TTX (mean 403.8 μg/kg). The immunohistochemistry analysis of their results showed TTX in the outer cells of the siphons, but also in the digestive system, foot, and gill tissue. They proposed that detecting TTX in organs participating in feeding gives preliminary proof to back the assumption of an exogenous source in *P. australis*. On the other hand, Boundy et al. (2020) reported that TTX levels were observed to be low and similar to those reported in shellfish from other countries, except for pipi (*Paphies australis*). They claimed that all pipi analyzed as part of their survey were observed to comprise obvious levels of TTX, and pipi from a sampling site in Hokianga Harbour contained constantly elevated levels. Their results also showed that the levels attained in shellfish were much lower than those recorded in other marine organisms known to contain TTX and cause human intoxication (e.g., pufferfish).

Regarding the presence of tetrodotoxin in Europe, the first report was on the digestive glands of *Mytilus galloprovincialis* mussels and *Venus verrucosa* clams gathered in 2006 and 2008, respectively, in Greece (Vlamis et al. 2015). Further records on the tetrodotoxin occurrence in bivalve mollusks collected between 2013 and 2019 from six more countries, that is, England, the Netherlands, China, Italy, Spain, and France, progressively present in the literature (Turner et al. 2015; Gerssen et al. 2018; Han et al. 2018; Leão et al. 2018; Dell'Aversano et al. 2019; Hort et al. 2020; Bacchiocchi et al. 2021; Bordin et al. 2021; Réveillon et al. 2021; Turner et al. 2021; Turner et al. 2017; Dhanji-Rapkova et al. 2021), along with additional TTX reports from other species in New Zealand (Biessy et al. 2020; Boundy et al. 2020; McNabb et al. 2014; Ogilvie et al. 2012). At the moment, among the highest TTX concentrations detected in bivalve mollusks were those of 1600 μg/kg in greenshell mussels (*Perna canaliculus*) (Boundy et al. 2020) and 800 μg/kg in "pipi" clams (*Paphies australis*) from New Zealand (McNabb et al. 2014), 541 μg/kg in Mediterranean mussels (*M. galloprovincialis*) from Italy (Bordin et al. 2021), and 253 μg/kg in Pacific oysters (*Magallana gigas*) from England and the Netherlands (Gerssen et al. 2018; Turner et al. 2017).

Tetrodotoxin in Gastropods and Echinoderms
Gastropods are the second most recurrent source of TTX poisoning, after pufferfish (Guardone et al. 2020), with frequent human intoxication incidences appearing in the literature, mostly in Asian countries, such as Taiwan, Japan, China, and Vietnam, where TTXs have been conventionally present and where such aquatic organisms are a popular food (Huang et al. 2008). The reports from across the world on the occurrence of gastropods accountable for human poisoning cases have been utterly reviewed by the works of Arakawa et al. (2010), Biessy et al. (2019a, b), and Guardone et al. (2020), while reports in echinoderms are scarce.

The first known occurrence of TTXs in gastropods was from temperate areas, which is associated with a nonlethal food poisoning case in Spain and the intricate trumpet shell species *Charonia lampas* (formerly *Charonia lampas lampas*) collected from the south coast of Portugal in 2007. The specimen's digestive glands comprised extremely high parent TTX and total TTX levels, with noticed concentrations of 315,000 and 1,319,000 μg/kg, respectively (Rodriguez et al. 2008). Additional studies in Portuguese and Moroccan gastropods showed the occurrence of TTXs at computable levels in three more species, *Steromphala umbilicalis* (formerly *Gibbula umbilicalis*), *Phorcus lineatus* (formerly *Monodonta lineata*), and *Onchidella celtica*, as well as in additional *C. lampas* specimens, with some at high concentrations (Silva et al. 2012). Generally speaking, it looks that higher TTX contents are perceived in the viscera of gastropods, possibly signifying a dietary route of accumulation; nevertheless, in some species, TTXs are wholly detected in the flesh instead, which proposes the possible existence of alternative TTX-binding mechanisms or sources (Biessy et al. 2019a, b).

The first appearance of TTXs in the southern hemisphere is found in New Zealand in 2009, which resulted from the investigation of poisoning cases in dogs after consuming washed-up native gray side-gilled sea slugs, *Pleurobranchaea maculata*, in Auckland (Biessy 2021). *P. macu-*

lata was found to contain TTX at levels as high as 850,000 μg/kg (McNabb et al. 2010). This discovery triggered more intensive TTX testing in the years to follow, where TTXs were found in additional *P. maculata* specimens, as well as in bivalve mollusks and gastropod and echinoderm species, such as *Arachnoides zelandiae* and *Lunella smaragda* (formerly *Turbo smaragdus*) (Ogilvie et al. 2012).

Tetrodotoxin in Other Mollusks and Crustaceans

The presence of tetrodotoxin in crustaceans and other mollusks (excluding bivalves, gastropods, and echinoderms) seems scarce. Throughout the past period, boosted levels of tetrodotoxin have been recorded in two blue-lined octopus species (*Hapalochlaena lunulata* and *H. fasciata*) in Taiwan and Japan (Wu et al. 2014). Regarding the body of the organism, the greatest concentrations were in the posterior salivary glands, and it is believed that TTX in these species performs as a hunting and defense mechanism against predators (Wu et al. 2014). These octopus species are not counted as safe to eat, though adequate recognition of the species could lead to severe illness and creates a substantial risk from a food hygiene point of view (Asakawa et al. 2019).

Recently, the occurrence of TTXs in other mollusks contains two starfish species, *Astrostole scabra* and *Ophidiaster ophidianus*, from New Zealand and Portugal, respectively, with the latter specimen containing TTXs at levels as high as 44,242 μg/kg (Silva et al. 2019). Regarding crustaceans, two latest reports were present in the literature. The first referred to a Chinese *Carcinoscorpius rotundicauda* horseshoe crab, containing much lower TTX levels of 162 μg/kg (Han et al. 2018), whereas the second record of TTX concentration is high as 78.225 μg/kg in the appendages of *Zosimus aeneus* xanthid crabs harvested in Japan (Zhang et al. 2021).

Lastly, it should be documented that TTXs at concentrations up to 500 mg/kg have been recorded in marine worms and flatworms. Even though these species are not eatable, they live inside other organisms, such as shellfish (Turner et al. 2018). Consequently, their occurrence inside eatable aquatic organisms could cause bad health effects to consumers, establishing a potential danger involving further investigation (Biessy et al. 2019a, b).

Causative Agent

Several species of marine bacteria are shown to be responsible for producing tetrodotoxin: *Pseudomonas* sp., *Vibrio alginolyticus*, *Alteromonas tetraodonis*, *Shewanella alga*, *S. putrefaciens*, *Microbacterium arabinogalactanolyticum*, and *Serratia marcescens* (Yasumoto et al. 1986; Yotsu-Yamashita et al. 2013; Noguchi et al. 1986). The toxin is transmitted to different aquatic organisms mainly through the food chain (Noguchi and Arakawa 2008; Hwang and Noguchi 2007; Fuchi et al. 1991). Tetrodotoxin and its derivatives have a heterocyclic guanidine structure (Chulanetra et al. 2011).

The mechanism of action of tetrodotoxin involves mainly selective blockade of voltage-gated neuronal channels in the cell by binding extracellularly at certain receptors. This results in the inhibition of the generation and propagation of nerve and muscle action potentials (Bentur et al. 2008). The blockade of neuromuscular transmission happens on motor nerve axons and muscle fiber membranes and not on motor end plates. Axonal block possibly also happens at sensory nerves (Kiernan et al. 2005). Blockade of sympathetic vasomotor nerves, direct influences on arterial smooth muscles, and medullary cardiovascular effects are thought to be the main mechanisms for vasodilation and hypotension (Cottrell et al. 1984; Chang et al. 1996).

Symptoms

A grading system was introduced by Fukuda and Tani (1941) in which the symptoms of tetrodotoxin poisoning can be evaluated. The first phase of the poisoning incidence, a perioral numbness and paresthesia, will appear with or without gastrointestinal symptoms such as nausea. In the second phase, the victim will begin to have lingual, facial numbness followed by early motor paralysis and incoordination. In this phase, the patient will suffer from denigrated speech. In the third phase, the patient will experience a generalized flaccid paralysis, respiratory failure, aphonia, and fixed or dilated pupils. Then the patient becomes sleepless. In the fourth phase, severe respiratory failure, hypoxia, hypotension, bradycardia, and cardiac dysrhythmias will be observed. In this phase, the patient will suffer from severe respiratory failure and hypoxia, hypotension, bradycardia, and cardiac dysrhythmias. The patient may enter an unconscious period. The degree of intoxication can be assessed according to three factors: the amount of tetrodotoxin ingested, time lag after ingestion until admission to hospital, and preexisting diseases (Leung et al. 2011). The first two phases are reasonably moderate; however, the third and fourth phases are considered severe instability.

Treatment

There is no specific treatment for tetrodotoxin poisoning; treatment is chiefly by helpful therapy and may include mechanical ventilation for oxygen supply, normal saline infusion for distending the intravascular volume, and gastric emptying actions (Leung et al. 2011). To decrease the level of mortality owed to tetrodotoxin poisoning, initial diagnosis and prompt clinical management are necessary.

To aid in the procedure of diagnosis of tetrodotoxin poisoning, inspecting the poison in leftover food is another way to make the judgement. On the other hand, the recognition of the toxin in the patient's urine or blood is important to verify the identification of poisoning (Leung et al. 2011). Death is principally owing to respiratory muscle paralysis and less frequently, intense hypotension (Halstead 1988; Yang et al. 1996; Yang and Deng 1996). Death usually happens within 6–24 h of consumption, sometimes as early as 17 min after ingestion of the toxin (Halstead 1988; Yang and Deng 1996).

Prevention

Generally, and if anybody wants to evade intoxication with tetrodotoxin, the best idea is to avoid scaleless fishes. In Japan, people encourage first-class restaurants to eat fugu. It is essential how the individuals cook the fish for food. They should have information about where in the fish body the toxin is found and keep away these parts. Tetrodotoxin is heat resistible and therefore frying, stewing, baking, boiling, and so on do not inactivate the poison. The only way to deactivate the toxin is by boiling the meat in a strong solution of sodium bicarbonate, "baking soda," for a prolonged period, but the fish will be rendered tasteless and not useful for eating. Immunity against the toxin cannot be gotten after a repetition of injections. Individuals may become poisoned several times and die after another incident of poisoning (Halstead 1967).

5.11.1 Species of Tetrodontoxic Fishes

Family: Tetraodontidae
Arothron stellatus (Anonymous, 1798) (Fig. 5.21)
Common name: stellate puffer

Diagnostic Features

The body is covered with a small spine except on the top of the snout, base of fins, and sides of the caudal peduncle. The caudal fin is rounded. The body is mainly white with black spots on the head.

Distribution

This species is distributed in the Indo-Pacific region from the Red Sea southward to the East African coasts (Smith and Heemstra 1986) and eastward to Japan and south to the Lord Howe Islands (Froese and Pauly 2023). It is also

Fig. 5.21 Arothron stellatus (Anonymous, 1798). Courtesy of Hamid Badar Osmany, Pakistan

reported from the south coast of South Africa (Smith and Heemstra 1986). It is reported from the eastern and southern coasts of the Arabian Peninsula (Manilo and Bogorodsky 2003).

Habitat, Biology, and Ecology

The stellate puffer is a marine species that enters brackish waters sometimes. It is a reef-associated species living at a depth range 3–58 m (Myers 1991). Individuals of this species of puffers reach a maximum total length of 1200 mm (Lieske and Myers 1994), with a common total length of 540 mm (Bouhlel 1988). This species is relatively uncommon and lives in patch reefs and coral slopes. The juveniles live on sandy and weedy inner reefs, while the adults live on clear lagoons and seaward reefs (Lieske and Myers 1994). The larvae are pelagic and may disperse over great distances (Kuiter and Tonozuka 2001). It was reported to enter estuaries (Talwar and Jhingran 1991).

Economic Value

The meat of this species is used for human consumption in some regions along its geographical distribution.

Conservation Status

Shao et al. (2014) evaluated this species as Least Concern for the Red List of the IUCN based on the following reasons. This species has a wide geographical distribution in the Indo-Pacific region and appears to be common. It is a secondary component of the marine aquarium trade; however, harvest levels are unidentified. This species is included in the artisanal fisheries and is by the way taken as bycatch in numerous trawl fisheries. There have been no documented population declines of this species.

Lagocephalus lagocephalus (Linnaeus, 1758) (Fig. 5.22)
Common name: oceanic puffer
Etymology: *Lagocephalus:* Greek, *lagos* = hare + Greek, *kephale* = head (Romero 2002)

Diagnostic Features

On the body of this species, there are two separate lines, the lower line being on a fold of the skin. The caudal peduncle is rather narrow, while the caudal fin is lunate in shape. There are two nostrils with developed papilla present on each side of the snout. There were no bony plates on the back. The body coloration is bluish gray above and white below. There are eight crossbands on the dorsal side of the body, which is a feature of juveniles. Dark spots on the front and middle of the belly and on the side near the pectoral base in the specimens with less than 330 mm are present.

Fig. 5.22 Lagocephalus lagocephalus (Linnaeus, 1758). Courtesy of Hamid Badar Osmany, Pakistan

Distribution

This species of puffer has a circumglobal type of distribution in the tropical to temperate seas.

Habitat, Biology, and Ecology

The oceanic puffer is a marine species that inhabits the benthopelagic area (Reide 2004) at a depth range of 10–476 m (Anonymous 2001b), but is usually found at a depth range of 10–100 m (Palomares et al. 1989). Individuals of this species reach a maximum total length of 610 mm (Eschmeyer et al. 1983) and a maximum reported weight of 3.2 kg (IGFA 2001). Individuals of this species live mainly in the open ocean (R Eschmeyer et al. 1983; Schneider 1990), but they may enter estuaries (Smith and Heemstra 1986). They feed on crustaceans and squids (Tortonese 1986).

Economic Value

The flesh of this species is reported to have tetrodotoxin that causes fatal poisoning (Smith and Heemstra 1986). Therefore, it should not be eaten (Bianchi et al. 1999).

Conservation Status

Hardy et al. (2014) evaluated this species of puffer as Least Concern for the Red List of the IUCN based on the following reasons. This species has wide geographical distribution. It is available in large quantities in certain regions of its range. There are no confirmed species-specific threats to this species.

Sphoeroides pachygaster (Müller & Troschel, 1848) (Fig. 5.23)
Common name: blunthead puffer

Fig. 5.23 Sphoeroides pachygaster (Müller & Troschel, 1848). Courtesy of Noble, Brandi NOAA

Etymology: *Sphoeroides:* Greek, *sphaira* = ball + Greek, suffix, *oides* = similar to (Romero 2002)

Diagnostic Features

The body is partially compressed and inflatable and has no scales or spines. There is one lateral line on the side of the body. The lips are papillose. It lacks pelvic fin. The dorsal and anal fins are single. The caudal fin is truncated in shape. The dorsal part of the body is brown, while the ventral part is whitish. The pigmentation on the body is mostly uniform, except for a few dark spots on the flanks.

Distribution

This species of puffer fish has a circumglobal distribution in the tropical and temperate seas.

Habitat, Biology, and Ecology

The blunthead puffer is a marine species that inhabits the benthopelagic region at a depth range of 50–480 m (Matsuura and Tyler 1997), but it is usually found at a depth range of 50–250 m (Ragonese et al. 1992). Individuals of this puffer fish reach a maximum standard length of 405 mm (Ragonese et al. 1997), with a common standard length of 260 mm (de Figueiredo et al. 2002). This species prefers sandy, muddy, and rocky bottoms (Schneider 1990). The young are pelagic (Robins and Ray 1986). The blunthead puffer feeds mainly on squid.

Economic Value

This species of puffer has a commercial value in some regions across its geographical distribution.

Conservation Status

Roberts (1996) evaluated this species as Vulnerable without given any reasons.

Other Fish Species Found in New Zealand Potentially Causing Tetrodotoxin Poisoning

Family: Tetraodontidae

1. *Arothron firmamentum*
2. *Canthigaster callisterna*

3. *Contusus richei*
4. *Lagocephalus cheesemanii*
5. *Pelagocephalus marki*
6. *Torquigener altipinnis*

Family: Diodontidae

1. *Allomycterus pilatus*
2. *Chilomycterus reticulatus*
3. *Diodon hystrix*

Family: Molidae

1. *Masturus lanceolatus*
2. *Mola ramsayi*
3. *Ranzania laevis*

5.12 Ichthyootoxic Fishes

This kind of intoxication is less acknowledged compared with the other well-known poisonings such as ciguatera and tetrodontoxism. In fish species causing ichthyootoxication, the poison is contained to the gonads, where there is a certain link between gonad activity and toxin production. Fresh water species are mainly the source of this poisoning, but there are some other anadromous, brackish water, or marine species that are included within this category.

5.13 Background

Even though ichthyootoxism is not a well-known intoxication, it is one of the oldest known intoxications. Since 1491 it has been stated in Europe that the ingestion of the roe of the cyprinid fish *Barbus barbus* will cause a case of intoxication (Halstead 1967). Francii (1683) give details on a case of a pregnant woman who became very ill after eating a large amount of gravid barbel. Incidences of intoxication after eating roe of American gar were reported by Brooks (1850), Coker (1929), and Taft (1945). As the intoxication happens during the breeding season of the barbels, this fish is forbidden from sale during the reproductive cycle (Meyer-Ahrens 1855). Toxic materials were obtained from the milt of some fish species such as salmon called "salmin," from *Acipenser sturio*, "sturin," and from *Cyprinus carpio*, "cyprinin" (Knox 1888). A test on the influence of pike and barbel roe was done on dogs and rabbits by Kossel (1896) and demonstrated sensory instability, respiratory and muscular paralysis, and finally death. In 1921, McGrudden (in Halstead and Schall 1955) wrote on the chemistry and toxicity of roe poisons, and he thought at that time it was "sapotoxins." Among the marine species whose roe was showed to be toxic is a species of groupers found at the coast of Ecuador (Halstead and Schall 1955). A spread of intoxication owing to eating marine fishes at Mori, Japan, during 1952–1953 was stated by Takayanagi et al. (1953; cited by Asano and Itoh 1962). Ichthyootoxin was discovered to harbor a lipoprotein called "dinogrunellin" as a result of studies done by Asano and Itoh (1962).

Causative Agent

Polysaturated fatty acids present in aquatic organisms derive mainly from the food chain, which in turn come from plankton-type organisms (Mancini et al. 2011). The polyunsaturated fatty acids are used by aquatic organisms to adapt themselves in cold environments by mounting membrane fluidity and avoiding any tendency to crystallize (Jütter 2001). In species of the genus *Barbus* (Family: Cyprinidae), there are high concentrations of polyunsaturated fatty acids (u3 and u6) initiating from the food chain (Aras et al. 2009). Irrespective of their useful input to human health, high concentrations of free unsaturated fatty acids have been reported to encourage toxic influences in higher invertebrates and humans as some of them cause hemolysis, which could be accountable for icthyotoxicity by accumulating in fish (Fu et al. 2004). It was learnt that the toxicity of unsaturated fatty acids was up to eightfold higher than that of saturated ones, the most toxic being linolenic acid (C18:3 u3), AA (20:4 u6), and EPA (C20:5. u3). Arachidonic acid, which is abundant in the extract of roe of the barbel, has hemolytic and cytotoxic activities (Mancini et al. 2011).

Symptoms
There are a number of symptoms given by Halstead (1967) that the patient showed after eating the roe of the causative fish species, including abdominal pain, nausea, vomiting, diarrhea, dizziness, headache, fever, bitter taste, dryness of the mouth, intense thirst, sensation of constriction of the chest, cold sweats, rapid irregular weak pulse, low blood pressure, cyanosis, pupillary dilatation, dysphagia, and tinnitus. In critical incidences and before death occurs, muscular cramps, paralysis, convulsions, and coma are usually suffered by victims.

Treatment
The main treatment step to be taken is to empty and wash the stomach of the patient. The remaining treatment steps are symptomatic. The treatment of ichthyootoxin is different from ciguatoxin or tetrodotoxin.

5.13.1 Ichthyootoxic Effect on Human in New Zealand

In New Zealand, Hallegraeff et al. (2021) reported on the case of the wide growth and spread of the algae *Karenia brevisulcata* (as *Gymnodinium brevisulcatum*, Chang 1999) that caused large, massive mortalities in summer 1998 in Wellington Harbour. During this event, over 500 cases of human respiratory distress were reported, although no food poisoning was documented. This species yields a range of ichthyotoxic complexes containing brevisulcatic acids (BSX) and brevisulcenals (KBT) which are however well-defined from brevetoxins (Holland et al. 2015). Laboratory experiments revealed that *K. brevisulcata* was toxic to juvenile snapper and Chinook salmon as well as greenshell mussel, paua, and sea urchin larvae, confirming it as a risk to aquaculture through seawater intake pipes. *Karenia selliformis* was linked to mass deaths of fish and shellfish in 1994 in Southland, New Zealand, with gymnodimine, a novel fast-acting marine biotoxin, identified as the possible cause (Seki et al., 1995). *Karenia mikimotoi* caused damage in fish farms in Marlborough Sounds in 1987. In January 1989, a *Heterosigma* bloom in Big Glory Bay, Stuart Island, killed cage-reared chinook salmon worth NZD 12 M (Chang et al., 1990), and in 2010 in the Marlborough Sounds, a major mortality of Chinook salmon, particularly those >3.5 kg, was associated with the dictyochophyte *Pseudochattonella verruculosa* (MacKenzie et al. 2011).

In 1993 a thorough marine biotoxin watching programme was begun around the entire New Zealand coastline, covering commercial and noncommercial shellfish harvesting areas (Hallegraeff et al. 2021). The greatest number of closures occurred in 1993 following an outbreak of neurotoxic shellfish poisoning (NSP) human shellfish poisonings, associated with the proof of brevetoxins in cockle, greenshell mussel, and oysters (Ishida et al., 2004). During 1993–1997, the mouse bioassay used for lipid-soluble toxins such as the NST brevetoxins and DST dinophysistoxin and okadaic acid were mystified by toxin extraction methods which were later revealed to include gymnodimine, pinnatoxins, pectenotoxins, free fatty acids, and yessotoxins. After some time, enhancement of both toxin extraction methods and understanding of mouse bioassay results, along with developed knowledge of co-extracted compounds and their oral toxicity, resulted in a decrease in "NSP" closures. Since 2001 with the establishment of LC-MS techniques for DST, no further anomalous closures have occurred. The area most insistently affected by unidentified "NSP" was Rangaunu Harbour, with almost constant oyster farm closures from 1993 to 1996, but since 2008 positive NSP results in acetone extraction mouse bioassay were identified due to pinnatoxins and positive ether extraction mouse bioassays possibly due to okadaic acid and analogues (McNabb et al., 2008). Paralytic shellfish toxin (PST) analyses between January and July 1993 used an extraction method (Hall 1991) that was later discovered to be faulty and have caused unnecessary closures.

5.13.2 Example of the Ichthyootoxic Effect on Human Health

Jawad (2021) reported on the case of ichthyootoxication in Iraq. This incidence was developed from eating roe of the cyprinid fish *Carasobarbus luteus*. This case has occurred in the southern part of Iraq at Basrah Province. The patients experienced abdominal pain, vomiting for 6 h, diarrhea, intense thirst, cyanosis, and faintness, among milder symptoms.

The incidence that Jawad (2021) reported includes six members of a family including parents and seven children who were taken to the emergency department of a hospital in Basrah, Iraq, showing symptoms of food poisoning. They informed the medical team at the hospital that they ate grilled roe of the cyprinid fish *Carasobarbus luteus* known by its local common name "Himri." All the family members were established to have the following symptoms: faintness, abdominal discomfort, nausea, diarrhea, bitter taste, dryness of mouth, and weakness. They were hospitalized for 2 days, and their stomachs were washed out. An hour after ingestion, the patients experienced cold sweats, chills, dizziness, headache, cyanosis, and vomiting for 6 h but recovered after 1 day's rest.

The habit of eating grilled roe of fish is common in Iraq in general and in its southern part in particular. Hence, people need to distinguish and learn about roe of what species they need to eat and which they need to keep away. The fish species responsible for the ichthyootoxism are greatly freshwater.

The polysaturated fatty acids found in aquatic organisms come mainly from the food chain that is related to the plankton-type organisms (Mancini et al. 2011). Such fatty acids are employed by aquatic organisms to acclimatize themselves in cold habitats by upsurging membrane fluidity and stopping any tendency to crystallize (Jütter 2001). In species of the genera *Barbus* and *Carasobarbus* (family: Cyprinidae), there are high dilutions of polyunsaturated fatty acids (u3 and u6) coming from the food chain (Aras et al. 2009). Regardless of their valuable input to human health, high additives of free unsaturated fatty acids are known to cause toxic influences in higher invertebrates and humans as some of them prompt hemolysis, which could be accountable for icthyotoxicity by loading in fish (Fu et al. 2004). It is known that the noxiousness of unsaturated fatty acids was observed to be eight times more than that of saturated ones, the most poisonous being linolenic acid (C18:3 u3), AA (20:4 u6), and EPA (C20:5. u3). Arachidonic acid, which is available in large quantities in extricate of roe of the barbel, has hemolytic and cytotoxic actions (Mancini et al. 2011).

5.13.3 Fish Species Causing Ichthyootoxic Poisoning for Humans in New Zealand

Family: Cyprinidae

Cyprinus carpio (Linnaeus 1758) (Figs. 5.24 and 5.25)

Common name: Common carp, Koi carp

Etymology: Cyprinus: Latin, *cyprinus* = carp; *carpio*: *carpio* is the Latinized form of carp (Scott and Crossman 1973). Cyprinus is the old-world name for carp

Diagnostic Features

Two pairs of barbels are present. The caudal fin is deeply emarginated. The pharyngeal teeth have a robust molar-like shape. The scales on the body are large. The last simple anal ray is bony and serrated (Kottelat and Freyhof 2007; Spillman 1961; Kottelat 2001).

Distribution

The common carp is distributed in the river systems of both Europe and Asia. It is also introduced throughout the world (Froese and Pauly 2016). In New Zealand, this species is common in most freshwater bodies (David et al. 2018; Pullan 2020).

Habitat, Biology, and Ecology

The common carp is a freshwater species that sometimes enters brackish water (Riede 2004). It lives at a depth range of 0–29 m (NYDEC 2019).

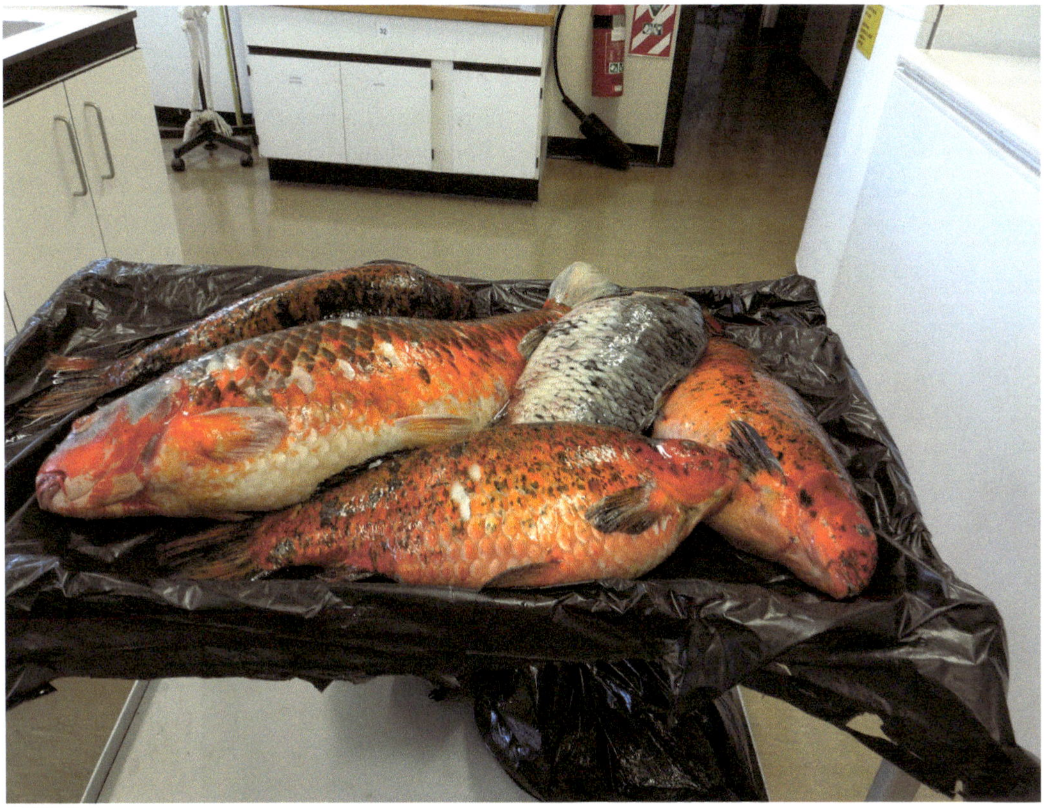

Fig. 5.24 *Cyprinus carpio* (Linnaeus 1758). Courtesy Laith Jawad

Individuals of this species reach maturity at a total length range of 250–360 mm with a maximum total length of 1200 mm (Murdy et al. 1997) and a common total length of 310 mm (Chugunova 1959). The maximum recorded weight of this species is 40.1 kg (Machacek 2007) and a maximum reported age is 38 years (Hinton 1962).

The common carp usually inhabits warm, deep, slow-flowing and still waters such as lowland rivers and large, well-vegetated lakes (Kottelat and Freyhof 2007). It can live and tolerate variable sorts of environments, but mostly choose large water bodies with slow flowing or standing water and soft bottom sediments. Individuals of this species flourish in large turbid rivers (Scott and Crossman 1973). They usually function at dusk and dawn. They feed on a wide range of benthic organisms and plant materials. The common carp breeds along shores or in backwaters. Adults often take on extensive spawning migration to appropriate backwaters and flooded fields.

Individuals of the common carp spawn in marginal, shallow, weed-infested areas. A polytypic plastic species is marked tendency to produce "varieties" and "races" in response to selective breeding and environmental influences. Carp is polygamous. A spawning female is usually followed by several males (Alikunhi 1966). Females are known to lay more than a million eggs in a season and breed at a temperature range of 15–20 °C; eggs hatch in 4 days (Pethiyagoda 1991). The sticky eggs are stuck to water plants or other submerged items. Reproductive success is limited to years when the water level starts rising and when high temperatures and flooding of terrestrial vegetation last for a long period (Kottelat and Freyhof 2007).

Economic Value

The meat of the common carp is consumed fresh and frozen (Frimodt 1995). It is also included in the aquarium industry, with groups of five or more individuals and requiring a minimum

Fig. 5.25 Common carp *Cyprinus carpio* (Linnaeus 1758) on offer for sale at one of the leading supermarkets in New Zealand. Courtesy Laith Jawad

aquarium size of >2000 mm. This species is not recommended for home aquariums (BMELF 1999).

Conservation

Freyhof and Kottelat (2008a) evaluated this species as Vulnerable in the Red List of the IUCN based on the following reasons. The populations of the common carp in some parts of the world are gradually but constantly declining due to river regulation and hybridization with domesticated introduced stocks. Nevertheless, externally pure carp (currently it is impossible to identify pure carp by genetic analysis) are still abundant in the lower parts of rivers within its native range. Most likely, only very few genetically pure stocks persist because of this long-lasting progression.

Tinca tinca (Linnaeus, 1758) (Fig. 5.26)
Common name: tench
Etymology: *Tinca:* Latin, *timica*; related to a predatory fish = timi, but the etymology of tench does not conform with this meaning (Romero 2002); *tinca: tinca* which is the Latin name for tench (Scott and Crossman 1973)

Diagnostic Features

The body is thick, heavy, and laterally compressed. The caudal peduncle is deep and short. The skin is thick and slimy. The scales are small and embedded. The body color is olive-green, sometimes dark green, or almost black, with golden reflections on the ventral surface. The head is triangular. The eye is orange-red and small. The snout is relatively long. The interorbital area is broad. The mouth is terminal and small, with thick lips. A pair of well-developed barbels, one at each corner of the mouth, is present. The caudal fin has 19 rays (Spillman 1961).

Distribution

This species is distributed in Eurasia, and it is considered as native in most Europe, naturally absent only in Ireland, Scandinavia north of 61°30'N, eastern Adriatic basin, and western and southern Greece where it is now introduced. In Asia, native eastward to western Yenisei drainage south of 60° N. Introduced elsewhere. It is reported from New Zealand (Roberts et al. 2015).

Habitat, Biology, and Ecology

The tench is a freshwater species that lives in the demersal habitat (Riede 2004) at a depth range from 1 m and reaches maturity at a length of about 2000 mm (Darwin 1877). Individuals of this species reach a maximum recorded standard length of 700 mm (Allen et al. 2002), with a common total length of 200 mm (Muus and Dahlström 1968). The maximum recorded weight of this species is 7.5 kg (Muus and Dahlström 1968).

The tench usually lives in shallow, densely vegetated lakes and backwaters. It regularly spent

Fig. 5.26 Tinca tinca (Linnaeus, 1758). Courtesy of McDowall, R.M., New Zealand

wintertime buried in mud. The larvae and juveniles are restricted to dense vegetation (Kottelat and Freyhof 2007). The adults reside in warm lakes and pools with weed and mud bottoms. Individuals of this species can stand low oxygen saturations (Billard 1997; Muus and Dahlström 1968). They usually feed on detritus, benthic animals, and plant materials. The adult frequently feed mainly on mollusks. The tench spawns among dense vegetation in still water (Kottelat and Freyhof 2007). The males reproduce at 2–3 years and females at 3–7 years.

Economic Value

It is used as a fodder fish for bass (Skelton 1993) and utilized fresh and frozen and eaten pan-fried, broiled, and baked (Frimodt 1995). It is popular with amateur sport fishers. Its flesh is highly esteemed (Billard 1997). But it is locally under threat due to river engineering (Kottelat and Freyhof 2007) and is reported to reach a maximum length of 84 cm TL (Page and Burr 1991).

Conservation

Freyhof and Kottelat (2008c) evaluated this species as Least Concern for the Red List of the IUCN based on the reason that this species is assumed to be native in most of Europe where it is locally threatened by river engineering.

Family: Salmonidae
Salmo salar Linnaeus, 1758 (Fig. 5.27)
Common name: Atlantic salmon
Etymology: *Salmo:* Latin, *salmo, Plinius* = salmon (Romero 2002); *salar:* From the Latin *salio* meaning to leap (Hart 1973)

Diagnostic Features

The Atlantic salmon can be separated from its closest species in having the following set of features: presence of 10–13 scales between ends of the adipose base and lateral line, there are 17–24 gill rakers (Kottelat and Freyhof 2007), the caudal fin is deeply forked in individuals smaller than 20 cm of standard length, hyaline or gray adipose margin is present, and the posterior part of the vomer is toothless (Kottelat and Freyhof 2007). The mouth reaches the area below the rear of the eye and has well-developed teeth (Rochard and Elie 1994). The vomerine teeth are weak (Robins and Ray 1986). The caudal fin has 19 rays (Spillman 1961). The scales are small (Rochard and Elie 1994). The juveniles have 8–12 blue-violet spots on the flanks with little red spots in-between (Rochard and Elie 1994). The

Fig. 5.27 Salmo salar Linnaeus, 1758. Courtesy of McDowall, R.M., New Zealand

adults at sea are bluish green dorsally, turning silvery along the sides and white ventrally, with a few black spots but none under the lateral line (Bigelow et al. 1963; Rochard and Elie 1994). The caudal fin normally has no spots and the adipose fin is not black bordered. The males are characterized in having an elongated hooked jaws that meet at the tips once they enter the spawning season, thicker fins, and slime covering their body.

Distribution

Salmo salar is distributed in the North Atlantic Ocean (Rochard and Elie 1994). It has been reported from the Western Atlantic. It is recorded from northern Quebec, Canada, to Connecticut and New York, USA, and inland to Lake Ontario where it is now extirpated (Page and Burr 1991). It is also reported from the Eastern Atlantic at the White and Barents Sea basins through northeastern Europe to the Baltic and North Sea basins, including Iceland (Kottelat and Freyhof 2007). It is introduced to New Zealand, Chile, southern Argentina (Kottelat and Freyhof 2007), and Australia (Kailola et al. 1993).

Habitat, Biology, and Ecology

The Atlantic salmon is a marine species that enters the fresh and brackish waters (Kangur et al. 2003). It lives at a depth range of 0–210 m, but it is usually found at a depth range 10–23 m (OBIS 2006). Individuals of this species reach a maximum total length of 1500 mm (Robins and Ray 1986), with a common total length of 380 mm (Chugunova 1959). The maximum recorded weight of this species is 46.8 kg (Daymond 1963) and the maximum recorded age is 13 years (Flower 1935).

Individuals of this species spend most of their life in freshwater (Rochard and Elie 1994). They are found in lakes and rocky runs and pools of small to large rivers (Page and Burr 1991). Some landlocked populations exist. They are found in all rivers where temperature rises above 10 °C for about 3 months per year and does not exceed 20 °C for more than a few weeks in summer (Kottelat and Freyhof 2007). The young endure freshwater for 1–6 years and then move around to coastal marine waters or even to open oceans where they stay for 1–4 years before returning to freshwater for spawning (Rochard and Elie 1994). The adults inhabit cooler waters with strong to moderate flow (Allen et al. 2002). The preferred food of the juveniles is aquatic insects, mollusks, crustaceans, and fish. At the sea, the adults feed on squids, shrimps, and fish (Rochard and Elie 1994).

At the start of sexual maturity, individuals of this species are realized to return from the ocean to the river where it was born and even to its precise natal site (Balon 1990; Rochard and Elie 1994). Lacustrine populations move to tributaries (Kottelat and Freyhof 2007). This species spawns in pairs. The male guards and defends the female against other males (Kottelat and Freyhof 2007). A female discharges between 8000 and 25,000 eggs during a spawning season (Balon 1990; Muus and Nielsen 1999), weighing 500–2000 per

Fig. 5.28 *Salmo trutta* Linnaeus, 1758. Courtesy of Karelj, Prague, Czech Republic

kg (Rochard and Elie 1994). The period of spawning lasts for 1–2 weeks. Most males die after spawning, while 10–40% of females may survive and return to the sea in autumn or overwinter in rivers, feed one summer, and migrate again (Rochard and Elie 1994).

Economic Value
The meat if this species is marketed fresh, dried, or salted, smoked, and frozen and eaten steamed, fried, broiled, cooked in microwave, and baked (Frimodt 1995).

Conservation
WCMC (1996) evaluated the Atlantic Salmon as Least Concern for the Red List of the IUCN without giving any reasons.

Salmo Trutta Linnaeus, 1758 (Fig. 5.28)
Common name: sea trout
Etymology: *Salmo:* Latin, *salmo*, *Plinius* = salmon (Romero 2002); *trutta: trutta* is a Latin name for trout (Etnier and Starnes 1993)

Diagnostic Features
The head is small and pointed (Rochard and Elie 1994). The mouth is large, reaching after the eye, and has well-developed teeth (Rochard and Elie 1994). The teeth are numerous and strongly developed located on the shaft of the vomer (Robins and Ray 1986). The caudal fin has 18–19 rays (Spillman 1961). The caudal peduncle is thick and rounded (Rochard and Elie 1994). The scales are small (Rochard and Elie 1994). The body coloration is gray-blue, with numerous spots; they are also found below the lateral line (Rochard and Elie 1994). The upper part of the body is blackish, usually orange on the sides, bounded by pale halos. The adipose fin has a red margin.

Distribution
The sea trout is distributed in both Europe and Asia. It is reported from the Atlantic, North, White, and Baltic Sea basins, from Spain to Chosha Bay (Russia). It is also recorded from Iceland and northernmost rivers of Great Britain and Scandinavia. It is introduced widely all over the world.

Habitat, Biology, and Ecology

Salmo trutta is a marine species that enters fresh and brackish water (Riede 2004). It lives at a depth range of 0–28 m (Rikardsen et al. 2007), but it is usually found at 1–2 m (Rikardsen et al. 2007). Individuals of this species reach a maximum standard length of 1400 mm (Muus and Dahlström 1967), with a common total length of 720 mm (Bauchot 1987). The maximum recorded weight of this species is 50.0 kg (Muus and Dahlström 1967) and the maximum reported age is 38 years (Svalastog 1991).

This species prefers streams, ponds, rivers, and lakes to live in (Scott and Scott 1988). It spends 1–5 years in freshwater and 6 months–5 years in salt water (Rochard and Elie 1994). The juveniles reach maturity in 3–4 years (Hart 1973). The lacustrine populations conduct migration to tributaries and lake outlets to spawn, seldom spawning on stone, wave-washed lake shores. Individuals of this species spawn in rivers and streams with swift currents, usually distinguished by downward movement of water into gravel (Kottelat and Freyhof 2007).

The fecundity is about 10,000 eggs (Muus and Nielsen 1999; Rochard and Elie 1994). This species is mainly diurnal (Muus and Dahlström 1967). The main food of the juveniles are aquatic and terrestrial insects, while adults feed on mollusks, crustaceans, and small fish (Cadwallader and Backhouse 1983.; Rochard and Elie 1994).

Economic Value

The meat of this species is marketed fresh and smoked and eaten fried, broiled, boiled, cooked in microwave, and baked (Frimodt 1995).

Conservation

Freyhof (2013) evaluated this species as Least Concern to the Red List of the IUCN based on the following reasons. The species is widely distributed around the world. Nevertheless, anadromous sections of its populations and many lacustrine stocks have significantly declined because of pollution (and possibly from impacts from salmon farming).

Family: Percidae
Perca fluviatilis Linnaeus, 1758 (Fig. 5.29)
Common name: European perch
Etymology: *Perca:* Greek, *perke* = perch, a fish without identification (Romero 2002)

Fig. 5.29 Perca fluviatilis Linnaeus, 1758. Courtesy of Citron, Wikipedia

Diagnostic Features

The European perch can be separated from the remaining species of the family Percidae in the following set of characteristics: the pelvic and anal fins are yellow to red; the posterior part of the first dorsal fin is with a dark blotch; the flank has 5–8 bold dark bars, usually Y-shaped; two dorsal fins are present, which are clearly separated from each other; and it has 56–77 scales along lateral line (Kottelat and Freyhof 2007). The body is greenish yellow; presence of 5–9 transverse black bands on the sides; the first dorsal fin is gray, with a black spot at the tip; the second dorsal fin is greenish yellow; the pectoral fins are yellow; and the remaining fins are red in color. The first dorsal fin is noticeably higher than the second dorsal fin. The caudal fin is emarginate in shape (Berg 1965).

Distribution

This species is distributed in Eurasia. It is reported from throughout Europe to the northernmost extremity of Scandinavia, except Iberian Peninsula, Central Italy, and Adriatic basin, Aegean Sea basin in Matriza and from Struma to Aliakmon drainages, Aral Sea basin, and Siberia in rivers draining the Arctic Ocean eastward to Kolyma. It is widely introduced around the world.

Habitat, Biology, and Ecology

Perca fluviatilis is a freshwater species that enters brackish waters sometimes (Riede 2004). It lives at a depth range 1–30 m (Frimodt 1995), but it is usually found at 3–4 m (Smolian 1920). The maximum standard length recorded of this species is 600 mm (Kottelat and Freyhof 2007), with a common total length of 250 mm (Muus and Dahlström 1968). The maximum reported weight is 4.8 kg (Berg 1965) and the maximum published age is 22 years (Beverton and Holt 1959).

Individuals of this species live in a very wide range of habitats from estuarine lagoons and lakes of all types to medium-sized streams. The larvae are found in the open water. This species is an opportunistic diurnal feeder which targets mainly during sunrise and sunset, using all available prey. The males reach their first sexual maturity at 1–2 years and females at 2–4 years of age (Kottelat and Freyhof 2007). Eggs clustered in long white ribbons (up to 1 m) are found over submerged bodies (Pinder 2001). Eggs hatch in about 8–16 days at normal temperatures (Maitland and Campbell 1992).

Economic Value

Its flesh is excellent and not so bony. It is utilized fresh and frozen and eaten pan-fried and baked (Frimodt 1995). It may be captured with natural or artificial bait (Billard 1997).

Conservation

Freyhof and Kottelat (2008b) evaluated this species as Least Concern in the Red List of the IUCN based on the following reason. The species is widely distributed and no recognized threats.

5.14 Ichthyogallotoxic Fishes

The species belonging to this group of poisonous fishes include several species belonging to the family Cyprinidae. In New Zealand, two of these species are present; these are the grass carp *Ctenopharyngodon idella* and the silver carp (silver bighead) *Hypophthalmichthys molitrix*. The grass and the silver carp are basically species of the aquaculture industry.

The habit of eating the gallbladder of the fish is well-known and many people from different ethnicities have this habit. In rural Asia, people eat gallbladders from various fish species for the treatment of various disorders, including arthritis and decreased visual acuity (Lin and Shih 1999). The famous fish species from which gallbladders are eaten are the species of carp, ostariophysan family Cyprinidae. When linked with toxic effects, it represents a form of ichthyotoxism resulting in acute renal failure. The toxin is believed to be cyprinol sulfate or cyprinol, a C27 bile acid (Hwang et al. 2000). Gallbladders of certain freshwater fish are increasingly recognized as an important source of ichthyosarcotoxism in Asian populations.

The raw carp bile has both nephrotoxic and hepatotoxic effects that are not well-known. Park et al. (1990) studied cases of such intoxication with raw carp bile. The patients experienced toxic acute renal failure and toxic hepatitis after eating raw bile of carp in three, grass carp in eight, and silver carp in two cases. The seriousness of the symptoms relies on the amount of bile ingested. All the patients recovered with conservative therapy and hemodialysis. Biopsy of the kidney showed findings similar with acute tubular necrosis like that produced by other nephrotoxins. Biopsy of the liver revealed a constant acute toxic hepatitis. Both suggest toxic effects of carp bile as a cause of toxic acute renal failure and hepatitis.

There are several reasons for acute renal failure after eating carp bile. First, and most notable, is the direct nephrotoxicity of some constituents of bile (Chen 1984). Second, bradycardia and hypotension can happen due to bile salts (Joubert 1978). Third, diarrhea arises because of the increased permeability of intestinal mucosa after inhibition of Na-K-dependent adenosine triphosphatase and because of reduced reabsorption of water and increased secretion due to mucosal damage (Chen 1984; Mekhjian and Phillips 1970; Chadwick et al. 1979). Fourth, volume reduction can happen owing to loss of salt and water via the kidneys (Topuzlu and Stahl 1960; Alon et al. 1982). Park et al. (1990) suggested that the urinary indices and pathologic findings in the kidney propose an acute tubular necrosis due to nephrotoxins rather than prerenal azotemia. The bile is toxic and has long been observed (Emerson 1929; Better and Bomzon 1988). The known toxin has not been identified yet. But the presence of such a toxin is confirmed by the fact that boiled bile is toxic in animals (Chan et al. 1985) and in man, as shown in the cases described by Park et al. (1990). The bile salts, which are the main constituents of bile, are believed to be toxic elements (Emerson 1929) since they make up about 90% of the dry weight of the bile salt fraction (Lin and Lin 1999). Some studies attempted to extract the toxic matter from grass carp bile (Yip et al. 1981; Lin et al. 1977) but were unsuccessful. Lately, it was shown that severe renal failure was established after eating of carp bile in rats, while it did not develop after raw hog bile (Lin and Lin 1999), so the nephrotoxicity of fish bile is specific for the carp family, but mammalian bile is relatively nonnephrotoxic. It was proposed that cyprinol, a C27 bile alcohol derived from cholesterol, which was found to be present in bile of freshwater fishes which belong to the family Cyprinidae denoted by the grass carp, may be the toxic component of carp bile (Chen et al. 1976; Tsuchiya 1962). Lin et al. (1977) isolated bile from the grass carp and showed that the toxic ingredient was recovered in the ethanol-soluble and ether-insoluble fraction of which bile acids and bile salts are the main elements. When it was given to rats orally, their livers displayed multiple focal necrosis and their kidneys showed congestion and cloudy swelling of the tubular epithelium. The question here is which element of bile acids is liable for the toxicity? What kinds of fish are toxic and what the toxin itself is? These questions remain undetermined and need more examination. Immediate renal and liver injury may result from exposure to a collection of toxins, containing carbon tetrachloride, trichloroethylene, chloroform, *Amanita phalloides*, arsenic, copper sulfate, and chromium (Levenson et al. 1988). But for the patient with toxic acute renal failure and toxic hepatitis without exposure to such toxins and drugs, the option of ingestion of raw carp bile should always be taken into attention, particularly in those areas of Asia where it is used to improve health. In some places in Asia, people do not know that carp bile is toxic, and it is even given as an oriental medication to control cough and hypertension (Ng and Kum 1977). It is proven that the toxic and healing doses are very close (Ng and Kum 1977). People working in public health must tell the people that carp bile is very toxic and that they should stop eating it extensively.

5.14.1 Reports on Cases of Ichthyogallotoxication from around the World

Park et al. (1990) reported on 13 intoxication cases from the Republic of Korea. These cases happened over 2 years of time. The patients in all these cases showed a severe renal failure and hepatitis after eating a raw carp bile. After admission to the hospital, it appeared that the patients have eaten raw bile of carp (*Cyprinus carpio*), grass carp (*Ctenopharyngodon idella*), and silver carp *Hypophthalmichthys molitrix*). Most fish were over 1 m long. The patients were ten men and three women, aged between 30 and 60 years. All patients did not experienced any renal or hepatic diseases before swallowing the raw bile, which was taken instantly after the fish was captured; only one patient ate it after boiling. The amount of bile ingested differed from 15 to 30 ml. Percutaneous renal biopsy was implemented in four and liver biopsy in three cases. A light and electron microscopic studies were applied to the specimens. The renal biopsies were also subjected to immunofluorescent studies.

All patients in the cases reported by Park et al. (1990) appeared at the start having a gastrointestinal upset, involving abdominal pain, nausea, vomiting, and diarrhea from 10 min to 12 h after eating. These indications were followed by oliguria heralding acute renal failure in seven patients (54%). Hematuria was noted in ten (77%) and jaundice in eight patients (62%). In addition to renal failure and hepatitis, one patient (8%) had severe congestive heart failure that was relieved by hemodialysis and digitalization. There were no coma, convulsions, or neurological abnormalities, excluding increased knee jerks in one case. The seriousness of the symptoms varied based on the amount of bile eaten.

The laboratory results showed that there were two cases with mild leukocytosis. Urine examination revealed microscopic hematuria in ten (77%) and proteinuria in seven (54%) patients, the urine specific gravity ranged from 1.003 to 1.015. Arterial blood gas analysis showed metabolic acidosis with a grown anion gap in seven (70%) of the ten patients checked.

The pathologic results of the light microscopic study of the kidneys showed that the main pathologic lesions were confined to the tubules and much less in the interstitium. The tubules showed focal necrosis and atrophy. A part of the tubular epithelium displayed a loss of nuclei and/or brush border. The cytoplasm was occasionally swollen. Some tubular lumina included protein casts. Light microscopic study of the liver revealed that the architecture was well conserved; the hepatocytes demonstrated focal ballooning degeneration and slight anisocytosis. Kupffer cell activity was slightly increased. Electron microscopic study showed massive degeneration or lysis of the hepatocytes, with a prominently decreased number of cytoplasmic organelles including mitochondria and lysosomes. An accumulation of lipid vacuoles in the cytoplasm was also observed.

Xuan et al. (2003) reported on 17 cases of ichthyogallotoxication from Ho Chi Minh City, Vietnam. All the patients consumed gallbladders from four fish species. The gallbladder was eaten to cure arthritis, urticaria, and eczema and to improve sexual vitality and vision. The patients were aged between 14 and 40 years. They had eaten between 3.7 and 5.4 gallbladders. The fish species that were eaten by seven patients were mainly the grass carp (*Ctenopharyngodon idella*) and the black shark (minnow) fish (*Morulius chrysophekadion*).

The results of the examination of the patients revealed that renal failure was present in all patients who presented initially to Cho Ray Hospital. Intravenous fluid (IVF) administered to 15 patients did not improve renal function and contributed to peripheral or pulmonary edema in four patients. Most patients were either anuric or oliguric. Serum creatinine concentrations increased to 14.7–3.9 mg/dL. Eleven patients underwent hemodialysis, most with anuria or continued oliguria despite fluids and diuretics. Six patients underwent a single hemodialysis treatment, three patients underwent two treatments, and two patients required four treatments. The mean peak creatinine level in this group was 15.6 mg/dL. Renal function recovered over a course of days to weeks in all patients, except for one patient who died. Depending on the rate of recovery, serum creatinine concentrations were tracked for a mean of 9.1–3.4 days after completion of the peak creatinine level. Patients were discharged while still recovering.

The light microscopic analysis showed features of acute tubular injury, including simplification of the tubular epithelium and focal karyorrhexis. Intratubular sloughed epithelium was present. Acute interstitial nephritis was unlikely to account for the severe impairment in renal function in the absence of significant interstitial inflammatory cells.

Jawad (2021) reported on three cases of eating gallbladder of carp species *Cyprinus carpio* and *Ctenopharyngodon idella*. These incidences indicate the decline level of education of the locals in the countryside villages located in Basrah and Mysan Provinces, south of Iraq.

In case no. 1, a 57-year-old woman living in a remote village at the edge of the southern marshes ate the grass carp gallbladders stewed with vinegar in a hope to improve general sickness. She experienced nausea and epigastric pain 2 hours after eating and arrived at an emergency department 8 hours after ingestion. Her liver showed severe parenchymal damage. Her alanine aminotransferase (ALT) (formerly glutamic pyruvic transaminase (GPT)) level was 5227 IU/L at 88 h and peaked at 7340 IU/L 13 hours after ingestion. Even with adequate hydration, she developed oliguria (2 mL/h) on day 3. The identification was the carp *Ctenopharyngodon idella* gallbladder poisoning. Hemodialysis was accomplished for continuing fluid retention. On day 26, her renal function showed continued improvement. She was stable and followed up with renal function monitoring.

In case no. 2, a 47-year-old housewife living in Mysan Province, south Iraq, ate the raw gallbladder of a grass carp *Ctenopharyngodon idella* and 6 hours later had central colicky abdominal pain, vomiting, and watery diarrhea for several hours. She was admitted to the clinic 3 days later due to oliguria. Physical inspection demonstrated only jaundice. There was no dehydration. Results of the medical examination were not normal. The renal tissue sample at the fourth week showed sign of new interstitial edema with interstitial fibrosis and acute tubular necrosis with regeneration. Both renal and liver functions go back to normal on discharge.

In case no. 3, a 50-year-old male living in a remote village north of Basrah was admitted in the emergency department. The patient established a severe kidney wound and changed sensorium with history of consumption of large amounts of gallbladder of carp (*Cyprinus carpio*) fish with bile, in dinner a few days ago. The patient started to experience vomiting and abdomen discomfort. He became unconscious later in the day. The patient was found to be nervous and twitchy suffering from pain in the abdomen. Medical investigations showed reasonable dehydration; icterus was present, but no pallor and edema. The case of the patient was steady except for tenderness over the upper abdomen. Further medical examinations showed an enhanced echo texture of both kidneys with normal collecting system. Later the patient was managed for all clinical problems and was cleared after 8 days of hospitalization.

5.14.2 The Fish Species Causing Ichthyogallotoxication

Family: Xenocyprididae

Ctenopharyngodon idella (Valenciennes, 1844) (Fig. 5.30)

Common name: Grass carp

Etymology: *Ctenopharyngodon:* Greek, *kteis*, *ktenos* = comb + Greek, *pharyngx* = pharynx + Greek, *odous* = teeth (Romero 2002); *idella: Cteno* = comb; *pharynx* = throat; *odon* = tooth (in reference to its comblike pharyngeal teeth); and *idella*: presumably derived from the Greek *idios*, distinctive or peculiar (Hassan-Williams et al. 2007)

Fig. 5.30 Ctenopharyngodon idella (Valenciennes, 1844). Courtesy of Courtesy of Hamid Badar Osmany, Pakistan

Diagnostic Features
This species of barb has no barbels. The snout is very short, and its length is less than or equal to the eye diameter. The postorbital length is more than half the head length (Eccles 1992). There are 18 soft rays in the caudal fin (Keith and Allardi 2001). It is separated from rather similar species, *Mylopharyngodon piceus,* by having the following characteristics: the body is olive to brassy green above and silvery white to yellow below; the body is cylindrical; the pharyngeal teeth are laterally compressed and serrated, with a groove along the grinding surface; and they are usually in two rows, 2,5–4,2 (Kottelat and Freyhof 2007).

Distribution
This species is distributed throughout Asia. It has been reported from Eastern China and Russia (Shireman and Smith 1983). It is also recorded in eastern Siberia, Amur River system (Berg 1964). It is widely transferred around the world (Skelton 1993). It was introduced into several countries around the world including New Zealand. Several countries report adverse ecological impact after introduction.

Habitat, Biology, and Ecology
The grass carp is a freshwater species that sometimes enter brackish water (Riede 2004). It lives at a depth range 0–30 m (Shao and Lim 1991). Individuals of this species reach a maximum total length of 1500 mm (Billard 1997). The maximum reported weight is 45.0 kg (Skelton 1993) and the maximum recorded age is 21 years (Shireman and Smith 1983). The adults live in lakes, ponds, pools, and backwaters of large rivers as they preferred large, slow-flowing, or standing water bodies with vegetation (Page and Burr 1991). They are standing to a wide range of temperatures from 0 to 38 °C, and salinities to as much as 10 ppt and oxygen levels down to 0.5 ppm (Frimodt 1995). They feed on higher aquatic plants and underwater grasses; they also eat detritus, insects, and other invertebrates (Page and Burr 1991). They spawn on riverbeds with very strong current (Billard 1997).

Economic Value
It is one of the world's most important aquaculture species and used for controlling weeds in rivers, fishponds, and reservoirs (Frimodt 1995). It is considered a pest in most countries because of the damages made to submerged vegetation (Kottelat 2001) and utilized fresh and eaten steamed, panfried, broiled, and baked (Frimodt 1995).

Conservation
The grass carp is not evaluated for the Red List of the IUCN.

Hypophthalmichthys molitrix (Valenciennes, 1844) (Fig. 5.31)
Common name: silver carp
Etymology: *Hypophthalmichthys:* Greek, *hypo* = under + Greek, *ophthalmos* = eye + Greek, *ichthys* = fish (Romero 2002); *molitrix: molitrix*, approximately grinder (referring to the pharyngeal grinding apparatus) (Etnier and Starnes 1993)

Diagnostic Features
The body is olivaceous to silvery and has no barbels. The keels extend from the isthmus to the anus. The edge of the last simple dorsal ray is not serrated. The branched anal rays are 12–13.5 (Kottelat et al. 1993). This species differs from *Hypophthalmichthys nobilis* by having sharp scaleless keel from the pectoral region to the anal origin, slender gill rakers, head length of 24–29% SL, and plain pale coloration, greenish gray above and whitish below (Kottelat and Freyhof 2007).

Fig. 5.31 Hypophthalmichthys molitrix (Valenciennes, 1844). Courtesy of Courtesy of Hamid Badar Osmany, Pakistan

Distribution

This species is distributed in Asia. It is native to most major Pacific drainage of East Asia from Amur to Xi Jiang, China (Kottelat and Freyhof 2007), and Hanoi, Vietnam (FAO 2019). It has been introduced around the world including New Zealand for aquaculture and control of algal blooms. Some countries report adverse ecological impact after introduction (Welcomme 1988). It is frequently puzzled with *Hypophthalmichthys nobilis* (Kottelat and Freyhof 2007).

Habitat, Biology, and Ecology

The silver carp is a freshwater species that enters brackish water sometimes (Riede 2004). It lives at depth range of 0–20 m (Shao and Lim 1991). Individuals of this species reach a maximum total length of 1200 mm (Page and Burr 1991). The maximum recorded weight is 50 kg (Page and Burr 1991) and the maximum recorded age is 20 years (Nico et al. 2020).

This species prefers to live in rivers with marked water-level fluctuations and overwinters in middle and lower stretches, swimming just beneath the surface (Page and Burr 1991). Individuals of this species feed in shallow (0.5–1.0 m deep) and warm (over 21 °C) backwaters, lakes, and flooded areas with slow current on phytoplankton and zooplankton (Etnier and Starnes 1993). The bigger individuals from about 1.5 cm SL feed only on phytoplankton, while larvae and small juveniles prey on zooplankton (Kottelat and Freyhof 2007). The adults breed in rivers or tributaries over shallow rapids with gravel or sand bottoms, in upper water layers, or even at surface during floods when the water level increases by 50–120 cm above the normal level. Larvae move downstream and settle in floodplain lakes, shallow shores, and backwaters with little or no current (Kottelat and Freyhof 2007). In aquaculture, it can survive brackish water (up to 7 ppt) when released into estuaries and coastal lakes (Kottelat and Freyhof 2007). Escape from fish farms is widely known even as they are stocked in large rivers and almost all still water bodies such as lakes and ponds.

Economic Value

The meat of this species is consumed by humans and introduced to many countries including New Zealand where its talent to clean reservoirs and other waters of clogging algae is appreciated even more than its food value (Frimodt 1995). It is one of the three or four species of cyprinids whose world production in aquaculture exceeds one million tons per year (Page and Burr 1991).

Conservation

Zhao (2011) evaluated this species as Nearly Threatened for the Red List of the IUCN based on the following reasons. The distribution of this species has been influenced by dams along its geographical distribution. In general, the populations of this species are declining around the world.

References

Akyol O, Ünal V, Ceyhan T, Bilecenoglu M. First confirmed record of *Lagocephalus sceleratus* (Gmelin, 1789) in the Mediterranean Sea. J Fish Biol. 2005;66:1183–6.

Albaret J-J. Mugilidae. In: Lévêque C, Paugy D, Teugels GG, editors. Faune des poissons d'eaux douce et saumâtres de l'Afrique de l'Ouest, Tome 2. Coll. Faune et Flore tropicales 40. Musée Royal de l'Afrique Centrale, Tervuren, Belgique, Museum National d'Histoire Naturalle, Paris, France and Institut de Recherche pour le Développement, Paris, France; 2003. p. 601–11, 815 p

Alexander J, Autrup H, Bard D, Carere A, Costa LG, Cravedi J-P, Di Domenico A, Fanelli R, Fink-Gremmels J, Gilbert J, Grandjean P, Johansson N, et al. Opinion of the scientific panel on contaminants in the food chain on a request from the commission related to the toxicity of fishery products belonging to the family of Gempylidae. Eur Food Saf Authority J. 2004;92:1–5.

Alikunhi KH. Synopsis of biological data on common carp *Cyprinus carpio* (Linnaeus), 1758 (Asia and the Far East). FAO Fish Synop. 1966;31(2):39 p.

Allen GR. Field guide to the freshwater fishes of New Guinea, Publication, no. 9. 268 p. Madang, Papua New Guinea: Christensen Research Institute; 1991.

Allen GR, Erdmann MV. Reef fishes of the East Indies. Perth: University of Hawai'i Press, Volumes I-III. Tropical Reef Research; 2012.

Allen GR, Midgley SH, Allen M. Field guide to the freshwater fishes of Australia. Perth: Western Australian Museum; 2002. 394 p.

Alon U, Berant M, Mordechovitz D, et al. The effect of intrarenal infusion of bile on the kidney function in the dog. Clin Sci. 1982;62:431.

Altman PL, Dittmer DS. Growth, including reproduction and morphological development. Federation of American Societies for Experimental Biology; 1962.

Anastasius M, Yiannikas J. Scombroid fish poisoning illness and coronary artery vasospasm. Australas Med J. 2015;8(3):96.

Anderson W. An account of some poisonous fish in the south seas. In a letter to Sir John Pringle, Bart. PRS from Mr. William Anderson, late surgeon's mate on board his Majesty's ship the resolution, now surgeon of that ship. Philos Trans R Soc Lond. 1776;66:544–74.

Anderson T. Isolation and characterization of wax esters from Calanus finmarchicus. M.Sc. thesis. Tromsø: University of Tromsø; 2010.

Anonymous. Increasing competition between fisheries and whales. In: Japan's Whale Research in the Western North Pacific (JARPN II). Fisheries Agency, Government of Japan; 2001a.

Anonymous. Fish collection database of the Institut de Ciencies del Mar (C.S.I.C.). In: Dept. Recursos Marins Renovables Grup d'Ictiologia Marina (G.I.M.) Psg. Joan de Borbo s/n 08039. Barcelona; 2001b.

Ansdell VE. Food poisoning from marine toxins. CDC Health Inf Int Travel 2016. 2015;100(8):91.

Arakawa O, Hwang DF, Taniyama S, Takayani T. Toxins of pufferfish that cause human intoxications. In: Lie A, Ishimatsu H-J, editors. Coastal environmental and ecosystem issues of the East China Sea. Nagasaki: Terrapub and Nagasaki University; 2010. p. 227–44.

Arakawa O, Takatani T, Taniyama S, Tatsuno R. Toxins of pufferfish—distribution, accumulation mechanism, and physiologic functions. Aqua-BioSci Monogr. 2017;10:41–80.

Aras NM, Hasiloglu MA, Haliloglu HI, Bayir A, Sirkecioglu AN. Comparison of fatty acid composition of some tissues and conversion ratios of stomach containing fatty acids to tissue fatty acids in *Barbus capito capito* Güldenstaed, 1773. Asian J Chem. 2009;21:6969–74.

Armitage RO, Payne DA, Lockley GJ, Currie HM, Colban RL, Lamb BG, Paul LJ, editors. Guidebook to New Zealand commercial fish species. Revised edition. Wellington: New Zealand Fishing Industry Board; 1994. 216 p.

Armstrong P, Murray P, Nesdale A, Peckler B. Ciguatera fish poisoning NZMJ. 2016;129:113–6.

Arnold SH, Brown WD. Histamine (?) toxicity from fish products. Adv Food Res. 1978;24:113–54.

Asakawa M, Matsumoto T, Umezaki K, Kaneko K, Yu X, Gomez-Delan G, Tomano S, Noguchi T, Ohtsuka S. Toxicity and toxin composition of the greater blue-ringed octopus *Hapalochlaena lunulata* from Ishigaki Island, Okinawa prefecture. Japan Toxins. 2019;11:245.

Asano M, Itoh M. Toxicity of a lipoprotein and lipids from the roe of a blenny, *Dinogunellus grigorjewi herzenstein*. Tohoku J Agric Res. 1962;13(2):151–67.

Ascione A, Barresi LS, Sarullo FM, De Silvestre G. Two cases of "scombroid syndrome" with severe cardiovascular compromise. Cardiologia. 1997;42:1285.

Bacchet P, Zysman T, Lefèvre Y. Guide des poissons de Tahiti et ses îles. Tahiti (Polynésie Francaise): Éditions Au Vent des Îles; 2006. 608 p.

Bacchiocchi S, Campacci D, Siracusa M, Dubbini A, Leoni F, Tavoloni T, Accoroni S, Gorbi S, Giuliani ME, Stramenga A, Piersanti A. Tetrodotoxins (Ttxs) and vibrio alginolyticus in mussels from Central Adriatic Sea (Italy): are they closely related? Mar Drugs. 2021;19(6):304.

Bagnis R. Ciguatera fish poisoning. In: Falconer I, editor. Algal toxins in seafood and drinking water. London: Academic; 1993. p. 105–15.

Bagnis R, Kuberski T, Laugier S. Clinical observations on 3,009 cases of ciguatera (fish poisoning) in the South Pacific. Am J trop Med Hygiene. 1979;28:1067.

Bagnis R, Spiegel A, Boutin JP, Burucoa C, Nguyen L, Cartel JL, Capdevielle P, Imbert P, Prigent D, Gras C. Evaluation of the efficacy of mannitol in the treatment of ciguatera in French Polynesia. Med Trop. 1992;52:67–73.

Baker AN. Reproduction, early life history, age growth relationships of the New Zealand pilchard, *Sardinops neopilchardus* (Steindachner). Fish Res Bull Fish Res N Z Minist Agricult Fish. 1972;5:1–64.

Balon EK. Epigenesis of an epigeneticist: the development of some alternative concepts on the early ontogeny and evolution of fishes. Guelph Ichthyol Rev. 1990;1:1–48.

Bane V, Lehane M, Dikshit M, O'Riordan A, Furey A. Tetrodotoxin: chemistry, toxicity, source, distribution and detection. Toxins. 2014;6:693–755.

Banner AH. Hallucinatory mullet poisoning. A case from Oahu. Hawaii Med J. 1973;32(5):330–1.

Banner AH, Boroughs H. Observations on toxins of poisonous fishes. Proc Soc Exp Biol Med. 1958;98:776–8.

Banner AH, Scheuer PJ, Sasaki S, Helfrich P, Alender CB. Observations on ciguatera-type toxins in fish. Ann N Y Acad Sci. 1960;90(3):770–87.

Barling PM, Foong YH. Oily fish, liquid wax esters and keriorrhoea–a review. Int e-J Sci Med Educ. 2015;9(1):21–5.

Bass AJ, Compagno LJV, Heemstra PC. Squalidae. In: Smith MM, Heemstra PC, editors. Smiths' sea fishes. Berlin: Springer; 1986. p. 49–62.

Bauchot, M.-L., 1987. Poissons osseux. p. 891-1421. In W. Fischer, M.L. Bauchot and M. Schneider (eds.) Fiches FAO d'identification pour les besoins de la pêche. (rev. 1). Méditerranée et mer Noire. Zone de pêche 37. Vol. II. Commission des Communautés Européennes and FAO, Rome.

Becke L. Poisonous fish of the Pacific Islands. 1901.

Bédry R, de Haro L, Bentur Y, Senechal N, Galil BS. Toxicological risks on the human health of populations living around the Mediterranean Sea linked to the invasion of non-indigenous marine species from the Red Sea: A review. Toxicon. 2021;191:69–82.

Benoit E, Juzans P, Legrand A, Molgo J. Nodal swelling produced by ciguatoxin-induced selective activation of sodium channels in myelinated nerve fibers. Neuroscience. 1996;71:1121–31.

Benoit E, Laurent D, Mattei C, Legrand AM, Molgo J. Reversal of Pacific ciguatoxin1B effects on myelinated axons by agents used in ciguatera treatment. Cybium. 2000;24:33–40.

Bentur Y, Ashkar J, Lurie Y, Levy Y, Azzam ZS, Litmanovich M, Golik M, Gurevych B, Golani D, Eisenman A. Lessepsian migration and tetrodotoxin poisoning due to *Lagocephalus sceleratus* in the eastern Mediterranean. Toxicon. 2008;52:964–8.

Berg LS. Freshwater fishes of the U.S.S.R. and adjacent countries, vol. 2. 4th ed. Jerusalem: Israel Program for Scientific Translations Ltd; 1964. (Russian version published 1949)

Berg, L.S., 1965. Freshwater fishes of the U.S.S.R. and adjacent countries. Volume 3, 4th edition. Israel program for scientific translations ltd, Jerusalem. (Russian version published 1949).

Berman P, Harley EH, Spark AA. Keriorrhea—the passage of oil per rectum—after ingestion of marine wax esters. S Afr Med J. 1981;59:791–2.

Better OS, Bomzon A. Effect of jaundice on the renal and cardiovascular systems. In: Epstein M, editor. The kidney in liver disease. 3rd ed. Baltimore: Williams & Wilkins; 1988. p. 508–34.

Beverton, R.J.H. and S.J. Holt, 1959. A review of the lifespans and mortality rates of fish in nature, and their relation to growth and other physiological characteristics. p. 142–180. In G.E.W. Wolstenholme and M. O'Connor (eds.) CIBA foundation colloquia on ageing: the lifespan of animals. volume 5. J & A Churchill Ltd, London.

Bianchi G, Carpenter KE, Roux J-P, Molloy FJ, Boyer D, Boyer HJ. FAO species identification guide for fishery purposes. Field guide to the living marine resources of Namibia. Rome: FAO; 1999. 265 p.

Biessy, L. Elucidating the Origin and Transmission of Tetrodotoxin in New Zealand Bivalves; The University of Waikato: Hamilton, New Zealand, 2021; Available online: https://hdl.handle.net/10289/14548. Accessed 30 Apr 2023.

Biessy L, Smith KF, Boundy MJ, Webb SC, Hawes I, Wood SA. Distribution of tetrodotoxin in the New Zealand clam, Paphies australis, established using immunohistochemistry and liquid chromatography-tandem quadrupole mass spectrometry. Toxins. 2018;10(7):282.

Biessy L, Boundy MJ, Smith KF, Harwood DT, Hawes I, Wood SA. Tetrodotoxin in marine bivalves and edible gastropods: a mini-review. Chemosphere. 2019a;236:124404.

Biessy L, Smith KF, Harwood DT, Boundy MJ, Hawes I, Wood SA. Spatial variability and depuration of tetrodotoxin in the bivalve Paphies australis from New Zealand. Toxicon X. 2019b;2:100008.

Biessy L, Pearman JK, Smith KF, Hawes I, Wood SA. Seasonal and spatial variations in bacterial communities from Tetrodotoxin-bearing and non-tetrodotoxin-bearing clams. Front Microbiol. 2020:11, 1860.

Bigelow HB, Bradbury MG, Dymond JR, Greeley JR, Hildebrand SF, Mead GW, Miller RR, Rivas LR, Schroeder WL, Suttkus RD, Vladykov VD. Fishes of the western North Atlantic. Part three: New Haven, Sears Found. Mar. Res., Yale Univ; 1963.

Billard R. Les poissons d'eau douce des rivières de France. In: Identification, inventaire et répartition des 83 espèces. Lausanne: Delachaux & Niestlé; 1997. 192 p.

Birinyi-Strachan LC, Davies MJ, Lewis RJ, Nicholson GM. Neuroprotectant effects of iso-osmolar d-mannitol to prevent Pacific ciguatoxin-1 induced alterations in neuronal excitability: a comparison with other osmotic agents and free radical scavengers. Neuropharmacology. 2005;49:669–86.

Bjeldanes LF, Schutz DE, Morris MM. On the aetiology of scombroid poisoning: cadaverine potentiation of histamine toxicity in the Guinea pig. Food Cosmet Toxicol. 1978;16(2):157–9.

Blaber SJM. The food and feeding ecology of Mugilidae in the St. Lucia lake systems. Biol J Linn Soc. 1976;8:267–77.

Blakesley ML. Scombroid poisoning: prompt resolution of symptoms with cimetidine. Ann Emerg Med. 1983;12(2):104–6.

Blythe DG, de Sylva DP. Mother's milk turns toxic following fish feast. JAMA J Am Med Assoc. 1990;264:2074.

Blythe DG, Hack E, Washington G, Fleming LE. The medical management of seafood poisoning. In: Hui YH, Kitts D, Stanfield PS, editors. Foodborne disease handbook. New York: Marcel Dekker; 2001. p. 311–20.

Boisier P, Ranaivoson G, Rasolofonirina N, Roux J, Chanteau S, Takeshi Y. Fatal mass poisoning in Madagascar following ingestion of a shark (Carcharhinus leucas): clinical and epidemiological aspects and isolation of toxins. Toxicon. 1995;33(10):1359–64.

Bordin P, Dall'Ara S, Tartaglione L, Antonelli P, Calfapietra A, Varriale F, Guiatti D, Milandri A, Dell'Aversano C, Arcangeli G, et al. First occurrence of tetrodotoxins in bivalve mollusks from Northern Adriatic Sea (Italy). Food Control. 2021;120:107510.

Bouder H, Cavallo A, Bouder MJ. Poissons veneneux et ichryosarcotoxisme. Monaco: Bull. de l'Ins tit. Ocean; 1962. p. 66.

Bouhlel M. Poissons de Djibouti. Placerville: RDA Internatiônal, Inc; 1988. 416 p.

Boundy MJ, Biessy L, Roughan B, Nicolas J, Harwood DT. Survey of tetrodotoxin in New Zealand bivalve molluscan shellfish over a 16-month period. Toxins. 2020;12(8):512.

Bourdy G, Cabalion P, Amade P, Laurent D. Traditional remedies used in the western Pacific for the treatment of ciguatera poisoning. J Ethnopharmacol. 1992;36:163–74.

Boydron-Le Garrec R, Benoit E, Sauviat MP, Lewis RJ, Molgó J, Laurent D. Ability of some plant extracts, traditionally used to treat ciguatera fish poisoning, to prevent the in vitro neurotoxicity produced by sodium channel activators. Toxicon. 2005;46(6):625–34.

Breder CM, Rosen DE. Modes of reproduction in fishes. Neptune City: T.F.H. Publications; 1966. 941 p.

Brooks W. A family poisoned by eating a gar. NW Med Surg J. 1850;7:437.

Bucciarelli GM, Lechner M, Fontes A, Kats LB, Eisthen HL, Shaffer HB. From poison to promise: the evolution of tetrodotoxin and its potential as a therapeutic. Toxins. 2021;13:517.

Bundesministerium für Ernährung, Landwirtschaft und Forsten (BMELF). Gutachten über Mindestanforderungen an die Haltung von Zierfischen (Süßwasser). Landwirtschaft und Forsten (BMELF), Bonn: Bundesministerium für Ernährung; 1999. 16 p.

Bykov VP. Marine fishes: chemical composition and processing properties. New Delhi: Amerind Publishing; 1983.

Cadwallader PL, Backhouse GN. A guide to the freshwater fish of Victoria. Melbourne: Government Printers; 1983. 249 p.

Caillaud A, De la Iglesia P, Darius HT, Pauillac S, Aligizaki K, Fraga S, Chinain M, Dioge'ne J. Update on methodologies available for ciguatoxin determination: perspectives to confront the onset of ciguatera fish poisoning in Europe. Mar Drugs. 2010;8(6):1838–907.

Calvert GM, Hryhorczuk DO, Leikin JB. Treatment of ciguatera fish poisoning with amitriptyline and nifedipine. J Toxicol Clin Toxicol. 1987;25:423–8.

Camara KD, Carpenter KE, Djiman R, Nunoo F, Sagna A, Sidibé A, Sylla M, de Morais L, Jelks H, Tolan J, Vega-Cendejas M, Espinosa-Perez H, Chao L, Moretzsohn F, Sparks JS. *Mugil cephalus*. The IUCN red list of threatened species 2019: e.T135567A127923853. 2019. https://doi.org/10.2305/IUCN.UK.2019-2.RLTS.T135567A127923853.en

Cameron J, Flowers AE, Capra MF. Electrophysiological studies on ciguatera poisoning in man (part II). J Neurol Sci. 1991;101:93–7.

Cardona L. Effects of salinity on the habitat selection and growth performance of Mediterranean flathead grey mullet *Mugil cephalus* (Osteichthyes, Mugilidae). Estuar Coast Shelf Sci. 2000;50:727–37.

Caro RJ, Cosculluela AM, Beltran LF, Rihuete HMA. Diarrea oleosa anaranjada. Keriorrhea inducida por pescado An Pediatr. 2011;74:67–8.

Cayré P, Amon Kothias JB, Diouf T, Stretta JM. Biology of tuna. In: Fonteneau A, Marcille J, editors. Resources, fishing, and biology of the tropical tunas of the Eastern Central Atlantic, FAO Fish. Tech. Pap. 292. Rome: FAO; 1993. p. 147, 354 p–244.

Cervigón F. Los peces marinos de Venezuela, vol. 3. Caracas: Fundación Científica Los Roques; 1994. 295 p.

Cervigón F.,Cipriani R, Fischer W, Garibaldi L, Hendrickx M, Lemus AJ, Marquez R, Poutiers JM, Robaina G and Rodriguez B. Fichas FAO de identificacio'n de especies para los fines de la pesca. Guı'a de campo de las especies comerciales marinas y de aquas salobres de la costa septentrional de Sur Ame'rica. Rome: FAO; 1992. 513 p. Preparado con el financiamento de la Comisio'n de Comunidades Europeas y de NORAD.

Chadwick VS, Gaginella TS, Carlson GL, et al. Effect of molecular structure on bile-acid-induced alterations in absorptive function, permeability, and morphology in the perfused rabbit colon. J Lab Clin Med. 1979;94:661–74.

Chan DW, Yeung CK, Chan MK. Acute renal failure after eating raw fish gall bladder. BMJ. 1985;290:897.

Chang FJ, Anderson C, Boustead NC. First record of a *Heterosigma* (Raphidophyceae) bloom with associated mortality of cage-reared salmon in Big Glory Bay, New Zealand. New Zealand J Marine Freshwater Res. 1990;24:461–9.

Chang FC, Bauer RM, Benton BJ, Keller SA, Capacio BR. 4-Aminopyridine antagonizes saxitoxin-and tetrodotoxin-induced cardiorespiratory depression. Toxicon. 1996;34(6):671–90.

Château-Degat ML. Les toxines marines: problèmes de santé en émergence. VertigO. 2003;4:1–11. https://doi.org/10.4000/vertigo.4698.

Chau R, Kalaitzis JA, Wood SA, Neilan BA. Diversity and biosynthetic potential of culturable microbes associated with toxic marine animals. Mar Drugs. 2013;11:2695–712.

Chen CF. The effect on gastrointestinal motility after ingestion of raw bile of grass carp (Ctenopharyngodon idellus) in the conscious rat. Proc Natl Sci Counc Repub China B. 1984;8:78–82.

Chen WY, Yen TS, Cheng JT, et al. Acute renal failure due to ingestion of raw bile of grass carp (Ctenopharyngodon idellus). J Formosan Med Assoc. 1976;75:149–57.

Chevaldonne P. Ciguatera and the saupe, Sarpa salpa, in the Mediterranean: a possible misinterpretation. J Fish Biol. 1990;37:503–4.

Chinain M, Faust M, Pauillac S. Morphology and molecular analyses of three toxic species of Gambierdiscus (Dinophyceae): G. pacificus, sp. nov., G. australes, sp. nov., and G. polynesiensis, sp. nov. J Phycol. 1999a;35:1282–96.

Chinain M, Germain M, Deparis X, Pauillac S, Legrand A. Seasonal abundance and toxicity of the dinoflagellate Gambierdiscus spp.(Dinophyceae), the causative agent of ciguatera in Tahiti, French Polynesia. Mar Biol. 1999b;135:259–67.

Chisholm C. On the poison fish. Edinburgh Med Surg Inst. 1808;4:393–422.

Chong W. Sale of oilfish to be curbed. The Standard 26 Jan 2007.

Chugunova NI. Age and growth studies in fish. Moskva: A systematic guide for ichthyologists. Israel Program for Scientific Translations. Izdatel'stvo Akademii Nauk SSSR; 1959. 132 p.

Chulanetra M, Sookrung N, Srimanote P, Indrawattana N, Thanongsaksrikul J, Sakolvaree Y, Chongsa-Nguan M, Kurazono H, Chaicumpa W. Toxic marine puffer fish in Thailand seas and tetrodotoxin they contained. Toxins. 2011;3(10):1249–62.

Chung C. Label mistake revealed in oilfish saga. The standard 29 Jan 2007a.

Chung C. Legislation to regulate fish products on the table. The Standard 3 Feb 2007b.

Ciguatera PF. Memoria sobre la enfermedad ocasionadapor la peces venenosos. Repert Fisico Natural Isla Cuba (Havana). 1966;2:1–39.

Clavijero FJ. Historia de la Antigua ó Baja California. Mexico: Juan R. Navarro; 1852. 60 p.

Coker RE. Studies of common fishes of the Mississippi River at Keokuk. Bull US Bur Fish. 1929;45:155.

Collette BB. Scombridae (including Thunnidae, Scomberomoridae, Gasterochismatidae and Sardidae). In: Whitehead PJP, Bauchot M-L, Hureau J-C, Nielsen J, Tortonese E, editors. Fishes of the North-Eastern Atlantic and the Mediterranean, vol. 2. Paris: UNESCO; 1986. p. 981–97.

Collette BB. Scombridae. Atunes, bacoretas, bonitos, caballas, estorninos, melva, etc. In: Fischer W, Krupp F, Schneider W, Sommer C, Carpenter KE, Niem V, editors. Guia FAO para Identification de Especies para lo Fines de la Pesca. Pacifico Centro-Oriental. 3 Vols. Rome: FAO; 1995. p. 1521–43.

Collette BB. Scombridae. Tunas (also, albacore, bonitos, mackerels, seerfishes, and wahoo). In: Carpenter KE, Niem V, editors. FAO species identification guide for fishery purposes. The living marine resources of the Western Central Pacific. Vol. 6. Bony fishes part 4 (Labridae to Latimeriidae), estuarine crocodiles. Rome: FAO; 2001. p. 3721–56.

Collette BB, Graves JE. Tunas and billfishes of the world. Baltimore: Johns Hopkins University Press; 2019.

Collette BB, Nauen, CE. FAO species catalogue. Vol. 2. Scombrids of the world. An annotated and illustrated catalogue of tunas, mackerels, bonitos and related species known to date. Rome: FAO. FAO Fish. Synop. 1983; 125(2):137 p.

Collette B, Acero A, Amorim AF, Boustany A, Canales Ramirez C, Cardenas G, Carpenter KE, de Oliveira Leite Jr. N, Di Natale A, Fox W, Fredou FL, Graves J, Guzman-Mora A, Viera Hazin FH, Juan Jorda M, Kada O, Minte Vera C, Miyabe N, Montano Cruz R, Nelson R, Oxenford H, Salas E, Schaefer K, Serra R, Sun C, Teixeira Lessa RP, Pires Ferreira Travassos PE, Uozumi Y, Yanez E. *Auxis thazard*. The IUCN red list of threatened species 2011: e.T170344A6757270. 2011b. https://doi.org/10.2305/IUCN.UK.2011-2.RLTS.T170344A6757270.en. Accessed 13 Apr 2023.

Collette B, Acero A, Canales Ramirez C, Cardenas G, Carpenter KE, Chang S-K, Chiang W, Di Natale A, Fox W, Guzman-Mora A, Juan Jorda M, Miyabe N, Montano Cruz R, Nelson R, Salas E, Schaefer K, Serra R, Sun C, Uozumi Y, Wang S, Wu J, Yanez E, Yeh S. *Scomber australasicus*. The IUCN red list of threatened species 2011: e.T170329A6750490. 2011c. https://doi.org/10.2305/IUCN.UK.2011-2.RLTS.T170329A6750490.en. Accessed 16 Apr 2023.

Collette B, Di Natale A, Fox W, Juan Jorda M, Nelson R. *Sarda australis*. The IUCN red list of threatened species 2011: e.T1703326752615. 2011d. https://doi.org/10.2305/IUCN.UK.2011-2.RLTS.T170332A6752615.en. Accessed 16 Apr 2023.

Collette B, Amorim AF, Boustany A, Carpenter KE, de Oliveira Leite Jr, N, Di Natale A, Fox W, Fredou FL, Graves J, Viera Hazin FH, Juan Jorda M, Minte Vera C, Miyabe N, Nelson R, Oxenford H, Sun C, Teixeira Lessa RP, Pires Ferreira Travassos PE, Uozumi Y. *Allothunnus fallai*. The IUCN red list of threatened species 2011e: e.T170349A6761139. 2011. https://doi.org/10.2305/IUCN.UK.2011-2.RLTS.T170349A6761139.en. Accessed 12 Apr 2023.

Collette, B., Smith-Vaniz, W.F., Williams, J.T., Pina Amargos, F., Curtis, M. & Grijalba Bendeck, L. 2015. *Gempylus serpens*. The IUCN red list of threatened species 2015: e.T16509126A16510882. 2005. https://doi.org/10.2305/IUCN.UK.2015-4.RLTS.T16509126A16510882.en

Collette BB, Boustany A, Fox W, Graves J, Juan Jorda M, Restrepo V . *Katsuwonus pelamis*. The IUCN red list of threatened species 2021: e.T170310A46644566. 2021a. https://doi.org/10.2305/IUCN.UK.2021-2.RLTS.T170310A46644566.en. Accessed 15 Apr 2023.

Collette BB, Boustany A, Fox W, Graves J, Juan Jorda M, Restrepo V. *Thunnus alalunga*. The IUCN red list of threatened species 2021: e.T21856A46911332. 2021b. https://doi.org/10.2305/IUCN.UK.2021-2.RLTS.T21856A46911332.en. Accessed 17 Apr 2023.

Compagno LJV. FAO Species Catalogue. Vol. 4. Sharks of the world. An annotated and illustrated catalogue of shark species known to date. Part 1 – Hexanchiformes to Lamniformes. FAO Fish. Synop. 125(4/1):1–249. Rome: FAO; 1984.

Compagno LJV, Niem VH. Hexanchidae. Cow sharks, sixgill, and seven gill sharks. In: Carpenter KE, Niem VH, editors. FAO identification guide for fishery purposes. The living marine resources of the Western Central Pacific. Rome: FAO; 1998. p. 1208–10.

Compagno LJV, Ebert DA, Smale MJ. Guide to the sharks and rays of southern Africa. London: New Holland (Publ.) Ltd.; 1989. 158 p.

Connolly N, Wong M, Fong L. ParknShop withdraws oilfish, cod as 14 fall ill The South China Morning Post 24 Jan 2007

Cook J. A voyage towards the south pole and round the world. London: Strahan & Cadell; 1777. pp. 39–40, 112–37

Cook SF, Compagno LJV. Hexanchus griseus. The IUCN red list of threatened species 2005: e.T10030A3155348. 2005. https://doi.org/10.2305/IUCN.UK.2005.RLTS. T10030A3155348.en. Downloaded on 5 November 2022.

Cooper MJ. Ciguatera and other marine poisoning in the Gilbert Islands. Pac Sci. 1964;18(4):411–40.

Cottrell JE, Hartung J, Giffin JP, Shwiry B. Intracranial pressure during tetrodotoxin-induced hypotension. Anesth Analg. 1984;63(11):1005–8.

Coutand H. Observations sur sept cas d'empoisonnement par le fois de requin a l'Ile des Pins (Nouvelle Caledonie) en 1878. These, Faculte de Medecine, Universite´ de Montpellier; 1879.

Cox G, Francis M. Sharks and rays of New Zealand. Canterbury Univ. Press, Univ. of Canterbury; 1997. 68 p.

Cox WM, Reid EE. The chemical composition of oil of Ruvettus pretiosus, the "castor oil fish". J Am Chem Soc. 1932;54:220–9.

D'Arras L. Essai sur les accidents cause's par les poissons. The'se no. 156, fac. Me'd. Paris. 1877. 69 p.

Darwin CR. The descent of man, selection in relation to sex. London: John Murray; 1877.

David BO, Özkundakci D, Pingram M, Bergin D, Bergin M. "The CarP-N neutral project": removal, processing and reuse of invasive fish in local terrestrial conservation projects. J Appl Ecol. 2018;55(4):1567–74.

Daymond JR. Family Salmonidae. Sears Found Mar Res Mem. 1963;1(3):457–546.

de Figueiredo JL, dos Santos AP, Yamaguti N, Bernardes RA, Del Bianco Rossi-Wongtschowski CL. Peixes da zona econômica exclusiva da Região Sudeste-Sul do Brasil: Levantamento com Rede de Meia-Água. São-Paulo: Editora da Universidade de São Paulo; Imprensa Oficial do Estado; 2002. 242 p.

De Haro L, Pommier P. Hallucinatory fish poisoning (Ichthyoallyeinotoxism): two case reports from the Western. Clin Toxicol. 2006;44:185–8. https://doi.org/10.1080/15563650500514590.

de Haro L, Treffot MJ, Jouglard J, Perringue C. Trois cas d'intoxication de type ciguate'rique apre's ingestion de sparidae de Me'diterrane'e. Ictyophysiologica Acta. 1993;16:133–46.

de Haro L, Pommier P, Valli M. Emergence of imported ciguatera in Europe: report of 18 cases at the poison control Centre of Marseille. J Toxicol Clin Toxicol. 2003;41(7):927–30.

De Rochas V. Essai sur la topographie hygie'nique et me'dicale de la Nouvelle-Cale'donie. M.D. these 250. Fac. Med. Paris. 1860. p. 35.

De Sauvages FB. De veneatis galline animalibus. M.D. Thesis, Rothomagensi Academia. Jo-annis Martel, Montspelii; 1758. p. 22.

De Sauvages FB. Les chefs-d'oeuvres de monsieur de Savages, vol. 2. Lyon: V. Reguilliat; 1770. p. 212.

Dell'Aversano C, Tartaglione L, Polito G, Dean K, Giacobbe M, Casabianca S, Capellacci S, Penna A, Turner AD. First detection of tetrodotoxin and high levels of paralytic shellfish poisoning toxins in shellfish from Sicily (Italy) by three different analytical methods. Chemosphere. 2019;215:881–95.

Dhanji-Rapkova M, Turner AD, Baker-Austin C, Huggett JF, Ritchie JM. Distribution of tetrodotoxin in Pacific oysters (*Crassostrea gigas*). Mar Drugs. 2021;19:84.

Doucette GJ, Kodama M, Franca S, Gallacher S. Bacterial interactions with harmful algal bloom species: bloom ecology, toxinogenesis, and cytology. In: Anderson DM, Cembella AD, Hallegraeff GM, editors. Physiological ecology of harmful algal blooms, vol. G41. Berlin: Springer; 1998. p. 619–49.

Ebert DA, Fowler S, Compagno L. Sharks of the world: a fully illustrated guide. Plymouth: Wild Nature Press; 2013. 528 p.

Eccles DH. FAO species identification sheets for fishery purposes. Field guide to the freshwater fishes of Tanzania. In: Prepared and published with the support of the United Nations Development Programme (project URT/87/016). Rome: FAO; 1992. 145 p.

Emerson WC. The toxic constituent of bile. J Lab Clin Med. 1929;14:635–43.

Engelsen H. Om giftfiskog giftigefisk Nord Hyg Tskr. 1922;3:316–25.

Eschmeyer WN, Herald ES, Hammann H. A field guide to Pacific coast fishes of North America. Boston: Houghton Mifflin Company; 1983. xii+336 p.

Etnier DA, Starnes WC. The fishes of Tennessee. AGRIS - International System for Agricultural Science and Technology; 1993.

ETYFish. 2022. https://etyfish.org/ETYFish_Somniosidae.pdf. (Downloaded on 16/11/2022).

European Union. Corrigendum to regulation (EC) no 853/2004 of the European Parliament and of the Council of 29 April 2004 laying down specific hygiene rules for food of animal origin. Off J Eur Union. 2004;226:22–82.

European Union. Commission Implementing Regulation (EU) 2019/627 of 15 March 2019. Laying down uniform practical arrangements for the performance of official controls on products of animal origin intended for human consumption in accordance with regulation (EU) 2017/625. Off J Eur Union 2019, 131, 51–100.

Fabricius O. Fauna Groenlandica, systematice sistens, animalia Groenlandiae occidentalis. Leipzig: Copenhagen; 1780.

Fadeev NS. Guide to biology and fisheries of fishes of the North Pacific Ocean. Vladivostok: TINRO-Center; 2005. 366 p.

FAO. *Hypophthalmichthys molitrix*. 2019. http://www.fao.org/fi/website/FISearchAction.do. Accessed 13 Sept 2019.

Faust M. Observation of sand-dwelling toxic dinoflagellates (Dinophyceae) from widely differing sites, including two new species. J Phycol. 1995;31:996–1003.

FDA (Food and Drug Administration). Scombrotoxin (histamine) formation. In: Fish and fishery products hazards and controls guidance. 3rd ed. Washington, DC: Center for Food Safety and Applied Nutrition,Office of Seafood; 2001. p. 83–102.

Feldman KA, Werner SB, Cronan S, Hernandez M, Horvath AR, Lea CS, Au AM, Vugia DJ. A large outbreak of scombroid fish poisoning associated with eating escolar fish (Lepidocybium flavobrunneum). Epidemiol Infect. 2005;133:29–33.

Finucci B. Ecology of New Zealand deep-sea Chondrichthyans. Victoria University of Wellington: Doctor of Philosophy in Marine Biology; 2017. 297 p.

Finucci B, Kyne PM. *Centroscymnus owstonii*. The IUCN red list of threatened species 2018: e.T41749A68615392. 2018. https://doi.org/10.2305/IUCN.UK.2018-2.RLTS.T41749A68615392.en

Finucci B, Barnett A, Bineesh KK, Cheok J, Cotton CF, Dharmadi, Graham KJ, Kulka DW, Neat FC, Pacoureau N, Rigby CL, Tanaka S, Walker TI. *Hexanchus griseus*. The IUCN red list of threatened species 2020: e.T10030A495630. 2020a. https://doi.org/10.2305/IUCN.UK.2020-3.RLTS.T10030A495630.en

Finucci B, Cheok J, Cotton CF, Kulka DW, Neat FC, Pacoureau N, Rigby CL, Tanaka S, Walker TI. *Centroselachus crepidater*. The IUCN red list of threatened species 2020: e.T46864A68615502. 2020b. https://doi.org/10.2305/IUCN.UK.2020-3.RLTS.T46864A68615502.en

Fischer W and Bianchi G (1984) Volume IV Bony fishes families: Scatophagidae to Trichiuridae. In FAO species identification sheets for fishery purposes—Western Indian Ocean (fishing area 51). Rome: Food and Agriculture Organization of the United Nations.

Fischer W, Sousa I, Silva C, de Freitas A, Poutiers JM, Schneider W, Borges TC, Feral JP, Massinga A. Fichas FAO de identificaçao de espécies para actividades de pesca. Guia de campo das espécies comerciais marinhas e de águas salobras de Moçambique. Publicaçao preparada em collaboraçao com o Instituto de Investigaçao Pesquiera de Moçambique, com financiamento do Projecto PNUD/FAO MOZ/86/030 e de NORAD. Roma: FAO; 1990. 424 p.

Fish CJ, Cobb MC. Noxious marine animals of the central and western Pacific Ocean. US Fish Wildl Res Rep. 1954;36:iii + 45.

Fitch JE. Toxicity and taxonomic notes on the squaretail. Tetragonurus cuvieri Calif Fish Game. 1952;38:251–2.

Fleming LE, Blythe DG, Baden DG. Ciguatera fish poisoning. Shoreman's Travel Medicine Monthly. 1997;1:1–5.

Fleming LE, Baden DG, Bean JA, Weisman R, Blythe DG. Seafood toxin diseases: issues in epidemiology and community outreach. In: Reguera B, Blanco J, Fernandez ML, Wyatt T, editors. Harmful Algae: Xunta de Galicia and Intergovernmental Oceanographic Commission of UNESCO; 1998. p. 245–8.

Flower SS. Further notes on the duration of life in animals. I. Fishes: as determined by otolith and scale-readings and direct observations on living individuals. Proc Zool Soc London. 1935;2:265–304.

Forster G. Voyage round the world, in his Britanic Majesty's sloop, resolution, commanded by Captain James Cook, during the years 1772–1775, 2 vols. London: G. Robinson; 1777.

Francii DG. Ova barbi comesta noxia. Acta Acad Nat Curios 1683;1:61–3.

Freyhof J. *Salmo trutta*. The IUCN red list of threatened species 2013: e.T19861A9050312. 2013. https://doi.org/10.2305/IUCN.UK.2008.RLTS.T19861A9050312.en

Freyhof J, Kottelat M. *Cyprinus carpio*. The IUCN red list of threatened species 2008: e.T6181A12559362. 2008a. https://doi.org/10.2305/IUCN.UK.2008.RLTS.T6181A12559362.en

Freyhof J, Kottelat M. *Perca fluviatilis*. The IUCN red list of threatened species 2008: e.T16580A6135168. 2008b. https://doi.org/10.2305/IUCN.UK.2008.RLTS.T16580A6135168.en

Freyhof J, Kottelat M. *Tinca tinca*. The IUCN red list of threatened species 2008: e.T21912A9339248. 2008c. https://doi.org/10.2305/IUCN.UK.2008.RLTS.T21912A9339248.en

Fricke R, Kulbicki M, Wantiez L. Checklist of the fishes of New Caledonia, and their distribution in the Southwest Pacific Ocean (Pisces). Stuttgarter Beiträge zur Naturkunde A, Neue Serie. 2011;4:341–463.

Friedman MA, Fleming LE, Fernandez M, et al. Ciguatera fish poisoning: treatment, prevention and management. Mar Drugs. 2008;6:456–79.

Friedman MA, Fernandez M, Backer LC, Dickey RW, Bernstein J, Schrank K, Kibler S, Stephan W, Gribble MO, Bienfang P, et al. An updated review of ciguatera fish poisoning: clinical, epidemiological, environmental, and public health management. Mar Drugs. 2017;15:72.

Frimodt C. Multilingual illustrated guide to the world's commercial Coldwater fish. Osney Mead, Oxford: Fishing News Books; 1995. 215 p.

Froese R. The science in Fishbase. In: Christensen V, Maclean J, editors. Ecosystem approaches to fisheries: a global perspective. Cambridge: Cambridge University Press; 2011. p. 47–54.

Froese R, Pauly D. Editors. FishBase. World Wide Web electronic publication. 2016. www.fishbase.org, version (06/2016).

Froese R, Pauly D, editors. FishBase. World Wide Web electronic publication; 2022. www.fishbase.org, version (011/2022).

Froese R, Pauly D, editors. FishBase. World Wide Web electronic publication; 2023. www.fishbase.org, version (01/2023).

Fu M, Koulman A, van Rijssel M, Lützen A, de Boer MK, Tyl MR, Liebezeit G. Chemical characterization of three hemolytic compounds from the microalgal species *Fibrocapsa japonica* (Raphidophyceae). Toxicon. 2004;43:355–63.

Fuchi Y, Narimatsa H, Nakama S, Kotobuki H, Hirakawa H, Torishima Y, Noguchi T, Ohtomo N. Tissue distribution of toxicity in a puffer fish, Arothron firmamentum ("hoshifugu"). J Food Hyg Soc Jpn. 1991;32:520–4.

Fukuda T. Violent increase of cases of puffer poisoning (in Japanese). Clin Stud. 1951;29(2):1762.

Fukuda A, Tani A. Records of Puffer poisonings. Nippon Igaku Oyobi Kenko Hoken. 1941;3528:7–13.

Gaillard C. Recherches sur les poissons représentés dans quelques tombeaux Egyptiens De l'ancien Empire. Mém Inst Français Arch Orient. 1923;51:97–100.

Gaughan D, Munroe TA, Hata H, Mohd Arshaad W. *Sprattus antipodum*. The IUCN red list of threatened species 2017: e.T98839225A98886993. 2017a. https://doi.org/10.2305/IUCN.UK.2017-3.RLTS.T98839225A98886993.en

Gaughan D, Munroe TA, Hata H, Mohd Arshaad W. *Sprattus muelleri*. The IUCN red list of threatened species 2017: e.T98839582A98887018. 2017b. https://doi.org/10.2305/IUCN.UK.2017-3.RLTS.T98839582A98887018.en

Gaughan D, Di Dario F, Hata H. Sardinops sagax (errata version published in 2019). The IUCN red list of threatened species 2018: e.T183347A143831586. 2018. https://doi.org/10.2305/IUCN.UK.2018-2.RLTS.T183347A143831586.en

Geiger E. Histamine content of unprocessed and canned fish. A tentative method of quantitative determination of spoilage. Food Res. 1944;9:293–7.

Gerssen A, Bovee THF, Klijnstra MD, Poelman M, Portier L, Hoogenboom RLAP. First report on the occurrence of tetrodotoxins in bivalve mollusks in The Netherlands. Toxins. 2018;10:450.

Gilbert RJ, Hoeas G, Murnwy CK, Cruickshank JG, Young SEJ. Scombrotoxic fish poisoning: features of the first SO incidents to be reported in Britain (1976–79). Br Med J 1980;281:71.

Gillespie NC. Possible origins of ciguatera. In: Covacevich J, Davie P, Pearn J, editors. Toxic plants and animals, a guide for Australia. Brisbane: Queensland Museum; 1987. p. 171–9.

Gillespie NC, Lewis RJ, Pearn JH, Bourke AT, Holmes MJ, Bourke JB, Shields WJ, Ciguatera in Australia. Occurrence, clinical features, pathophysiology, and management. Med J Aust. 1986;145:584–90.

Givney RC. Illness associated with rudderfish/ escolar in South Australia Communicable Disease Intelligence. 2002;26:440.

Glibert PM, Anderson DM, Gentien P, Graneli E, Sellner KG. The global, complex phenomena of harmful algal blooms. Oceanography. 2005;18(2):136–47.

Goh L. Importers can use fast test for oilfish, say researchers The South China Morning Post 7 Feb 2007.

Gómez F, Quit D, Lopes RM, Lin S. Fukuyoa paulensis gen. et sp. nov., a new genus for the globular species of the dinoflagellate Gambierdiscus (Dinophyceae). PLoS One. 2015;10:e0119676.

Gorreta F, Bernasconi R, Galliani G, Salmona M, Tacconi MT, Bianchi R. Wax esters of o-3 polyunsaturated fatty acids: A new formulation as a potential food supplement—digestion and absorption in rats. Lebensm Wiss Technol. 2002;35:458–65.

Grant WS, Clark AM, Bowen BW. Why restriction fragment length polymorphism analysis of mitochondrial DNA failed to resolve sardine (*Sardinops*) biogeography: insights from mitochondrial DNA cytochrome b sequences. Can J Fish Aquat Sci. 1998;55(12):2539–47.

Gregory J. Outbreaks of diarrhoea associated with butterfish in Victoria. Communicable Disease Intelligence. 2002;26:439–44.

Guardone L, Maneschi A, Meucci V, Gasperetti L, Nucera D, Armani A. A global retrospective study on human cases of Tetrodotoxin (TTX) poisoning after seafood consumption. Food Rev Int. 2020;36:645–67.

Gudger EW. A new purgative, the oil of the "castor oil fish,". Ruvettus Boston Med Surg J. 1925;192:107–11.

Gudger EW, Mowbray LL. The oilfish, Ruvettus pretiosus, at Bermuda. Science. 1927;65:145–6.

Guss DA. Scombroid fish poisoning: successful treatment with cimetidine. Undersea Hyperb Med. 1998;25(2):123.

Hall, S., 1991. Interim revision to the PSP mouse bioassay protocol. United States Food & Drug Administration, Washington 4th January 1991.

Hallegraeff GM. Ocean climate change, phytoplankton community responses, and harmful algal blooms: a formidable predictive challenge. J Phycol. 2010;46:220–35.

Hallegraeff GM, Schweibold L, Jaffrezic E, Rhodes L, MacKenzie L, Hay B, Farrell H. Overview of Australian and New Zealand harmful algal species occurrences and their societal impacts in the period 1985 to 2018, including a compilation of historic records. Harmful Algae. 2021;102:101848.

Halstead BW. Poisonous fishes. Public Health Rep. 1958;73(4):302.

Halstead BW. Clin Pharmacol Ther. 1964;5:615–27.

Halstead BW. Poisonous and venomous marine animals of the world, invertebrates, vol. 1. Washington, DC: US Government Printing Office; 1965.

Halstead BW. Poisonous and venomous marine animals of the world, vertebrates, vol. 2. Washington, DC: U.S. Government Printing Office; 1967.

Halstead B. Poisonous and venomous marine animals of the world. Princeton: Darwin Press; 1978.

Halstead BW, editor. Poisonous and venomous marine animals of the world. 2nd ed. Princeton: Darwin Press; 1988.

Halstead BW, Bunker NC. A survey of poisonous fish of Johnston Island. Zoologica. 1954;39:61–7.

Halstead BW, Carscallen LJ. Marine organisms. In: Altman PL, Dittmer DS, editors. Biology data Book. Washington, DC: Federal American Society of Experimental Biology; 1964. p. 336–43.

Halstead BW, Lively WM Jr. Poisonous fishes, and ichthyosarcotoxism. Their relationship to the armed forces. US Armed Forces Med J. 1954;5(2):157–75.

Halstead BW, Schall DW. Report on the poisonous fishes captured during the Woodrow G. Krieger expedition to the Galapagos Islands. In: Essays in the natural sciences in honor of Captain Allan Hancock. Los Angeles: University of Southern California Press; 1955. p. 147–72.

Halstead BW, Schall DW. A report on the poisonous fishes captured during the Woodrow G. Krieger expedition to Cocos Island; 1956.

Hamada J. The effect of tetrodotoxin crystals on the smooth and ganglionic active drugs in the isolated intestine and bronchial muscle (in Japanese, English summary). J Chiba Med Soc. 1960;36:1358–68.

Hamilton B, Hurbungs M, Jones A, Lewis RJ. Multiple ciguatoxins present in Indian Ocean reef fish. Toxicon. 2002a;40(9):1347–53.

Hamilton B, Hurbungs M, Vernoux JP, Jones A, Lewis RJ. Isolation and characterization of Indian Ocean ciguatoxin. Toxicon. 2002b;40(6):685–93.

Han C, Zhang X, Li L, Chen S, Yan Z, Gao X, Chang J. Analysis and evaluation of Tetrodotoxin in coastal aquatic products of Zhejiang Province. J Coast Res. 2018;83:380–5.

Hardy, G., Jing, L., Leis, J.L., Liu, M., Matsuura, K. & Shao, K. 2014. Lagocephalus lagocephalus. The IUCN red list of threatened species 2014: e.T190464A19929421. https://doi.org/10.2305/IUCN.UK.2014-3.RLTS.T190464A19929421.en. Accessed 15 May 2023.

Harrison IJ. Mugilidae. Lisas. In: Fischer W, Krupp F, Schneider W, Sommer C, Carpenter KE, Niem V, editors. Guia FAO para Identification de Especies para lo Fines de la Pesca. Pacifico Centro-Oriental. 3 Vols. Rome: FAO; 1995. p. 1293–8.

Harrison IJ. Mugilidae. In: Stiassny MLJ, Teugels GG, Hopkins CD, editors. The fresh and brackish water fishes of lower Guinea, west-Central Africa. Volume II. Collection Faune et Flore tropicales 42. Institut de Recherche pour le Développement, Paris, France, Muséum national d'Histoire Naturelle, Paris, France, and Musée royal de l'Afrique Centrale, Tervuren, Belgium; 2008. p. 450–71, 603 p.

Harrison IJ, Senou H. Order Mugiliformes. Mugilidae. Mullets. In: Carpenter KE, Niem VH, editors. FAO species identification guide for fishery purposes. The living marine resources of the Western Central Pacific. Volume 4. Bony fishes part 2 (Mugilidae to Carangidae). Rome: FAO; 1997. p. 2069–108.

Hart JL. Pacific fishes of Canada. Bull Fish Res Board Can. 1973;180:740 p.

Hassan-Williams C, Bonner TH, Thomas C (photogr.). Texas freshwater fishes. Texas State University-San Marcos: Biology Department/ Aquatic Station; 2007.

Helfman GS, Collette BB, Facey DE. The diversity of fishes. Malden: Blackwell Science; 1999.

Helfrich P. Fish poisoning in Hawaii. Hawaii Med J. 1963;22:361–72.

Helfrich P, Banner A. Hallucinatory mullet poisoning. J Trop Med Hyg. 1960;1(86–89):20.

Helfrich P, Banner AH. Experimental induction of ciguatera toxicity in fish through diet. Nature. 1963;197(4871):1025–6.

Hennemann RM. Sharks & rays: elasmobranch guide of the world. Frankfurt: IKAN-Unterwasserarchiv; 2001. 304 p.

Hidalgo J, Liberona J, Molgó J, Jaimovich E. Pacific ciguatoxin-1b effect over Na+ and K+ currents, inositol 1,4,5-triphosphate content and intracellular Ca2+ signals in cultured rat myotubes. Br J Pharmacol. 2002;137:1055–62.

Hinton. Horned shark, gar, mormyrid, characin, carp, armored catfish, arowana, upside down catfish. 1962.

Holmes M. Gambierdiscus yasumotoi sp. nov. (Dinophyceae), a toxic benthic dinoflagellate from Southeastern Asia. J Phycol. 1998;34:661–8.

Holland ND, Holland LZ, Holland PWH. Scenarios for the making of vertebrates. Nature. 2015;520:450–5.

Hort V, Arnich N, Guérin T, Lavison-Bompard G, Nicolas M. First detection of tetrodotoxin in bivalves and gastropods from the French mainland coasts. Toxins. 2020;12:599.

Huang HN, Lin J, Lin HL. Identification, and quantification of tetrodotoxin in the marine gastropod *Nassarius* by LC-MS. Toxicon. 2008;51:774–9.

Humann P. Reef fish identification: Florida, Caribbean. Bahamas: New World Publications, Jacksonville, Florida; 1994. 426 p.

Humann P, Deloach N. Reef fish identification. Galápagos. Florida: New World Publications, Inc.; 1993. 267 p.

Hungerford JM. Scombroid poisoning: a review. Toxicon. 2010;56(2):231–43.

Hwang DF, Noguchi T. Tetrodotoxin poisoning. Adv Food Nutr Res. 2007;52:142–236.

Hwang DF, Chang SH, Shiau CY, Chai T. High-performance liquid chromatographic determination of biogenic amines in fish implicated in food poisoning. J Chromatogr B. 1997;693:23–30.

Hwang DF, Yeh YH, Lai YS, Deng JF. Identification of cyprinol and cyprinol sulfate from grass carp bile and their toxic effects in rats. Toxicon. 2000;39:411–4.

Hwang CC, Lin CM, Huang CY, Huang YL, Kang FC, Hwang DF, et al. Chemical characterization, biogenic amines contents, and identification of fish species in cod and escolar steaks, and salted escolar roe products. Food Control. 2012;25:415–20.

Igarashi H. The pungent principles of fishes produced by decrease in freshness, part II. Bull Jpn Soc Sci Fish. 1939;8:158–60. In Annales: series Historia Naturalis (Vol. 21, no. 2, p. 167). Scientific and Research Center of the Republic of Slovenia.

IGFA. Database of IGFA angling records until 2001. Fort Lauderdale: IGFA; 2001.

International Association for Medical Assistance to Travellers (IAMAT). Available from: https://www.iamat.org/risks/ciguatera-fish-poisoning. Downloaded 23/11/2022.

Ishida H, Nozawa A, Nukaya H, Tsuji K. Comparative concentrations of brevetoxins PbTx-2, PbTx-3, BTX-B1, and BTX-B5 in cockle *Austrovenus stuchburyi*, Greenshell mussel, *Perna canaliculus*, and Pacific oyster, *Crassostera gigas*, involved in neurotoxic shellfish poisoning in New Zealand. Toxicon. 2004;43:779–89.

Ishihara F. The physiological study of puffer poison (in Japanese). Tokyo Med Assoc Manag. 1917;31:276–9.

Jacquet JL, Pauly D. Trade secrets: renaming and mislabeling of seafood. Mar Policy. 2008;32:309–18.

Jal S, Khora SS. An overview on the origin and production of tetrodotoxin, a potent neurotoxin. J Appl Microbiol. 2015;119:907–16.

Jawad LA. Fish species of the order Cypriniformes as a source of Ichthyootoxin and Ichthyogallotoxin in Iraq: cases reports. Tigris and Euphrates Rivers: Their Environment from Headwaters to Mouth; 2021. p. 1493–8.

Jensen AS. Mindeskr Japetus Steenstrups Fods. 1914;30:12–6.

Jordan DS, Evermann BW, Tanaka S. Notes on new or rare fishes from Hawaii. Proc Calif Acad Sci. 1927;16(20):649–80.

Joubert P. In vitro investigation of the negative chronotropic effect of cholic acid in the rat. Clin Exp Pharmacol Physiol. 1978;5:1.

Jütter F. Liberation of 5,8,11,14,17-eicosapentaenoic acid and other polyunsaturated fatty acids from lipids as a grazer defense reaction in epilithicdiatom biofilms. J Phycol. 2001;37:744–55.

Kailola PJ, Williams MJ, Stewart PC, Reichelt RE, McNee A, Grieve C. Australian fisheries resources. Canberra: Bureau of Resource Sciences; 1993. 422 p.

Kan K, Ushiyama H, Shindo T, Uehara S, Yasuda K. Outbreak of histamine poisoning due to ingestion of fish (Lepidocybium flavobrunneum). J Food Hyg Soc Jpn (Shokuhin Eiseigaku Zasshi). 2000;41:116–21.

Kangur M, Paaver T, Drevs T. Salmon, *Salmo salar* L. In: Ojaveer E, Pihu E, Saat T, editors. Fishes of Estonia. Tallinn: Estonian Academy Publishers; 2003. p. 91, 416 p–7.

Katikou P. Public health risks associated with tetrodotoxin and its analogues in European waters: recent advances after the EFSA scientific opinion. Toxins. 2019;11(5):240.

Katikou P, Vlamis A. Tetrodotoxins: recent advances in analysis methods and prevalence in European waters. Curr Opin Food Sci. 2017;18:1–6.

Katikou P, Georgantelis D, Sinouris N, Petsi A, Fotaras T. First report on toxicity assessment of the Lessepsian migrant pufferfish *Lagocephalus sceleratus* (Gmelin, 1789) from European waters (Aegean Sea, Greece). Toxicon. 2009;54:50–5.

Katikou P, Gokbulut C, Kosker AR, Campàs M, Ozogul F. An updated review of tetrodotoxin and its peculiarities. Mar Drugs. 2022;20(1):47.

Katzung BG. Histamine, serotonin, and ergot alkaloids. In: Katsung BG, editor. Basic and clinical pharmacology. New York: McGraw Hill; 2007. p. 255–64.

Kawataba T. Problems involved in the research on fish and shellfish poisonings. Jpn J Med Sci Biol. 1962;15:141–3.

Kawataba T, Ishizaka K, Miura T. Studies on the food poisoning associated with putrefaction of marine products I outbreaks of allergy-like food poisoning caused by "sama sakuraboshi" (dried seasoned saury) and canned seasoned mackerel (In Japanese, English summary). Bull Jpn Soc Sci Fish. 1955;21:335–40.

Keith P, Allardi J. Atlas des poissons d'eau douce de France. Muséum national d'Histoire naturelle, Paris. Patrimoines naturels. 2001;47:1–387.

Kiernan MC, Isbister GK, Lin CS, Burke D, Bostock H. Acute tetrodotoxin-induced neurotoxicity after ingestion of puffer fish. Ann Neurol. 2005;57(3):339–48.

Kim R. Flushing syndrome due to mahimahi (scombroid fish) poisoning. Arch Dermatol. 1979;115(8):963–5.

Kiraly SJ, Moore JA, Jasinski PH. Deepwater and other sharks of the U.S. Atlantic Ocean exclusive economic zone. Mar Fish Rev. 2003;65(4):1–64.

Knox G. Poisonous fishes and methods of preventing cases of poisoning by them. Voen Med Zh. 1888;161(399):464.

Knudsen S. *Kyphosus vaigiensis*. The IUCN red list of threatened species 2015: e.T46086190A46664689. 2015. https://doi.org/10.2305/IUCN.UK.2015-4.RLTS.T46086190A46664689.en.

Knudsen SW, Clements KD. Revision of the fish family Kyphosidae (Teleostei: Perciformes). Zootaxa. 2013;3751(1):001–101.

Knutsen HK, Alexander J, Barregård L, Bignami M, Brüschweiler B, Ceccatelli S, Cottrill B, Dinovi M, Edler L, Grasl-Kraupp B, et al. Risks for public health related to the presence of tetrodotoxin (TTX) and TTX analogues in marine bivalves and gastropods. EFSA J. 2017;15:e04752.

Kodama AM, Hokama Y, Yasumoto T, Fukui M, Manea SJ, Sutherland N. Clinical and laboratory findings implicating palytoxin as the cause of ciguatera poisoning due to Decapterus macrosoma (mackerel). Toxicon. 1989;27(9):1051–3.

Kodama M, Sato S, Ogata T. *Alexandrium tamarense* as a source of tetrodotoxin in the scallop *Patinopecten yessoensis*. In toxic phytoplankton blooms in the sea; Smayda, T.J., Shimizu, Y., Eds.; Elsevier B.V.: Amsterdam. The Netherlands. 1993;3:401–6.

Kodama M, Sato S, Sakamoto S, Ogata T. Occurrence of tetrodotoxin in *Alexandrium tamarense*, a causative dinoflagellate of paralytic shellfish poisoning. Toxicon. 1996;34:1101–5.

Kosker AR, Özogul F, Ayas D, Durmus M, Ucar Y, Regenstein JM, Özogul Y. Tetrodotoxin levels of three pufferfish species (*Lagocephalus* sp.) caught in the north-eastern Mediterranean Sea. Chemosphere. 2019;219:95–9.

Kossel A. Yber die basischen Stoffe des Zellkerns. S.-B. Kgl. preuss. Akad Wiss. 1896;18:403.

Kottelat M. Fishes of Laos. Colombo: WHT Publications Ltd; 2001. 198 p.

Kottelat M, Freyhof J. Handbook of European freshwater fishes. Cornol and Freyhof, Berlin: Publications Kottelat; 2007. 646 p.

Kottelat M, Whitten AJ, Kartikasari SN, Wirjoatmodjo S. Freshwater fishes of Western Indonesia and Sulawesi = Ikan air tawar Indonesia Bagian Barat dan Sulawesi. Hong Kong: Periplus Editions; 1993. 293 p.

Kounis NG. Coronary hypersensitivity disorder: the Kounis syndrome. Clin Ther. 2013;35(5):563–71.

Kuiter RH, Tonozuka T. Pictorial guide to Indonesian reef fishes. Part 3. Jawfishes - sunfishes, Opistognathidae - Molidae. Zoonetics; 2001. p. 623–893.

Kusche J, Bieganski T, Hesterberg R. The influence of carcinoma growth on diamine oxidase activity in human gastrointestinal tract. Agents Actions. 1980;10:110–3.

Lacono JM, Dougherty RM. Effects of polyunsaturated fats on blood pressure. Ann Rev Nutr. 1993;13:243–60.
Lago J, Rodriguez LP, Blanco L, Vieites JM, Cabado AG. Tetrodotoxin, an extremely potent marine neurotoxin: distribution, toxicity, origin and therapeutical uses. Mar Drugs. 2015;13:6384–406.
Lange WR, Lipkin KM, Yang GC. Can ciguatera be a sexually transmitted disease? J Toxicol Clin Toxicol. 1989;27:193–7.
Lange WR, Snyder FR, Fudala PJ. Travel and ciguatera fish poisoning. Arch Intern Med. 1992;152:2049–53.
Lartigue J, Jester E, Dickey R, Villareal T. Nitrogen source effects on the growth and toxicity of two strains of the ciguatera-causing dinoflagellate Gambierdiscus toxicus. Harmful Algae. 2009;8:781–91.
Last PR, Stevens JD. Sharks and rays of Australia, vol. 1994. Australia: CSIRO; 1994. 513 p.
Laurent D, Kerbrat AS, Darius HT, Girard E, Golubic S, Benoit E, Sauviat MP, Chinain M, Molgo J, Pauillac S. Are cyanobacteria involved in Ciguatera fish poisoning-like outbreaks in New Caledonia? Harmful Algae. 2008;7(6):827–38.
Leão JM, Lozano-Leon A, Giráldez J, Vilariño Ó, Gago-Martínez A. Preliminary results on the evaluation of the occurrence of tetrodotoxin associated to marine Vibrio spp. in bivalves from the Galician rias (northwest of Spain). Mar Drugs. 2018;16:81.
Leask A, Yankos P, Ferson MJ. Fish, so foul! Foodborne illness caused by combined fish histamine and wax ester poisoning. Commun Dis Intell. 2004;28:83–5.
Lebeau A. La ciguatera dans l'Oce'an Indien: e'tude des poissons ve'ne'neux des bancs de l'archipel des Mascareignes et de la cre'te centrale de l'Oce'an Indien. Rev Trav Inst Pe^ches Marit. 1979;42(4):325–45.
Lee RF, Hagen W, Kattner G. Lipid storage in marine zooplankton. Mar Ecol Prog Ser. 2006;307:273–306.
Legrand A, Fukui M, Cruchet P, Ishibashi Y, Yasumoto T. Characterization of ciguatoxins from different fish species and wild Gambierdiscus toxicus. In: Tosteson TR, editor. Proceeding of the 3rd international conference on Ciguatera, Puerto Rico, 1992. Quebec: Polysciences Publications; 1992. p. 25–32.
Legroux R, Levaditi JC, Boudin G, Bovet D. Pre'sence d'histamine dans la chair d'un thon responsible d'une intoxication collective. Ann Inst Pasteur. 1947;73:101–4.
Lehane L, Olley J. Histamine fish poisoning revisited. Int J Food Microbiol. 2000;58(1–2):1–37.
Lerke PA, Werner SB, Taylor SL, Guthertz LS, Scombroid poisoning. Report of an outbreak. West J Med. 1978;129:381–6.
Leung KSY, Fong BMW, Tsoi YK. Analytical challenges: determination of tetrodotoxin in human urine and plasma by LC-MS/MS. Mar Drugs. 2011;9(11):2291–303.
Levenson DJ, Shorecki KL, Newell GC, et al. Acute renal failure associated with hepatobiliary diseases. In: Brenner BM, Nazarus JM, editors. Acute renal failure. 2nd ed. London: Churchill Livingstone; 1988. p. 535–80.
Lewis R. Ciguatera management. SPC Live Reef Fish Inf Bull. 2000;7:11–3.
Lewis RJ. The changing face of ciguatera. Toxicon. 2001;39:97–106.
Lewis R, King G. In: Williamson JA, Fenner PJ, Burnett JW, Rifkin JF, editors. Ciguatera (fish poisoning). Sydney: University of New South Wales Press; 1996.
Li JL. Taiwanese Fish Expert's first lecture to seafood. Taipei: IF Publishing; 2007.
Lieske E, Myers R. Collins Pocket Guide. Coral reef fishes. Indo-Pacific & Caribbean including the Red Sea: Haper Collins Publishers; 1994. 400 p.
Lin YF, Lin SH. Simultaneous acute renal and hepatic failure after ingesting raw carp gall bladder. Nephrol Dial Transplant. 1999;1999(14):2011–2.
Lin YF, Shih HL. Simultaneous acute renal and hepatic failure after ingesting raw carp gall bladder. Nephrol Dial Transplant. 1999;14:2011–2.
Lin CT, Huang PC, Yen TS. Partial purification and some characteristic nature of a toxic fraction of the grass carp bile. J Chin Biochem Soc. 1977;1977(6):1–5.
Ling KH, Cheung CW, Cheng SW, Cheng L, Li SL, Nichols PD, Ward RD, Graham A, But PPH. Rapid detection of oilfish and escolar in fish steaks: a tool to prevent keriorrhea episodes. Food Chem. 2008a;110:538–46.
Ling KH, Cheung CW, Cheng SW, Cheng L, Li SL, Nichols PD, Ward RD, Graham A, But PPH. Erratum to "rapid detection of oilfish and escolar in fish steaks: A tool to prevent keriorrhea episodes". Food Chem. 2008b;111:795.
Ling HK, Nichols PD, But PP-H. Fish-induced Keriorrhea. In: Advances in Food and nutrition research. Volume 57, edited by Taylor, S. Academic Press is an imprint of Elsevier 30 corporate drive, suite 400, Burlington, MA 01803, USA; 2009.
Litaker RW, Vandersea MW, Faust MA, Kibler SR, Chinain M, Holmes MJ, Holland WC, Tester PA. Taxonomy of Gambierdiscus including four new species, Gambierdiscus caribaeus, Gambierdiscus carolinianus, Gambierdiscus carpenteri and Gambierdiscus ruetzleri (Gonyaulacales, Dinophyceae). Phycologia. 2009;48:344–90.
Llewellyn LE. Revisiting the association between sea surface temperature and the epidemiology of fish poisoning in the South Pacific: reassessing the link between ciguatera and climate change. Toxicon. 2009;56:691. https://doi.org/10.1016/j.toxicon.2009.08.011.79.
Lowe RT. A synopsis of the fishes of Madeira. Trans Zool Soc London. 1841;2:180–1.
Lukton A, Olcott HS. Content of free imidazole compounds in the muscle tissue of aquatic animals. Food Res. 1958;23:611–8.
Lum K. Safety of fish and seafood products. In: Oyarzabal OA, Backert S, editors. Microbial food safety. New York: Springer; 2012. p. 159–72.
Lyons DE, Beery JT, Lyons SA, Taylor SL. Cadaverine and aminoguanidine potentiate the uptake of histamine in vitro in perfused intestinal segments of rats. Toxicol Appl Pharmacol. 1983;70:445–58.

Machacek, H. (ed.), 2007. World records freshwater fishing. wwwfishing-worldrecordscom, November 2007.

Macht DI, Barba-Gose J. Pharmacology of Ruvettus pretiosus, or "castor-oil fish". Proc Soc Exp Biol Med. 1931a;28:772–4.

Macht DI, Barba-Gose J. Two new methods for pharmacological comparison of insoluble purgatives. J Am Pharmacol Assoc. 1931b;20:556–64.

Macht DI, Spencer EC. Physiological and toxicological effects of some fish muscle extracts. Proc Soc Exp Biol Med. 1941;46(2):228–33.

MacKenzie LA, Smith KF, Rhodes LL, Brown A, Langi V, Edgar M, Lovell G, Preece M. Mortalities of sea-cage salmon (Oncorhynchus tshawytscha) due to a bloom of Pseudochattonella verruculosa (Dictyochophyceae) in Queen Charlotte Sound, New Zealand. Harmful Algae. 2011;2011(11):45–53.

Magarlamov TY, Melnikova DI, Chernyshev AV. Tetrodotoxin-producing bacteria: detection, distribution and migration of the toxin in aquatic systems. Toxins. 2017;9:166.

Maintz L, Novak N. Histamine and histamine intolerance. Am J Clin Nutr. 2007;85(5):1185–96.

Maitland PS, Campbell RN. Freshwater fishes of the British Isles. London: HarperCollins Publishers; 1992. 368 p.

Mancini I, Defant A, Mesaricˇ T, Potocˇnik F, Batista U, Guella G, Turk T, Sepčić K. Fatty acid composition of common barbel (*Barbus barbus*) roe and evaluation of its haemolytic and cytotoxic activities. Toxicon. 2011;57(7):1017–22.

Manilo LG, Bogorodsky SV. Taxonomic composition, diversity and distribution of coastal fishes of the Arabian Sea. J Ichthyol. 2003;43(1):S75.

Maracacci A. Sur le pouvoir toxique du sang de thon. Arch Ital Biol. 1891;16:1.

Maralit BA, Aguila RD, Ventolero MFH, Perez SWL, Willette DA, Santo MD. Detection of mislabeled commercial fishery by-products in The Philippines using DNA barcodes and its implications to food traceability and safety. Food Control. 2013;33:119–25.

Markov S. Seefischvergiftung Wien Med Woschr. 1943;93(26–27):388.

Martin W, Banner AH. The effect of poisonous fish upon intestinal parasites. Trans Am Microsc Soc. 1958;77(3):304–6.

Masuda H, Amaoka K, Araga C, Uyeno T, Yoshino T. The fishes of the Japanese archipelago, vol. 1. Tokyo: Tokai University Press; 1984. 437 p. (text).

Matarese AC, Kendall AW, Blood DM, Vinter MV. Laboratory guide to early life history stages of Northeast Pacific fishes. NOAA Tech Rep NMFS. 1989;80:1–652.

Matsuura K, Tyler JC. Tetraodontiform fishes, mostly from deep waters, of New Caledonia. In: Séret B, editor. Résultats des Campagnes MUSORSTOM; 1997. p. 173–208. vol. 17. no.9. Mem. Mus. Natl. Hist. Nat. 174.

Mattei C, Dechraoui M, Molgó J, Meunier F, Legrand A, Benoit E. Neurotoxins targeting receptor site 5 of voltage-dependent sodium channels increase the nodal volume of myelinated axons. J Neurosci Res. 1999;55:666–73.

May JL, Maxwell JGH. Trawl fish from temperate waters of Australia. Tasmania: CSIRO Division of Fisheries Research; 1986. 492 p.

McCosker JE, Randall JE. Notes on the snake eels of the genera Apterichtus and Ichthyapus (Anguilliformes: Ophichthidae) of the centeral and south Pacific, with the description of a new species. Zootaxa. 2005;800:1–11.

McMillan PJ, Griggs LH, Francis MP, Marriott PJ, Paul LJ, Mackay E, Wood BA, Sui H, Wei F. New Zealand fishes. Volume 3: A field guide to common species caught by surface fishing. New Zealand Aquatic Environment and Biodiversity Report No. 69. 2011; 145 p.

McNabb P, Rhodes L, Selwood A. Results of analyses for brevetoxins and pinnatoxins in Rangaunu harbour oysters, 1993–2008. Cawthron Report No: Prepared for NZ Food Safety Authority; 2008. p. 1453.

McNabb P, Selwood AI, Munday R, Wood SA, Taylor DI, MacKenzie LA, van Ginkel R, Rhodes LL, Cornelisen C, Heasman K, et al. Detection of tetrodotoxin from the grey side-gilled sea slug—*Pleurobranchaea maculata*, and associated dog neurotoxicosis on beaches adjacent to the Hauraki Gulf, Auckland. New Zealand Toxicon. 2010;56:466–73.

McNabb PS, Taylor DI, Ogilvie SC, Wilkinson L, Anderson A, Hamon D, Wood SA, Peake BM. First detection of tetrodotoxin in the bivalve *Paphies australis* by liquid chromatography coupled to triple quadrupole mass spectrometry with and without precolumn reaction. J AOAC Int. 2014;97:325–33.

Mekhjian HS, Phillips SF. Perfusion of the canine colon with unconjugated bile acids. Effect on water and electrolyte transport, morphology and bile acid absorption. Gastroenterology. 1970;59:120–9.

Melnikova DI, Vlasenko AE, Magarlamov TY. Stable tetrodotoxin production by *Bacillus* sp. Strain 1839. Mar Drugs. 2019;17:704.

Melton RJ, Randall JE, Fusetani N, Weiner RS, Couch RD, Sims JK. Fatal sardine poisoning. Hawaii Med J. 1984;43(4):114–20.

Meyer FD. On the poisonous quality of a species of West Indian fish. Phila Med Phys J. 1805;1:43–8.

Meyer-Ahrens KM. Von den giftigen fischen. Schweiz Z Med Chir Geburtsh. 1855;3:188–230.

Michailidis, N. Study on the lessepsian migrant Lagocephalus sceleratus in Cyprus. In Proceedings of the EastMed, 2010. Report of the Sub-Regional Technical Meeting on the Lessepsian Migration and Its Impact on Eastern Mediterranean Fishery, Nicosia, Cyprus, 7–9 December 2010; GCP/INT/041/EC-GRE-ITA/TD-04; FAO: Athens, Greece, 2010; pp. 74–87. Available online: http://www.faoeastmed.org/pdf/publications/EastMed_TD04.pdf. Accessed 30 Apr 2023.

Mills AR. Poisonous fish in the South Pacific. J Trop Med Hyg. 1956;59:99–103.

Ministry of Primary Industries (MPI). ESR. Risk Profile: Ciguatoxins in seafood. New Zealand Food Safety Technical Paper No. 2019/06. Available from: https://www.mpi.govt.nz/dmsdocument/37859/direct

Mok D. Wellcome sold sushi and sashimi made from oilfish. The South Morning Post 2 Feb 2007.

Molgó J, Comella J, Legrand A. Ciguatoxin enhances quantal transmitter release from frog motor nerve terminals. Br J Pharmacol. 1990;99:695.

Molgó J, Shimahara T, Legrand A. Ciguatoxin, extracted from poisonous morays eels, causes sodium-dependent calcium mobilization in NG108-15 neuroblastoma x glioma hybrid cells. Neurosci Lett. 1993;158:147–50.

Mori M, Saito T, Nakanishi Y, Miyazawa K, Hashimoto Y. The composition and toxicity of wax in the flesh of castor oil fishes. Bull Jpn Soc Sci Fisheries. 1966;32:137–45.

Morrow JD, Margolies GR, Rowland J, Roberts LJ. Evidence that histamine is the causative toxin of scombroid-fish poisoning. N Engl J Med. 1991;324(11):716–20.

Mosher H, Fuhrman FA, Buchwald HD, Fischer HG. Tarichatoxin-tetrodotoxin: A potent neurotoxin. Science. 1964;144:1100–10.

Motil KJ, Scrimshaw NS. The role of exogenous histamine in scombroid poisoning. Toxicol Lett. 1979;3:219–23.

Munday R, Murray S, Rhodes LL, Larsson ME, Harwood DT. Ciguatoxins and maitotoxins in extracts of sixteen Gambierdiscus isolates and one Fukuyoa isolate from the South Pacific and their toxicity to mice by intraperitoneal and oral administration. Mar Drugs. 2017;15(7):208.

Mundy BC. Checklist of the fishes of the Hawaiian Archipelago. Bish Mus Bull Zool. 2005;6:1–704.

Munir S, Siddiqui PJA, Morton SL. The occurrence of the ciguatera fish poisoning producing dinoflagellate genus Gambierdiscus in Pakistan waters. Algae. 2011;26(4):317–25.

Murata M, Legrand A, Ishibashi Y, Fukui M, Yasumoto T. Structures and configurations of ciguatoxin from the moray eel Gymnothorax javanicus and its likely precursor from the dinoflagellate Gambierdiscus toxicus. J Am Chem Soc. 1990;112:4380–6.

Murata M, Naoki H, Iwashita T, Matsunaga S, Sasaki M, Yokoyama A, Yasumoto T. Structure of maitotoxin. J Am Chem Soc. 1993;115:2060–2.

Murdy EO, Birdsong RS, Musick JA. Fishes of Chesapeake Bay. Smithsonian Institution Press Washington and London; 1997. 324 p.

Murray JS, Harwood DT, Rhodes LL. Ciguatera fish poisoning event in New Zealand from imported tropical reef fish and confirmation of Pacific ciguatoxins by LC-MS/MS. Harmful Algae News. 2020;66:4–5.

Murray JS, Harwood DT, Rhodes L. Ciguatera poisoning and confirmation of ciguatoxins in fish imported into New Zealand. The New Zealand Medical Journal (Online). 2021;134(1536):100–4.

Muus BJ, Dahlström P. Guide des poissons d'eau douce et pêche. Copenhague: GEC Gads Forlag; 1967. 242 p.

Muus BJ, Dahlström P. Süßwasserfische. München: BLV Verlagsgesellschaft; 1968. 224 p.

Muus BJ, Nielsen JG. Die Meeresfische Europas in Nordsee, Ostsee und Atlantik. Stuttgart. 1999;Kosmos-Naturführer:336 p.

Myers RF. Micronesian reef fishes. 2nd ed. Barrigada: Coral Graphics; 1991. 298 p.

Nader, M.; Indary, S.; Boustany, L. FAO EastMed the Puffer Fish *Lagocephalus sceleratus* (Gmelin, 1789) in the Eastern Mediterranean. GCP/INT/041/EC–GRE –ITA/TD-10. Available online.: http://www.fao.org/3/ap967e/ap967e.pdf. Accessed 30 Nov 2021.

Nakamura I. Gempylidae. In: Fischer W, Bianchi G, editors. FAO species identification sheets for fishery purposes. Western Indian Ocean fishing area 51; 1984. Vol. II.

Nakamura I. Gempylidae. Escolares. In: Fischer W, Krupp F, Schneider W, Sommer C, Carpenter KE, Niem V, editors. Guia FAO para Identification de Especies para lo Fines de la Pesca. Pacifico Centro-Oriental. 3 Vols. Rome: FAO; 1995. p. 1106–13.

Nakamura I, Parin NV. FAO species catalogue. Vol. 15. Snake mackerels and cutlassfishes of the world (families Gempylidae and Trichiuridae). An annotated and illustrated catalogue of the snake mackerels, snoeks, escolars, gemfishes, sackfishes, domine, oilfish, cutlassfishes, scabbardfishes, hairtails, and frostfishes known to date. FAO Fish Synop. 1993;125(15):136 p.

Nakamura I, Parin NV. Gempylidae. Snake mackerels. In: Carpenter KE, Niem V, editors. FAO species identification guide for fishery purposes. The living marine resources of the Western Central Pacific. Vol. 6. Bony fishes part 4 (Labridae to Latimeriidae), estuarine crocodiles. Rome: FAO; 2001. p. 3698–708.

New York Department of Environmental Conservation (NYDEC), 2019. Larger unusual fish species that anglers may encounter. https://www.dec.ny.gov/animals/7014.html. Accessed 18 May 2023.

Ng HL, Kum KP. Common ichthyogallotoxic fishes in China. Donwuxue Zazhi. 1977;3:28–9.

Nichols JT, Bartsch P. Fish and shells of the Pacific world. New York: Macmillan Co.; 1945. p. 201.

Nichols PD, Mooney BD, Elliott NG. Unusually high levels of nonsaponifiable lipids in fishes escolar and rudderfish: identification by gas and thin-layer chromatography. J Chromatogr A. 2001;936:183–91.

Nico LG, Fuller P, Li J. *Hypophthalmichthys molitrix* (Valenciennes in Cuvier and Valenciennes, 1844): U.S. Geological survey, nonindigenous aquatic species database, Gainesville. 2020. https://nas.er.usgs.gov/queries/factsheet.aspx?speciesID=549. Revision Date: 12/3/2019, Peer Review Date: 4/1/2016, Access Date: 2/1/2020.

Noguchi T, Arakawa O. Tetrodotoxin—distribution and accumulation in aquatic organisms, and cases of human intoxication. Mar Drugs. 2008;6:220–42.

Noguchi T, Arakawa O, Daigo K, Hashimoto K. Local differences in toxin composition of a xanthid crab Atergatis floridus inhabiting Ishigaki Island, Okinawa. Toxicon. 1986;24:705–11.

Noguchi T, Arakawa O, Takatani T. TTX accumulation in pufferfish. Comp Biochem Physiol-Part D Genom Proteom. 2006;1:145–52.

Ocean Biogeographic Information System (OBIS). 2006. OBIS-extracted Depth Data. Harvested by E.Agbayani July 2006 at www.iobis.org

Ogilvie, S.; Taylor, D.; McNabb, P.; Hamon, D.; Nathan, P.; Anderson, A. Tetrodotoxin in Kaimoana: Science and Mataurangā Mitigating Health Risks from a Lethal Neurotoxin. Prepared for Nga Pae O Te M ¯ aramatanga (Contract No. 10RF-18); Cawthron ¯ Report No. 2219. 22 p. plus appendices; 2012. Available online: http://www.maramatanga.ac.nz/sites/default/files/10RF-18%20Final%20contract%20report.pdf. Accessed 30 Mar 2023.

Ogura Y. The biological estimation of crystalline tetrodotoxin. III. On the isolated stomach vagal nerve preparation of rat (in Japanese, English summary). Annu Rep Food Microbiol. 1963;12:105–9.

Orfila M. A general system of toxicology. Philadelphia: Carey and Sons; 1817.

Orsolini L, Ciccarese M, Papanti D, De Berardis D, Guirguis A, Corkery JM, Schifano F. Psychedelic fauna for psychonaut hunters: a mini-review. Front Psych. 2018;9:153.

Oshiro N, Yogi K, Asato S, Sasaki T, Tamanaha K, Hirama M, Yasumoto T, Inafuku Y. Ciguatera incidence and fish toxicity in Okinawa. Japan Toxicon. 2009;56:656. https://doi.org/10.1016/j.toxicon.2009.05.036.

Otani Y, Hasegawa K, Hanaki K. Comparison of aerobic denitrifying activity among three cultural species with various carbon sources. Water Sci Technol. 2004;50:5–22.

Page LM, Burr BM. A field guide to freshwater fishes of North America north of Mexico. Boston: Houghton Mifflin Company; 1991. 432 p.

Paik JH-Y, Bjeldanes LF. Effects of cadaverine on histaminc transport and metabolism in isolated gut sections of the Guinea-pig. Food Cosmet Toxicol. 1979;17:629–32.

Palafox NA, Jain LG, Pinano AZ, Gulick TM, Williams RK, Schatz IJ. Successful treatment of ciguatera fish poisoning with intravenous mannitol. JAMA J Am Med Assoc. 1988;259:2740–2.

Pall M. Common etiology of posttraumatic stress disorder, fibromyalgia, chronic fatigue syndrome and multiple chemical sensitivity via elevated nitric oxide/peroxynitrite. Med Hypotheses. 2001;57:139–45.

Palomares MLD, Jarre A, Sambilay V Jr. Appendix I. Documentation of available 5 1/4' MSDOS data diskettes on the Peruvian upwelling ecosystem. In: Pauly D, Muck P, Mendo J, Tsukayama I, editors. The Peruvian upwelling ecosystem: dynamics and interactions. ICLARM Conf. Proc. 18; 1989. p. 408, 483 p–16.

Parin NV. Gempylidae. In: Whitehead PJP, Bauchot M-L, Hureau J-C, Nielsen J, Tortonese E, editors. Fishes of the North-Eastern Atlantic and the Mediterranean, vol. 2. Paris: UNESCO; 1986. p. 967–73.

Park SK, Kim DG, Kang SK, Han JS, Kim SG, Lee JS, Kim MC. Toxic acute renal failure and hepatitis after ingestion of raw carp bile. Nephron. 1990;56(2):188–93.

Parrish JD, Serra R, Grant WS. The monotypic sardines, *Sardina* and *Sardinops*: their taxonomy, distribution, stock structure, and zoogeography. Can J Fish Aquat Sci. 1989;46:2019–36.

Parsons M, Aligizaki A, Bottein M-YD, Fraga S, Morton S, Penna A, Rhodes L. *Gambierdiscus* and *Ostreopsis*: reassessment of the state of knowledge of their taxonomy, geography, ecophysiology and toxicology. Harmful Algae. 2012;14:107–29.

Pattaravivat J, Morioka K, Shirosaki M, Itoh Y. Effect of washing conditions on the removal of lipid from the fatty fish escolar (Lepidocybium flavobrunneum) meat. J Biol Sci. 2008;8:34–42.

Paul L, Fowler S (SSG Australia & Oceania Regional Workshop, March 2003). Heptranchias perlo. The IUCN red list of threatened species 2003: e. T41823A10572878. 2003. https://doi.org/10.2305/IUCN.UK.2003.RLTS.T41823A10572878.en. Downloaded on 11/2022.

Pearn J. Neurology of ciguatera. J Neurol Neurosurg Psychiatry. 2001;70(1):4–8.

Perez-Martin R, Franco J, Aubourg S, Gallardo J. Changes in free amino acids content in albacore (*Thunnus alalunga*) muscle during thermal processing. Z Lebensm Unters Forsch. 1988;187:432–5.

Pergola M. Bacte'ries toxige'nes chez le poisson frais avec ou sans la presence de l'ichthyovenin. Boll Sez Ital Soc Intern Microbiol. 1937;9:105–8.

Pergola M. L'ictioveleno e istamina. Rend Ist Sup Sanit. 1956;19:419–32.

Pethiyagoda R. Freshwater fishes of Sri Lanka. Colombo: The Wildlife Heritage Trust of Sri Lanka; 1991. 362 p.

Phisalix M. Animaux venimeux et venins, 2 vols. Paris: Masson et Cie; 1922.

Pinder, A.C., 2001. Keys to larval and juvenile stages of coarse fishes from fresh waters in the British Isles. Freshwater biological association. The ferry house, far Sawrey, Ambleside, Cumbria, UK. Scientific publication no. 60. 136 p.

Pottier I, Vernoux J, Jones A, Lewis R. Analysis of toxin profiles in three different fish species causing ciguatera fish poisoning in Guadeloupe. French West Indies Food Addit Contam. 2002a;19:1034–42.

Pottier I, Vernoux J, Jones A, Lewis R. Characterization of multiple Caribbean ciguatoxins and congeners in individual specimens of horse-eye jack (Caranx latus) by high-performance liquid chromatography/mass spectrometry. Toxicon. 2002b;40:929–39.

Pullan, S. (2020). Koi netting survey in Lake Waikare, Waikato first exercise, 4-6 September 2020 project №: KOI-WAIK-63. Report prepared for NZWR by Pullan environment and aquatic consulting. Steve Pullan con-

sultant H: 09 827 4058 I M: 021 262 0279 I E: s.pullan@xtra.co.nz 2 Rickards place New Lynn, Auckland 0600, New Zealand.
Quod JP, Turquet J. Ciguatera in Re'union Island: epidemiology and clinical patterns. Toxicon. 1996;34(7):779–85.
Racciatti D, Vecchiet J, Ceccomancini A, Ricci F, Pizzigallo E. Chronic fatigue syndrome following a toxic exposure. Sci Total Environ. 2001;270:27–31.
Ragonese S, Rivas G, Jereb P. Spreading of puffer *Sphoeroides cutaneus* Günther, 1870 (Pisces, Tetraodontidae) in the Sicilian Channel. Is it an exploding population? Rapp. Comm Int Mer Méditerr. 1992;33:308.
Ragonese S, Jereb P, Morara U. Morphometric relationships of *Sphoeroides pachygaster* (Pisces - Tetraodontidae) of the Strait of Sicily (Mediterranean Sea). Cah Biol Mar. 1997;38:283–9.
Raikhlin-Eisenkraft B, Rabbitfish ("Aras"), BY. An unusual source of ciguatera poisoning. Isr Med Assoc J. 2002;4:28–30.
Raisfeld R, Patronite R. The "it" fish. New York Magazine. http://www.nymagcom/restaurants/features/16899. 2006. Accessed 5 Mar 2023.
Raithel M, Ulrich P, Hochberger J, Hahn EG. Measurement of gut diamine oxidase activity: diamine oxidase as a new biologic marker of colorectal proliferation? Ann N Y Acad Sci. 1998;859(1):262–6.
Rambla-Alegre M, Reverté L, del Río V, de la Iglesia P, Palacios O, Flores C, Caixach J, Campbell K, Elliott CT, Izquierdo-Muñoz A, et al. Evaluation of tetrodotoxins in puffer fish caught along the Mediterranean coast of Spain. Toxin profile of Lagocephalus sceleratus. Environ Res. 2017;158:1–6.
Randall JE. A review of ciguatera, tropical fish poisoning, with a tentative explanation of its cause. Bull Mar Sci. 1958;8(3):236–67.
Rawles DD, Flick GJ, Martin RE. Biogenic amines in fish and shellfish. Adv Food Nutr Res. 1996;39:329–65.
Read BE. Chinese materia medica. Fish drugs Peiping Nat Hist Bull. 1939:136.
Reide K. Global register of migratory species—from global to regional scales. Final report of the R&DProjekt 808 05 081. Bonn: Federal Agency for Nature Conservation; 2004. 329 p.
Rémy C. Sur les poissons toxique du Japon. Comp Rend Soc Biol. 1883;7:3–28.
Réveillon D, Savar V, Schaefer E, Chevé J, Halm-Lemeille M-P, Hervio-Heath D, Travers M-A, Abadie E, Rolland J-L, Hess P. Tetrodotoxins in French bivalve mollusks—analytical methodology, environmental dynamics and screening of bacterial strain collections. Toxins. 2021;13:740.
Reverté L, De La Iglesia P, Del Río V, Campbell K, Elliott CT, Kawatsu K, Katikou P, Diogène J, Campàs M. Detection of Tetrodotoxins in puffer Fish by a self-assembled monolayer-based immunoassay and comparison with surface Plasmon resonance, LC-MS/MS, and mouse bioassay. Anal Chem. 2015;87:10839–47.
Rhodes L, Smith K. A checklist of the benthic and epiphytic marine dinoflagellates of New Zealand, including Rangitahuā/Kermadec Islands. N Z J Mar Freshw Res. 2019;53:258–77.
Rhodes L, Gimenez Papiol G, Smith K, Harwood T. Gambierdiscus cf. yasumotoi (Dinophyceae) isolated from New Zealand's sub-tropical northern coastal waters. N Z J Mar Freshw Res. 2014a;48(2):303–10.
Rhodes L, Smith K, Harwood T, Selwood A, Argyle P, Bedford C, Munday R. *Gambierdiscus* and *Ostreopsis* from New Zealand, the Kermadec Islands and the Cook Islands and the risk of ciguatera fish poisoning to New Zealand. In Marine and Freshwater Algae 2014, Proceeding the 16th International Conference Harmful Algae, Wellington, New Zealand, 27–31 August 2014; Cawthron Institute: Nelson, New Zealand, 2014b; pp. 180–183.
Rhodes LL, Smith KF, Murray S, Harwood DT, Trnski T, Munday R. The epiphytic genus Gambierdiscus (Dinophyceae) in the Kermadec Islands and Zealandia regions of the southwestern Pacific and the associated risk of ciguatera fish poisoning. Mar Drugs. 2017;15(7):219.
Rhodes LL, Smith KF, Murray JS, Nishimura T, Finch SC. Ciguatera fish poisoning: the risk from an Aotearoa/New Zealand perspective. Toxins. 2020;12(1):50.
Riede K. Global register of migratory species - from global to regional scales. In: Final report of the R&D-Projekt 808 05 081. Bonn: Federal Agency for Nature Conservation; 2004. 329 p.
Rikardsen AH, Diserud OH, Elliott JM, Dempson JB, Sturlaugsson J, Jensen AJ. The marine temperature and depth preferences of Arctic charr (Salvelinus alpinus) and sea trout (Salmo trutta), as recorded by data storage tags. Fish Oceanogr. 2007;16:436–47.
Ritchie AH, Mackie IM. The formation of diamines and polyamines during storage of mackerel (Scomber scombrus). In: Connell JJ, editor. Advances in fish science and technology. Surrey: Fishing News Books; 1979. p. 489–94.
Roberts C. *Liacos pachygaster*. The IUCN red list of threatened species 1996: e.T12103A3321801. 1996. https://doi.org/10.2305/IUCN.UK.1996.RLTS.T12103A3321801.en
Roberts CD, Stewart AL, Struthers CD, editors. The fishes of New Zealand. Te Papa: Museum of New Zealand; 2015. 2,000 pages over 4 volumes.
Robins CR, Ray GC. A field guide to Atlantic coast fishes of North America. Boston: Houghton Mifflin Company; 1986. 354 p.
Rochard E, Elie P. La macrofaune aquatique de l'estuaire de la Gironde. Contribution au livre blanc de l'Agence de l'Eau Adour Garonne. p. 1-56. In: Mauvais J-L, Guillaud J-F, editors. État des connaissances sur l'estuaire de la Gironde. Éditions Bergeret, Bordeaux, France: Agence de l'Eau Adour-Garonne; 1994. 115 p.
Rodriguez P, Alfonso A, Vale C, Alfonso C, Vale P, Tellez A, Botana LM. First toxicity report of tetrodotoxin and 5,6,11-trideoxyTTX in the trumpet shell

Charonia *Lampas lampas* in Europe. Anal Chem. 2008;80:5622–9.
Rodríguez I, Alfonso A, Alonso E, Rubiolo JA, Roel M, Vlamis A, Katikou P, Jackson SA, Menon ML, Dobson A, et al. The association of bacterial C9-based TTX-like compounds with *Prorocentrum minimum* opens new uncertainties about shellfish seafood safety. Sci Rep. 2017;7:40880.
Rolton A, Rhodes L, Hutson KS, Biessy L, Bui T, MacKenzie L, Symonds JE, Smith KF. Effects of harmful algal blooms on fish and shellfish species: a case study of New Zealand in a changing environment. Toxins. 2022;14(5):341.
Romero P. An etymological dictionary of taxonomy. Madrid, unpublished; 2002.
Rontani J. Production of Wax Esters by bacteria. In: Timmis KN, editor. Handbook of hydrocarbon and lipid microbiology. New York: Springer; 2010. p. 459–70.
Rosenberg LJ, Wrathall JR. Time course studies on the effectiveness of tetrodotoxin in reducing consequences of spinal cord contusion. J Neurosci Res. 2001;66:191–202.
Roughly TC, Roberts BJ. Bounty descendant live on remote Norfolk Island. Natl Geogr Mag. 1960;116(6):575.
Ruello, N. (2004). Report on the oil content, composition, and the consumption of escolar prepared for the Australian Government Department of Health and ageing. Ruello and Associates Pty Ltd, Australia (unpublished report).
Ruff RL, Lewis J. Clinical aspects of ciguatera: an overview. Mem Qld Mus 1994 34:609–619.
Ruiz-Capillas C, Moral A. Free amino acids and biogenic amines in red and white muscle of tuna stored in controlled atmospheres. Amino Acids. 2004;26:125–32.
Ruiz-Gutierrez V, Perez-Zarza MC, Muriana FJG, Bravo L. Lipid and fatty acid composition of muscle and internal organs from Ruvettus pretiosus. J Fish Biol. 1997;50:1353–7.
Russell FE. Marine foams and venomous and poisonous marine animals. In: Advances to marine biology, vol. 3. London: Academic; 1965. p. 235.
Russell FE, Maretić Z. Scombroid poisoning: mini-review with case histories. Toxicon. 1986;24(10):967–73.
Sabroe RA, Kobza Black A. Scombrotoxic fish poisoning. Clin Exp Dermatol. 1998;23(6):258–9.
Sakami T, Nakahara H, Chinain M, Ishida Y. Effects of epiphytic bacteria on the growth of the toxic dinoflagellate Gambierdiscus toxicus (Dinophyceae). J Exp Mar Biol Ecol. 1999;233:231–46.
Sanchez-Guerrero IM, Vidal JB, Escudero AI. Scombroid fish poisoning: a potentially life-threatening allergic-like reaction. J Allergy Clin Immunol. 1997;100:433–4.
Saoudi M, Rabeh FB, Jammoussi K, Abdelmouleh A, Belbahri L, Feki AE. Biochemical and physiological responses in Wistar rat after administration of puffer fish (*Lagocephalus lagocephalus*) flesh. J Food Agric Environ. 2007;5:107–11.
Satake M, Murata M, Yasumoto T. The structure of CTX3C, a ciguatoxin congener isolated from cultured Gambierdiscus toxicus. Tetrahedron Lett. 1993;34:1975–8.
Savtschenko P. Atlas of the poisonous fishes. Description of the ravages produced by them in the human organism, and antidotes which may be employed (in Russian). St Petersburg; 1886. p. 53.
Schep LJ, Slaughter RJ, Temple WA, Beasley DMG. Ciguatera poisoning: an interesting occurrence in New Zealand. N Z Med J. 2010;123:100–2.
Schneider W. FAO species identification sheets for fishery purposes. Field guide to the commercial marine resources of the Gulf of Guinea. Prepared and published with the support of the FAO regional Office for Africa. Rome: FAO; 1990. 268 p.
Schnorf H, Taurarii M, Cundy T. Ciguatera fish poisoning: a double-blind randomized trial of mannitol therapy. Neurology. 2002;58:873–80.
Scott WB, Crossman EJ. Freshwater fishes of Canada. Bull. Fish. Res. Board Can. 1973;184:xi+1-966.
Scott WB, Scott MG. Atlantic fishes of Canada. Can Bull Fish Aquat Sci. 1988;219:731 p.
Seki T, Satake M, MacKenzie L, Kaspar H, Yasumoto T. Gymnodimine, a new marine toxin of unprecedented structure isolated from New Zealand oysters and the dinoflagellate, *Gymnodinium* sp. Tetrahedron Lett. 1995;36:7093–6.
Shadbolt C, Kirk M, Roche P. Editorial: diarrhea associated with consumption of escolar (rudderfish). Commun Dis Intell. 2002;26:436–8.
Shao KT, Lim PL. Fishes of freshwater and estuary. Encyclopedia of field guide in Taiwan, vol. 31. Taipei: Recreation Press, Co., Ltd.; 1991. 240 p. (in Chinese).
Shao K, Liu M, Jing L, Hardy G, Leis JL, Matsuura K. Arothron stellatus. The IUCN red list of threatened species 2014: e.T193712A2264205. 2014. https://doi.org/10.2305/IUCN.UK.2014-3.RLTS.T193712A2264205.en.
Shcherbachev YN. Preliminary list of thalassobathyal fishes of the tropical and subtropical waters of the Indian Ocean. J Ichthyol. 1987;27(2):37–46.
Shellfish Toxin Update (STU). Available online: https://www.seafoodnewzealand.org.nz. Accessed 23 Nov 2022.
Shireman JV, Smith CR. Synopsis of biological data on the grass carp, *Ctenopharyngodon idella* (Cuvier and Valenciennes, 1884). FAO Fish Synop. 1983;135:86 p.
Silva M, Azevedo J, Rodriguez P, Alfonso A, Botana LM, Vasconcelos V. New gastropod vectors and tetrodotoxin potential expansion in temperate waters of the Atlantic Ocean. Mar Drugs. 2012;10:712–26.
Silva M, Rodríguez I, Barreiro A, Kaufmann M, Neto AI, Hassouani M, Sabour B, Alfonso A, Botana LM, Vasconcelos V. Tetrodotoxins occurrence in non-traditional vectors of the North Atlantic waters (Portuguese maritime territory, and Morocco coast). Toxins. 2019;11:306.

Sims JK. A theoretical discourse on the pharmacology of toxic marine ingestions. Ann Emergency Med. 1987;16:1006.

Skelton PH. A complete guide to the freshwater fishes of southern Africa. Southern Book Publishers; 1993. 388 p.

Skinner MP, Brewer TD, Johnstone R, Fleming LE, Lewis RJ. Ciguatera fish poisoning in the Pacific islands. PLoS Negl Trop Dis. 2011;5:e1416.

Smith CL. National Audubon Society field guide to tropical marine fishes of the Caribbean, the Gulf of Mexico, Florida, The Bahamas, and Bermuda. New York: Alfred A. Knopf, Inc.; 1997. 720 p.

Smith MM, Heemstra PC. Tetraodontidae. In: Smith MM, Heemstra PC, editors. Smiths' sea fishes. Berlin: Springer; 1986. p. 894–903.

Smith MM, Smith JLB. Mugilidae. In: Smith MM, Heemstra PC, editors. Smiths' sea fishes. Berlin: Springer; 1986. p. 714–20.

Smith KF, Rhodes L, Verma A, Curley BG, Harwood DT, Kohli GS, Solomona D, Rongo T, Munday R, Murray SA. A new Gambierdiscus species (Dinophyceae) from Rarotonga, Cook Islands: *Gambierdiscus cheloniae* sp. nov. Harmful Algae. 2016;60:45–56.

Smith KF, Biessy L, Argyle PA, Trnski T, Halafihi T, Rhodes LL. Molecular identification of Gambierdiscus and Fukuyoa (Dinophyceae) from environmental samples. Mar Drugs. 2017;15(8):243.

Smith-Vaniz WF, Williams J, Pina Amargos F, Curtis M, Grijalba Bendeck L. Lepidocybium flavobrunneum. The IUCN red list of threatened species 2015: e.T190287A16510672. 2015. https://doi.org/10.2305/IUCN.UK.2015-4.RLTS.T190287A16510672.en

Smolian, K., 1920. Merkbuch der Binnenfischerei. Fischereiförderung GmbH, Berlin, Germany, p. 449, XXV.

Sommer C, Schneider W, Poutiers J-M. FAO species identification field guide for fishery purposes. The living marine resources of Somalia. Rome: FAO; 1996. 376 p.

Sparrman A. A voyage round the world with Captain James Cook in H.M.S. Resolution. London: Golden Cockerell Press; 1944. p. 218.

Sparrow L. Key factors influencing the occurrence and frequency of Ciguatera. Ph.D. thesis, College of Science and Engineering, James Cook University, Townsville, 2018.

Specht D. Scombroid fish poisoning. J Emerg Nurs. 1998;24(2):118–9.

Spillman, C.-J., 1961. Faune de France: Poissons d'eau douce. Fédération Française des Sociétés Naturelles,Tome 65. Paris. 303 p .

Stenhouse F. "Market Access Advice: Prohibition of Oilfish and Escolar Exports to South Korea (Aug 13)." Australian Quarantine and Inspection Service, Australia. 2007. http://www.daff.gov.au/aqis/export/fish/fish-notices/2007/FISH0708_Oilfish_escolar_market_ban_to_South_Korea.pdf. Accessed 5 Mar 2023.

Stevenson ML. Trawl survey of the west coast of the South Island and Tasman and Golden Bays, March-April 2003 (KAH0304). New Zealand fisheries assessment report 2004/4. 69 p. 2004.

Stobbs RE, Bruton MN. The fishery of the Comoros, with comments on its possible impact on coelacanth survival. Environ Biol Fishery. 1991;32:341–59.

Stommel EW. Scombroid fish poisoning. In: Simjee S, editor. Infectious disease: foodborne diseases. Totowa: Humana Press; 2007. p. 375–38.

Stratton JE, Taylor SL. Scombroid poisoning. In: Ward D, Hackney C, editors. Microbiology of marine food products. New York: Spectrum; 1991. p. 331–51.

Suehiro Y. Poison of globe-fish. In: Suehiro Y, editor. Practice of fish physiology (in Japanese). Tokyo: Takeuchi Bookstore; 1947. p. 140–59.

Suyama M, Yoshizawa Y. Free amino acid composition of the skeletal muscle of migratory fish. Bull Jpn Soc Sci Fish. 1973;39:1339–43.

Svalastog D. A note on maximum age of brown trout, *Salmo trutta* L. J Fish Biol. 1991;38(6):967–8.

Taft C. Poisonous marine animals. Tex Rep Biol Med. 1945;3:399.

Tahara Y. Reports of discovery of puffer toxin (in Japanese). Chugai Iji Shimpo. 1894;344:4–9.

Tahara Y. Ueber die giftigen bestandtheile des tetrodon. Congr Internat Hyg Demg, C.R. 1894 (Budapest). 1896;8:198–207.

Tahara Y. Discover of puffer poison (in Japanese). Dobutsugaku Zasshi. 1897;6:268–75.

Tahara Y. Uber das tetrodongift. Biochem € Z. 1910;30:255–75.

Tahara, Y. Über das Tetrodongift. Biochem. Z. 1911, 30, 255–275. Available online: https://archive.org/details/sim_biochemischezeitschrift-beitraege-zur-chemischen_1911_30/page/274/mode/2up. Accessed 30 Apr 2023.

Tahara Y, Hirata Y. Studies on the puffer fish toxin. J Pharm Soc Japan (Yakugaku Zasshi). 1909;328:587–625. (In Japanese)

Talwar PK, Jhingran AG. Inland fishes of India and adjacent countries, vol. 2. Rotterdam: A.A. Balkema; 1991.

Tamele IJ, Silva M, Vasconcelos V. The incidence of tetrodotoxin and its analogs in the Indian Ocean and the Red Sea. Mar Drugs. 2019;17:28.

Tanew A. Severe scombroid fish poisoning: an under recognized dermatological emergency. J Am Acad Dermatol. 2009;65:246–7.

Tang W-C. Chinese medicinal materials from the sea. Abstr Chin Med. 1987;1(4):571–600.

Taylor SL. Histamine food poisoning: toxicology and clinical aspects. Crit Rev Toxicol. 1986;17(2):91–128.

Taylor SL, Lieber ER. In vivo inhibition of rat intestinal histamine-metabolizing enzymes. Food Cosmet Toxicol. 1979;17:237–40.

Taylor SL, Speckard M. Isolation of histamine producing bacteria from frozen tuna. Mar Fish Rev. 1983;45:35–9.

Taylor SL, Stratton JE, Nordlee JA. Histamine poisoning (scombroid fish poisoning): an allergy-like intoxication. J Toxicol Clin Toxicol. 1989;27(4–5):225–40.

Thomas E. On the poison of fish. Mem Med Soc Lond 1799;5:94–111.

Thomson JM. Mugilidae. In: Daget J, Gosse J-P, Thys van den Audenaerde DFE, editors. Check-list of the freshwater fishes of Africa (CLOFFA), vol. 2. Brussels/Tervuren/Paris: ISNB/MRAC/ORSTOM; 1986. p. 344–9.

Thomson JM. Mugilidae. In: Quero JC, Hureau JC, Karrer C, Post A, Saldanha L, editors. Check-list of the fishes of the eastern tropical Atlantic (CLOFETA), vol. 2. Lisbon/Paris/Paris: JNICT/SEI/UNESCO; 1990. p. 855–9.

Ting JY, Brown AF. Ciguatera poisoning: a global issue with common management problems. Eur J Emerg Med. 2001;8:295–300.

Tinker SW. Hawaiian fishes. Honolulu, HI: Tongg Pub. Co.; 1944. p. 262–72.

Titcomb M. Native use of fish in Hawaii. Honolulu: The University Press of Hawaii; 1972. 175 p.

Titcomb M, Pukui MK. Native use offish in Hawai'i; 1952.

Topuzlu C, Stahl WM. Effect of bile infusion on the dog kidney. N Engl J Med. 1960;274:760–3.

Torres FSBJ, Pauly D. Tabular data on marine fishes from southern Africa. Part II: Growth parameters Fishbyte. 1991;9(2):37–8.

Tortonese E. Tetraodontidae. In: Whitehead PJP, Bauchot M-L, Hureau J-C, Nielsen J, Tortonese E, editors. Fishes of the North-Eastern Atlantic and the Mediterranean, vol. 3. Paris: UNESCO; 1986. p. 1341–7.

Trnski T, Schlumpf HA, editors. Kermadec biodiscovery expedition. Bull Auckland Museum, 2011; Volume 20, Available online: http://www.aucklandmuseum.com/research/pub/bulletin/20/1 Accessed 23 Nov 2022.

Tsuchiya K. Suisangaku-Zenshu, Suisankagaku. Tokyo: Koseisha-Koseikaku; 1962. p. 183–4.

Tung I-H. On the fishery biology of the grey mullet, *Mugil cephalus* Linnaeus, in Taiwan. Rep. Inst. Fish. Biol. Minist. Econ. Aff. Natl Taiwan Univ. 1981;3:38–102.

Turner AD, Higgins C, Higman W, Hungerford J. Potential threats posed by tetrodotoxins in UK waters: examination of detection methodology used in their control. Mar Drugs. 2015;13:7357–76.

Turner AD, Dhanji-Rapkova M, Coates L, Bickerstaff L, Milligan S, O'Neill A, Faulkner D, McEneny H, Baker-Austin C, Lees DN, et al. Detection of Tetrodotoxin shellfish poisoning (TSP) toxins and causative factors in bivalve molluscs from the UK. Mar Drugs. 2017;15:277.

Turner AD, Fenwick D, Powell A, Dhanji-Rapkova M, Ford C, Hatfield RG, Santos A, Martinez-Urtaza J, Bean TP, Baker-Austin C, et al. New invasive nemertean species (*Cephalothrix simula*) in England with high levels of tetrodotoxin and a microbiome linked to toxin metabolism. Mar Drugs. 2018;16:452.

Turner A, Dean K, Lewis A, Jenkins Z, Bear B, Batista F, Ryder D, Maskrey B, Later A, House C, et al. ScillyHAB: A Multi-Disciplinary Survey of Harmful Marine Phytoplankton and Shellfish Toxins in the Isles of Scilly, Utilizing Citizen Science in a Remote Offshore U.K. Territory during the COVID-19 Pandemia, 2021. Presentation no. SM-O-15_221 in the 19th International Conference on Harmful Algal Blooms, La Paz, Mexico, 10–15 October 2021. Available online: https://imsvirtualcenter.com.mx/icha. Accessed 30 Mar 2023.

Ueyama N, Sugimoto K, Kudo Y, Onodera KI, Cho Y, Konoki K, Nishikawa T, Yotsu-Yamashita M. Spiro bicyclic Guanidino compounds from pufferfish: possible biosynthetic intermediates of Tetrodotoxin in marine environments. Chem-A Eur J. 2018;24:7250–8.

Ujević I, Roje-Busatto R, Dragičević B, Dulčić J. Tetrodotoxin in Invasive Silver-cheeked Toadfish Lagocephalus sceleratus (Gmelin, 1789) in the Adriatic Sea. In The Handbook of Environmental Chemistry. In: Joksimović D, Đurović M, Zonn IS, Kostianoy AG, Semenov AV, editors. The montenegrin adriatic coast, vol. 110. Cham: Springer; 2020. p. 141–9.

Ukishima Y, Masui T, Matsubara S, Goto R, Okada S, Tsuji K, Kosuge T. Wax components of escolar (Lepidocybium flavobrunneum) and its application to base of medicine and cosmetics. Yakugaku Zasshi. 1987;107:883–90.

Vernoux J, Lewis R. Isolation and characterization of Caribbean ciguatoxins from the horse-eye jack (Caranx latus). Toxicon. 1997;35:889–900.

Vernoux J, Lahlou N, Abbad A, Riyeche N, Magras L. A study of the distribution of ciguatoxin in individual Caribbean fish. Acta Trop. 1985;42:225–33.

Villareal T, Hanson S, Qualia S, Jester E, Granade H, Dickey R. Petroleum production platforms as sites for the expansion of ciguatera in the northwestern Gulf of Mexico. Harmful Algae. 2007;6:253–9.

Vincent L. Les Poissons ve'ne'neus du Japon Arch Me'divcale Nav Paris. 1883;39:392–4.

Vinson LP. Ele'ments d'une topographie me'dicale de la Nouvelle-Cale'donie et de I'lle des Pins. Thesis No. 59, faculty of Medicine, Paris. 1858. p. 95.

Vision Dive. 2022. https://www.visiondive.com/toxic--fish-meat/. Downloaded 30/10/2022.

Vlamis A, Katikou P, Rodriguez I, Rey V, Alfonso A, Papazachariou A, Zacharaki T, Botana AM, Botana LM. First detection of tetrodotoxin in Greek shellfish by UPLC-MS/MS potentially linked to the presence of the dinoflagellate *Prorocentrum minimum*. Toxins. 2015;7:1779–807.

Waqanivavalagi S, Murphy M, Jones P. Two cases of ciguatera fish poisoning following eel consumption. NZMJ. 2019;29:21–2.

Welcomme RL. International introductions of inland aquatic species. FAO Fish Tech Pap. 1988;294:318 p.

Wheeler JFG. Report on the Mauritius-Seychelles fisheries surveys 1948–1949. Part I. The bottom fishes of economic importance. Fish Publ Lond. 1953;1(3):1–57.

White MV. The role of histamine in allergic diseases. J Allergy Clin Immunol. 1990;86:599–605.

Whitehead PJP. FAO Species Catalogue. Vol. 7. Clupeoid fishes of the world (suborder Clupeoidei). An annotated and illustrated catalogue of the herrings,

sardines, pilchards, sprats, shads, anchovies and wolf-herrings. FAO Fish. Synop. 125(7/1):1–303. Rome: FAO; 1985.
Whitehead PJP, Rodriguez-Sánchez YR. Clupeidae. Sardinas, sardinetas, machuelos, sábalos, piquitingas. In: Fischer W, Krupp F, Schneider W, Sommer C, Carpenter KE, Niem V, editors. Guia FAO para Identification de Especies para lo Fines de la Pesca. Pacifico Centro-Oriental. 3 Vols. Rome: FAO; 1995. p. 1015–25.
World Conservation Monitoring Centre (WCMC). Salmo salar. The IUCN Red List of Threatened Species 1996: e.T19855A9026693. 1996. https://doi.org/10.2305/IUCN.UK.1996.RLTS.T19855A9026693.en
Wu Z, Yang Y, Xie L, Xia G, Hu J, Wang S, Zhang R. Toxicity and distribution of tetrodotoxin-producing bacteria in puffer fish *Fugu rubripes* collected from the Bohai Sea of China. Toxicon. 2005;46:471–6.
Wu YJ, Lin CL, Chen CH, Hsieh CH, Jen HC, Jian SJ, Hwang DF. Toxin and species identification of toxic octopus implicated into food poisoning in Taiwan. Toxicon. 2014;91:96–102.
Xuan BHN, Thi TXN, Nguyen ST, Goldfarb DS, Stokes MB, Rabenou RA. Ichthyotoxic ARF after fish gallbladder ingestion: a large case series from Vietnam. Am J Kidney Dis. 2003;41(1):220–4.
Yamada U, Shirai S, Irie T, Tokimura M, Deng S, Zheng Y, Li C, Kim YU, Kim YS. Names and illustrations of fishes from the East China Sea and the Yellow Sea. Tokyo: Overseas Fishery Cooperation Foundation; 1995. 288 p.
Yang CC, Deng JF. Overview of marine toxins. I Tetrodotoxin Clin Med. 1996;38:125–35.
Yang CC, Liao SC, Deng JF. Tetrodotoxin poisoning in Taiwan: an analysis of poison center data. Vet Hum Toxicol. 1996;38(4):282–6.
Yano I. An experimental study on the globefish (fugu) Jpn Soc of Int Med 33rd Annu Meet. 1938;5:99–101.
Yasumoto T, Nakajima I, Bagnis R, Adachi R. Finding of a dinoflagellate as a likely culprit of ciguatera. Bull Jpn Soc Sci Fish. 1977;43:1021–6.
Yasumoto T, Inoue A, Bagnis R, Garcon M. Ecological survey on a dinoflagellate possibly responsible for the induction of ciguatera. Bull Jpn Soc Sci Fish. 1979;45:395–9.
Yasumoto T, Yasumura D, Yotsu M, Michishita T, Endo A, Kotaki Y. Bacterial production of tetrodotoxin and anhydrotetrodotoxin. Agric Biol Chem. 1986;50(3):793–5.
Yip LL, Chow CL, Yung KH, Chu KW. Toxic material from the gallbladder of the grass carp (Ctenopharyngodon idellus). Toxicon. 1981;19:567–9.
Yohannes K, Dalton CB, Halliday L, Unicomb LE, Kirk M. An outbreak of gastrointestinal illness associated with the consumption of escolar fish. Commun Dis Intell. 2002;26:441–5.
Yotsu-Yamashita M, Abe Y, Kudo Y, Ritson-Williams R, Paul VJ, Konoki K, Cho Y, Adachi M, Imazu T, Nishikawa T, Isobe M. First identification of 5,11-dideoxytetrodotoxin in marine animals, and characterization of major fragment ions of tetrodotoxin and its analogs by high resolution ESI-MS/MS. Mar Drugs. 2013;11(8):2799–813.
Zhang Y, Tsutsui H, Yamawaki N, Morii Y, Nishihara GN, Itoi S, Arakawa O, Takatani T. Geographic variations in the toxin profile of the xanthid crab *Zosimus aeneus* in a single reef on Ishigaki Island, Okinawa. Japan Mar Drugs. 2021;19:670.
Zhao H. *Hypophthalmichthys molitrix*. The IUCN red list of threatened species 2011: e.T166081A6168056. 2011; https://doi.org/10.2305/IUCN.UK.2011-2.RLTS.T166081A6168056.en.
Zidowitz H. Bericht des Workshops 'Erhaltung und Management von Tiefwasser-Knorpelfischen' 27.-29. November 2003, Dunedin. Neuseeland Elasmoskop. 2005;2005:28–32.

Part III

Venomous Fishes

6 Venomous Fishes

6.1 Ichthyocrinotoxic Fishes

This class of poisonous fishes has poisonous glands, but they do not have the venomous tools such as spines and teeth to convey the toxin to the body of the victim. With no venomous tool, the poisonous fishes discharge their toxins into the water. These fishes are classified under different phylogenetic units and form a wide spectrum of species variety. They use their toxins as a shield mechanism and as a repellent (Halstead 1967).

Poisonous fish are essential in the ecosystem due to toxins that they secrete to their habitat. These toxins are intricate in a series of processes that happened in the ecosystem and influence the occurrence or absence of other organisms, a process called the "exclusion effect." Toxins are metabolic products of poisonous fishes, and these complexes intrude with the vertical migration of animals in relation to the concentration of plants (Hardy 1956). Alteration in the input of organic materials in the environment has a direct outcome on chemical, physical, and biological factors, which in turn influence the group of animals present (Pearson and Rosenberg 1978). The variety of the organisms in any such habitat and through time will confront predictable fluctuations in the abundance and species richness. The outcome of the role of toxins secreted in the inspected water exhibited that the toxins of one group might impact the number and distribution of associated plants and animals (Halstead 1967).

Background

Much of the data about poisonous fishes are linked to toxins originating from the skin of these fishes. In the late nineteenth century, Cavazzani (1892), Kobert (1894, 1902), and Coutiere (1899) were the first to investigate about this poison in the skin of marine fishes. Afterward in the early twentieth century, researchers have trailed the same theories of the earlier writers in this field (Englesen 1922; Fredericq 1924; Maass 1937). Clarke (1918) was the first to publish on the poisonous characteristics of the jelly secretions of the labial glands of boxfishes. The toxicology of the Japanese puffers was the spotlight of scientists (Tani 1945), while other researchers became concerned in the toxic skin secretions such as Hashimoto (1950), Macober (1956), and Flaschentrager and Abdallah (1957) who found that the poison present in the internal organs of the fish is alike to that located in their skin, but the toxin of the skin is more dangerous (Larson et al. 1960). Scholars such as Collette (1966) and Nigrelli (1958, 1962) have published about the pectoral fin glands and their poison in some fish species. Subsequently, studies were appeared on various issues of the poisonous glands and the ichthyocrinotoxic fish species by Halstead (1988a, b), Church and Hodgson (2002), and Haddad et al. (2003a, b).

Poison Organs

The glandular organs that comprise toxin have the same main structure with some variation in

L. A. Jawad, *Dangerous Fishes of New Zealand*, https://doi.org/10.1007/978-3-031-69698-5_6

the different species. Boxfish is among the fish species that has poisonous glands in both the skin and mouth. Here the brief account of the histology of these two structures is dealt with grounded on the original researchers such as Thomson (1963, cited in Halstead 1967).

The skin of boxfish is different from that of other teleost fish species in having hard and soft parts. The hard bony carapace encloses almost the body, whereas the soft part is confined to the lips, the base of the fins, and the caudal peduncle. When the fish is alarmed, a foamy material is discharged at the base of the soft skin (i.e., the lips, base of the fins, and caudal peduncle areas). Pocket formations that planned to contain foamy toxic material are found in the soft areas of the skin.

The skin is constituted of three layers: the outer epidermis, the dermis, and subcutaneous tissue. The epidermis is the significant part of the skin, where the glandular secretion occurs. This layer contains of stratified squamous epithelium separated from the dermis by a layer of melanophores. Among the definite cells in this layer are the undifferentiated basal cells, mucous cells, and club cells. The undistinguishable cells have large nuclei and form the matrix of the epidermis. They modify their shape corresponding to their position in the epidermis. The mucous cells are lesser in size than the indistinguishable cells with nuclei of various shapes, and they are many at the epithelium surface. The third type of cells is the club cells, which are large and with inconsistent shapes. These cells can discharge their content while they are in the middle part of the epidermis and therefore have no need to come to the surface. Both mucous and club cells contain mucins, but of different type in each of them.

Poisonous glands are found in numerous sites in the mouth region of the boxfish. They are found in the lips, in the epithelial lining of the buccal cavity, and in the esophagus. The labial villi are an obvious configuration in the buccal cavity. These villi are additions of the mucous membrane and known as "labial glands" that extend deep in the epidermis and form a duct to empty outside the buccal cavity. The villi are held in pockets located above the dorsal row and below the ventral row of the teeth. Furthermore, there is an isolated pocket opposing each tooth. The three types of cells found in the skin are also found in the epidermis of the lips and with similar structure. These cells can secrete toxin and are contemplated poison glands, but with ichthyocrinotoxic construction, as the venomous tool is absent.

In the pufferfish, *Arthron hispidus*, the entire body is hidden under a numerous cartilaginous spine and concentrated on the ventral side of the body. The histological picture is dependent on the investigation made by Rosen (1913a, b) and Eger (1963) and given by Halstead (1967). The skin in this species is comprised of three layers, the epidermis, dermis, and subcutaneous. The poisonous secretions are in the epidermis layer only. The spines or prickles have no direct connection with the epidermis or the poisonous glands. These spines are enclosed with an epidermal layer that has contact with a poisonous gland deep in the epidermis. When the fish is disturbed, the stomach expands by taking in air or water. The ventral body wall of the fish inflates, and the prickles extend. This causes a stretching of the glandular tissue mass covering the prickles and the poison becomes ready to be expelled outside the body. The basic matter of the epidermis layer is like that of the boxfish with a slight variation. Here the toxin is formed in the club cells of the epidermis and flows around the prickles. In no way can the prickles be a venomous apparatus as their structure is very simple and differs greatly from the spines of catfishes, scorpionfishes, and weevers, but it is an ichthyocrinotoxic structure.

In toadfish, there are two types of poisonous glands, a pectoral and axillary. The description given below is centered on the study by Collette (1966) and stated by Halstead (1967). The glands of the pectoral fin are found on the surface of the upper ray of this fin, and the axillary glands are in a pouch present in the pectoral axilla. In the case of excitement, the pectoral fin expands, and the foramen of the gland opens wide, so the poisonous content is ready to be excreted outside in the water.

Catfish species may have smooth skin or their whole body is covered with bony plates (Sire

1993). There have been many studies linked to the structure of the skin of catfishes. (For review, see Arratia (2003).) The skin of catfishes is recognized as a paraneuron because of the presence of many receptor-secretory cells previously considered to be endocrine and sensory cells (Fujita 1994). The number of the various cells present in the epidermis differs with the body part of the individual (Arratia 2003).

The habitat that the fish lives in is full of microbes of different types. To evade the damaging action of these microbes, the fish established a typical physical barrier included of skin and skin mucus, which act as a first line of defense against these microorganisms. There are numerous immune ingredients found in this defense mechanism of the skin such as lysozyme, immunoglobulin, carbonic anhydrase, lectins, crinotoxins, calmodulin, C-reactive protein, proteolytic enzymes, and peptides, which have the capability to kill bacteria (Whyte 2007; Ramos et al. 2012). Fish skin mucus lectins demonstrate modification in their structural make-up and can be divided into five distinct compounds: galectin (Tasumi et al. 2004; Muramoto and Kamiya 1992; Muramoto et al. 1999), C-type (Tsutsui et al. 2003, 2007), B-type (Tsutsui et al. 2003), rhamnose-binding lectin (RBL; Okamoto et al. 2005), and pentraxin (Tasumi et al. 2004). Since fish possess a long evolutionary history and therefore demonstrate huge species variety, it may not be unexpected to find other types of lectins in their skin secretions.

6.1.1 Species of Ichthyocrinotoxic Fishes

Family: Diodontidae.

Diodon hystrix (Linnaeus 1758) (Figs. 6.1 and 6.2).

Common name: spot-fin porcupinefish.

Etymology: *Diodon:* Greek, *di* = two + Greek, *odous* = teeth (Romero 2002).

Diagnostic Features

The body is short, with a reasonable oval shape, and expandible. A long, sharp, erectile spines shield both the head and body, with the length of spines on the forehead 1–1.5 times the eye diameter and longer than spines on the dorsal and ventral areas of body. The lateral line is not clear. The eye is very large and rounded. The snout is short, with a broad interorbital width. The mouth is small, with the upper jaw longer than the lower. The teeth are in the outer surfaces of the upper and lower jaw fused in each jaw without a median suture. A nasal organ is present in front of the eye in the shape of a short tube with two openings. The gill opening is very small and looks like a vertical slot directly in front of the pectoral fin base. The dorsal and anal fins are positioned posteriorly with a symmetric shape and no fin spines and with vertical, slightly rounded general shape. The pectoral fins are truncated in shape and not reaching the anterior base of the dorsal fin. There are no pelvic fins. The caudal fin is rounded. The body is dark brown in color on the dorsal surfaces of both the body and head, with white ventrally. Black or dark brown spots are present on the dorsolateral side of the body. Presence of several black spots on all fins except for anal fin.

Fig. 6.1 Diodon hystrix (Linnaeus 1758). (Courtesy of Laith Jawad)

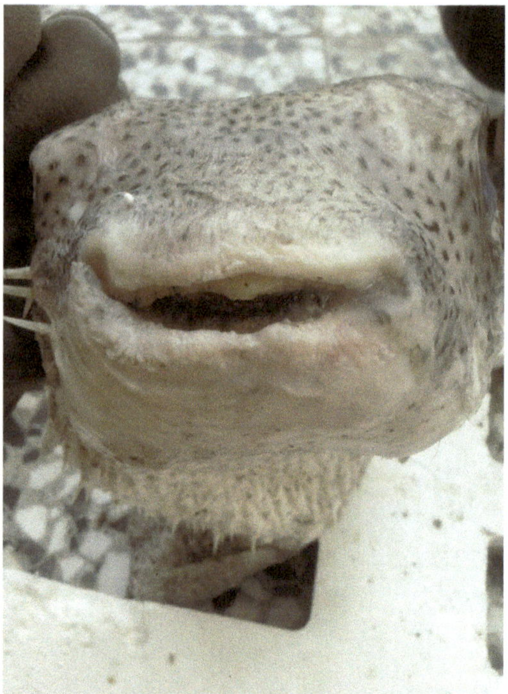

Fig. 6.2 Front view of Diodon hystrix (Linnaeus 1758). (Courtesy of Laith Jawad)

Distribution

This species of porcupinefish is circumtropical in distribution. It is reported in the Eastern Pacific from San Diego, California, USA, to Chile, including the Galapagos Islands (De la Cruz Agüero et al. 1997). It is also recorded from the Western Atlantic at Bermuda, Massachusetts (USA), and northern Gulf of Mexico to Brazil (Robins and Ray 1986). Its presence is also reported from the Eastern Atlantic: 30° N to 23° S (Duron and Queró 1990). It is recorded from the Western Indian Ocean: Red Sea to Madagascar, Reunion, and Mauritius (Fricke 1999.). Recently, it is reported from Khor Al Zubair area, north-west Arabian Gulf, Iraq (Jawad et al. 2019).

Habitat, Biology, and Ecology

Diodone hystrix is a marine species living in association with reefs at a depth range of 2–50 m (Leis 2001), but it is usually found at 3–20 m (Gasparini and Floeter 2001). It reaches a maximum total length of 910 mm (Eschmeyer et al. 1983), with a common total length of 400 mm (Sanches 1991). The maximum recorded weight is 2.8 kg (IGFA 2001). Individuals of this species prefer living in in lagoon and seaward reefs to at least 50 m. They usually present in caves and holes in shallow reefs (Smith 1997; Kuiter and Tonozuka 2001a, b). The juveniles to about 20 cm are pelagic. The adults are benthic (Sommer et al. 1996). This species is solitary and nocturnal and feeds on hard-shelled invertebrates such as sea urchins, gastropods, and hermit crabs (Leis 2001).

Economic Value

This species has no commercial value, and its meat is reported to be poisonous (Halstead et al. 1990).

Conservation Status

Liu et al. (2015) evaluated this species as Least Concern to the Red List of the IUCN with no reasons given.

Family: Ostraciidae.

Lactoria diaphana (Bloch and Schneider 1801) (Fig. 6.3).

Common name: roundbelly cowfish.

Etymology: *Lactoria:* Latin, concerned with milk (Romero 2002).

Diagnostic Features

The color of this species of cowfish is pale greenish gray, with a fine honeycomb reticulated pattern and brown blotches. The juveniles are semitransparent. The carapace of adults has a short spine projecting from above each eye, a broadly rounded lower surface with a posteriorly directed spine on each side just before the tail base, and a single small spine on the middle of the back. As its common name suggests, the belly region of the carapace is rounded.

Distribution

The roundbelly cowfish is reported from the Southeast Atlantic at the point off Swakopmund, Namibia (Heemstra 1995). It is also recorded from the Indo-Pacific and Eastern Pacific at the South Africa east through Indonesia to the Easter

Fig. 6.3 Lactoria diaphana (Bloch and Schneider 1801). (Courtesy of Izuzuki, Japan)

Island and Peru, north to southern Japan, Hawaii, and southern California, south to New Caledonia, New South Wales, and Kermadec Islands (Francis 1993).

Habitat, Biology, and Ecology

Lactoria diaphana is a marine species that enters brackish waters and lives in association with reef at a depth range of 8–50 m (Matsuura 2001). It reaches a maximum total length of 340 mm (Tinker 1978). Individuals of this species inhabit coral and rocky reefs of coastal areas (Matsuura 2001; Kuiter and Tonozuka 2001a, b). They are also found on offshore sand and mudflats (Kuiter and Tonozuka 2001a, b). The young fish live in oceanic waters, at or near the surface (Eschmeyer et al. 1983); semi-transparent young may found in estuaries (Smith 1986). This species feeds on benthic invertebrates (Matsuura 2001).

Economic Value

This species has no economic value and reports to be a venomous species (Halstead et al. 1990).

Conservation Status

This species is not evaluated for the Red List of the IUCN.

Lactoria fornasini (Bianconi 1846) (Fig. 6.4).
Common name: thornback cowfish.
Etymology: *Lactoria:* Latin, concerned with milk (Romero 2002).

Diagnostic Features

The body is covered with hexagonal, plate-like scales that are fused together to form a solid, box-like carapace, from which the eyes, mouth, fins, and tail protrude. The mouth is small and has protruding lips, the upper profile of the snout is straight, the back is somewhat convex, the flanks are concave, and the belly is rounded, causing the fish to resemble a purse. There are small preocular spines on either side of the head, a large spine projects from the dorsal ridge, and a further moderate-sized spine projects from each of the pelvic ridges near the anal fin. Neither the dorsal fin nor the anal fin has any spines, but both have nine soft rays, while the caudal fin is fan-shaped and has ten rays. The color of the body is variable and changes to match the fish's surroundings; it is generally pale tan with spots or wiggly markings in yellow, mauve, or blue.

Distribution

This species of cowfish is distributed in the Indo-West Pacific region from the East Africa to the

Fig. 6.4 Lactoria fornasini (Bianconi 1846). (Courtesy of Nicolas Bailly)

Hawaiian and Rapa Islands, north to southern Japan, south to Lord Howe Island. It is also reported from the Southeast Atlantic. It sometimes reaches the southeast coast of South Africa (Smith 1986).

Habitat, Biology, and Ecology
The thornback cowfish is a marine species that lives in association with reef at a depth range of 5–132 m (Randall and Cea 2011), but it is usually found at a range of 6–30 m (Lieske and Myers 1994). Individuals of this species reach a maximum total length of 230 mm (Smith 1986). They inhabit clear outer lagoon and seaward reefs (Myers 1991). They are also found in the benthopelagic over rubble and sand at 5–132 m (Mundy 2005). They are solitary (Allen and Erdmann 2012). The males are highly territorial (Lieske and Myers 1994).

Economic Value
This species has no economic value and reports to be a venomous species (Halstead et al. 1990).

Conservation Status
This species is not evaluated for the Red List of the IUCN.

6.2 Acanthotoxic (Venomous) Fishes

Venomous and poisonous organisms are dispersed in both land and aquatic environments. These creatures are dangerous to human life. There are about 2000 aquatic organisms that can be grouped under these two categories (Fitzgerald 2000). Most of the severe envenomation happens in the temperate or tropical waters of the Indo-Pacific region, with a trivial number of venomous fishes living in North American and European waters (Auerbach 1991; Russell 1958).

Envenomation in the aquatic habitats can be either surface stings (erythema, vesicles, and urticaria) or puncture wounds (bites, stings; Atkinson et al. 2006).

Four orders of teleost fishes are venomous (Church and Hodgson 2002; Haddad et al. 2003a, b; Halstead 1988a, b; Smith-Vaniz et al. 2001; Vetrano et al. 2002). They live in various territories extending from mountain streams to coral reefs and oceanic midwaters (Nelson 1994). Numerous kinds of signs result from cases of envenomation varying from blisters to intense pain, fever, and death (Haddad et al. 2003a, b; Halstead 1970, 1988a, b; Vetrano et al. 2002).

It is feasible to determine the venomous fish groups from around the world. They are presently disseminated among the catfishes (Siluriformes) and six groups of "acanthomorphs" or spiny-rayed fishes (Church and Hodgson 2002; Halstead 1970, 1988a, b; Nelson 1994; Smith-Vaniz et al. 2001): toadfishes (Batrachoidiformes); scorpionfishes (Scorpaeniformes: Scorpaenoidei); surgeonfishes, scats, and rabbitfishes (Perciformes: Acanthuroidei); saber-toothed blennies (Perciformes: Blennioidei); jacks (Perciformes: Percoidei); and stargazers and weever fishes (Perciformes: Trachinoidei).

Venomous creatures can produce poison in specific tissues or glands that are linked with application structures (e.g., stings), different from the poisonous organisms that usually produce poisons in nonspecialized tissues or accumulate them after ingestion of prey or algae and may be dangerous to people who consume them (Spanier 1987; Russell 1996). Antivenom is not obtainable for all types of envenomation, but antivenom for the treatment of stonefish and box stings is available. Moreover, conventional medications are used in some countries involving vinegar, fig juice, boiled cactus, heated stones, hot urine, hot water, and ice (Auerbach 1991; Russell 1958) (Fig. 6.5).

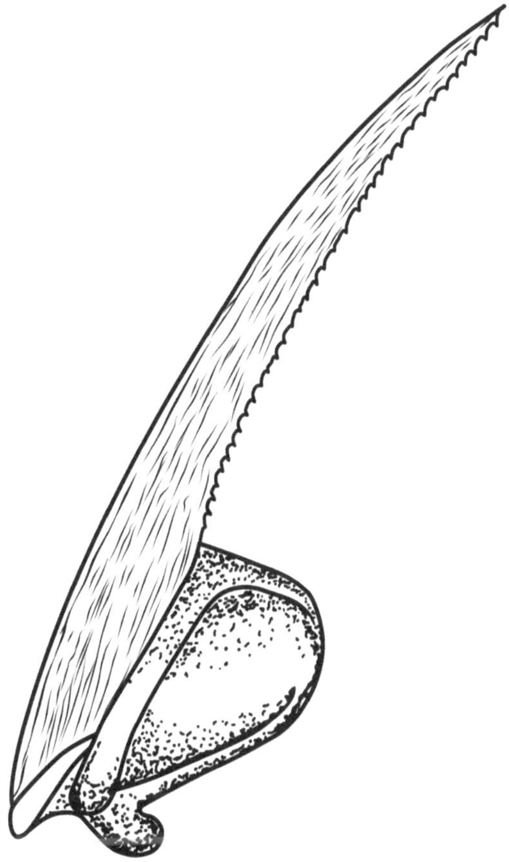

Fig. 6.5 Lateral view of the dorsal spine of the chimaera genus *Hydrolagus*

6.2.1 Venomous Chimaeras

The subclass Holocephali includes the chimaera, elephant fishes, or ratfishes, as they are diversely termed. They are a group of cartilaginous fishes having a single external opening on either side enclosed by an opercular skin fold with cartilaginous supports leading to a common branchial chamber unit which the true gill clefts open. Externally, chimaeras are compressed laterally, tapering posteriorly to a slender tail. The snout is rounded or conical, continued as a long-pointed beak, or bearing a curious shoe-shaped, proboscis. There are two dorsal fins. The first dorsal fin is triangular, normally higher than the second, and bordered anteriorly by a strong sharp-pointed bony spine which serves as a venom apparatus.

The information related to the venom of the chimaeras is very few and is restricted to two species, *Chimaera monstrata* of the North Atlantic Ocean and *Hydrolagus colliei* of the Pacific Ocean. Evans (1923, 1943) has made short reference to the dorsal spine of *H. affinis* but did not describe the venom apparatus that may present.

6.2.2 History of Research

In his review of the early on writings on venomous fishes, Autenrieth (1833) mentioned that a statement issued by Richter in 1764 in his *Ichthyologie*, namely, that the spines of the "sea cat" were greatly dreaded by Spanish fishermen and wounds formed by this fish may be fatal.

Coutière (1899) later implied to Richter's work and proposed that possibly Richter's seacat was *Chimaera monstrata* but misidentified he was not certain. Nonetheless, Richter's fish was confident to have three fins equipped with venomous spines. The fish under discussion was certainly a siluroid, or catfish, and not a chimaera. This matter has been considered by Phisalix (1922) and Halstead and Bunker (1952). A survey of the literature related to the medical ichthyology shows that Bottard (1889) was one of the first to propose the likelihood that the dorsal spine of chimaeras acts as a venom apparatus. However, in his investigations on *C. monstrata*, he was incapable to find what he believed to be a venom organ. Later, Evans (1923, 1943, 1944) acquired two specimens of *C. monstrata*, one from Lousy Bank in the North Sea and the other from the Mediterranean Sea. Through his histological work of the dorsal spines of these fishes, he was capable of discovering a layer of the glandular epithelium lying in the posterior groove of the dorsal spine. It was his estimation that this glandular tissue denoted a true venom organ. Unluckily, Evans did not have ready access to fresh material and was incapable of finishing his experimentations. The only other paper to appear thus far on the venom organ of chimaeras is by Halstead and Bunker (1952) on the Pacific ratfish *Hydrolagus colliei*. General references to venomous chimaeras have been published in the works of Halstead (1956, 1959), Halstead and Mitchell (1963), and Russell (1965a, b).

6.2.3 Morphology of the Venom Organ

The venom organ of the members of the genus *Hydrolagus* typically consists of the dorsal spine, the glandular epithelium of the spine and connecting membrane, and the enveloping integumentary sheath.

The developed spine is elongate, tapers to a sharp point, and comprises a cartilaginous core enveloped by a sheath of vasodentine. This construction is usually deemed to be an altered denticle (Goodrich 1909). A lateral view of the spine shows that it curves elegantly backward with the bend becoming vaguely more marked in the terminal one-fifth (Fig. 6.6). The anterior margin of the spine is keeled and smooth, while the two posterolateral margins are serrate with the teeth pointing in a downward direction. The anterior keel of the spine is translucent and colorless; the remainder is dark brown except for light areas where the pigment layer has been worn away. The anterolateral surface of the spine is marked by several fine ridges diminishing in size as they approached the posterior margin.

In cross section, the spine is roughly trigonal in outline with a pronounced anterior keel and a shallow posterior depression, named the interdentate depression. Resting within the depression is a strip of soft, grayish, sporadically pigmented tissue. Directing in a vertical path along the midline of this tissue is a dark line that marks the junction where the spine was previously linked with the dorsal fin. The most dorsal spine is attached with the anterior margin of the dorsal fin only at the proximal one-half of the spine, the terminal part having been torn away. The rest of the connecting membrane between the dorsal spine and fin is thickened and shiny white in appearance. When the dorsal fin is depressed and the spine is laying against the fin, the connecting membrane forms a fold that covers most of the dentate margin of the proximal one-half of the spine. Concluding from the anatomical verification presented, it seems that this membrane is a fundamental part of the venom organ of the dorsal spine. Seemingly, one of the functions of this membrane is to bath the dentate margins of the spine with secretions.

The base of the dorsal spine is located exactly above the origin of the pectoral fin. If the flesh at the base of the spine is removed away, it will be seen that the spine is contained of two portions, a hard dentinal part—continuing almost the entire length of the spine—and a short basal portion comprising the cartilage. The vasodentine portion is only an outer sheath covering the inner

Fig. 6.6 Callorhinchus milii (Bory de Saint-Vincent 1823). (Courtesy of Erik Tvskova)

core of the cartilage. This core is an elongate, superior extension of the cartilaginous base of the spine which continues up into the sheath for approximately six-sevenths of its total length. These anatomical relations can be best displayed by drying the spine. In the dried state, the dentical sheath can be taken off the rest of the spine, thus revealing the core.

The dorsal spine is reinforced by a highly modified, cartilaginous, neural spine of the anterior portion of the spinal column. The neural spine is greatly lifted and strongly compressed medially to form a moderately expanded anterior process and a broadly expanded, wing-like, posterior process. These two processes are disconnected above by a deep notch. The posterior side of the posterior neural process is depressed to form a fossa of moderate depth. This fossa is split into two lateral depressions by a low, vertical median ridge. The depressor muscles of the spine, one on each side, have their origin in these lateral depressions. Situated on the dorsum of the posterior neural process is the fossa in which the condyle of the dorsal spine articulates.

6.2.4 Mechanism of Intoxication

Envenomations from chimaeras are the result of wounds imposed by their single, sharp, dorsal spine.

6.2.5 Clinical Characteristics

Injuries caused by chimaeras are mostly of the puncture wound variety. There is very barely irritation existing as to the clinical features. Evans (1943) have stated that the wounds can be hurting and dangerous. Chimaera lacerations are mostly less severe than those produced by stingrays.

6.2.6 Treatment

Chimaera injuries should be handled in a manner like those produced by a stingray. Nevertheless, in the case of chimaera lacerations, there is less chance for the venomous glandular epithelium to become inserted in the wound of the victim due to the anatomical pattern of the venom organ. One must take extreme caution in handling live chimaeras. They can render a nasty bite, and the sting of the dorsal spine is extremely sharp.

6.2.7 Toxicology

Halstead and Bunker (1952) narrated on the toxicology of the chimaera venom saying that an intradermal scratch test achieved with the use of the dorsal spine of *Hydrolagus colliei* on the palmar surface of the left forearm of Halstead himself resulted in a zone of ischemia having raised borders bounded by a 2 cm zone of erythema, which exhibited in 3–4 min. In about 5 min, there was a mild dull ache in the area about the scratch which took for about 10 min. The eruption vanished in about 30 min. A comparable test was operated on the same arm using the dorsal spine of *Squalus acanthias*, but no reaction was delivered. It is assumed that the cause for this difference is that the glandular epithelium in *S. acanthias* is located further down the shaft of the spine, whereas in *H. colliei*, the glandular epithelium is sited nearer the top. In other words, to have successful analogous results, one would have to insert the *S. acanthias* spine much deeper into the skin.

Aqueous extracts were made from the glandular epithelium of the inter-dentate depression of the dorsal spine of *H. colliei*. A control test was made by making similar extracts from the skin taken from the back of the same fish. The extracts were injected intraperitoneally with laboratory mice. The extracts from the glandular epithelium of the dorsal spine proved to be toxic, resulting in immediate distress, pulling of the hind-limbs, jerking motions, motionlessness, and death of the mice on the third day. The control remained healthy.

6.2.8 Human Interaction

Hayes and Sim (2011) reported on two cases on envenomation by ratfish in the Western Isles Hospital (WIH), Stornoway, on the Isle of Lewis off Northwest Scotland. These cases are related to two foreign deep-sea fishermen who showed wounds of the spine of *Chimaera* species. The spine penetrated the lower limb and caused injuries.

In the first incidence, the victim is a 40-year-old French fisherman admitted to WIH after a penetrating *Chimaera* spine injury to his left foot. The fisherman had been doing his job onboard the fishing boat when, at around midnight, a fish spine of a *Chimaera* went through his leather boot into his left foot. He suffered serious pain and given two doses of co-amoxiclav 625 mg and diclofenac 100 mg on the way to hospital. After 21 h of his injury, the fisherman reached to the emergency department with a discolored left foot and the *Chimaera* spine-end protruding from a wound on the lateral side of his left fifth toe. It was not feasible to withdraw the spine under local anesthesia and therefore underwent general anesthesia. The spine had penetrated through the soft tissues of the fifth toe and into the plantar surface of the base of the fourth toe. The spine was eradicated completely, and the tract was irrigated with saline and cefuroxime. Postoperative recovery was uneventful and the victim was sent out back to France on oral cefuroxime, with plans for the drain be removed within 3–4 days.

In the second case, a 40-year-old Portuguese fisherman admitted to WIH with a piercing *Chimaera* spine injury to his left thigh. While handling a *Chimaera* aboard the vessel, the fisherman was stabbed by the fish's dorsal fin spine, and it inserted into the anterior part of his left thigh. The victim was transported from the offshore fishing boat by a coastguard helicopter to WIH. During the journey, Entonox gas was used for analgesia. On arrival at the hospital, the victim had intense peripheral pulses and was generally comfortable. A plain radiograph showed a whole spine deep in the left thigh soft tissues. The victim was given intravenous cefuroxime and metronidazole before operation. Under general anesthesia, the wound was surgically look at, and the *Chimaera* spine was found to have penetrated the subsartorial canal with the sharp tip adjacent to the femoral artery. The spine was safely taken. Instantly after operation, healing started slowly, with intravenous cefuroxime and metronidazole given prior to discharge home.

Hayes and Sim (2011) concluded that the incidences of the two victims discussed above demonstrate the distressing case of injury from defensive fish spines; it is obvious that the dorsal fin spines (8–9 cm in length) can pierce deep into soft tissues. For the second victim, where the sharp spine was located nearer to the femoral artery, it highlights the prospective for damage to major blood vessels, and they recommend cautious wound investigation where important structures may be involved.

6.2.9 The Toxicity of Chimaeras

Accounts on the toxicity of the chimaeras are contradictory. Therefore, Smith (1950) has confirmed that the flesh is wholesome, whereas Halstead and Bunker (1952) discovered that the oviduct was toxic and that the dorsal spine comprised a substance that produced a painful irritation on the surface of the skin (Hardy and Mackie 1971). The original investigation on the toxicity is that of Buniva (1803) (cited in Hardy and Mackie 1971) who at the same time showed some skepticisms on his information concerning the liver toxicity. Numerous researchers cite Lacepede who asserts that the Norwegians eat the liver and eggs without harmful consequence (Hardy and Mackie 1971). The only published laboratory result on this topic is the work of Halstead and Bunker (1952) who used only one fish, the oviduct of which was found to be toxic. Reports of De Clereq (1964) (cited in Hardy and Mackie 1971) and Halstead (1959, 1964) on this issue since the paper of Halstead and Bunker (1952) appeared are scanty and have simply been statements of claimed toxicity using their work as the main or sole reference. In an industrial fishery, separation of the various organs of the fish preceding to treatment is not economically practical; accordingly, if some of the organs have toxic constituents, they could, if they were not heat sensitive, introduce a toxin into the finished meal.

Hardy and Mackie (1971) reported on some deaths that did happen, but little implication is connected to them since a similar case of "non-specific" deaths is normally found with extracts of many edible fish species and indeed with many other protein materials (Hardy and Mackie 1971). In these cases, death occurs rapidly, usually within 30 min.

The iced storage tests show that postmortem changes in the flesh of the chimaeras at this temperature did not yield toxic materials. It was not conceivable, though, to set up studies that would repeat all possible types of storage, and it is practical that the past observations on the toxicity of ratfish were made on samples that had been differently stored after catching.

No verification of toxic products in the oviduct or the eggs themselves could be found. Halstead and Bunker (1952) perceived that the dorsal spine of ratfish caused painful lesions when drawn across his skin and observed that extracts prepared from this spine were fatal to mice after intraperitoneal injection. Throughout our fishing activities, the hands of the crew, some ten men, who were in constant contact with the fish, were repeatedly stabbed by the spines of these fish with no problematic effect. The crew were more fearful of similar cuts from redfish (*Sebastes marinus*). Ratfish are not new to the fishermen and there is no history in this area (north-east Scotland) of any toxic or unpleasant effects arising from handling these fish. It was felt in the light of these notes that the influences yielded by the spines of *Hydrolagus colliei* were due to substances not found in fish caught in the north-east Atlantic.

6.2.10 The Species of Chimaeras Living in the New Zealand Waters

In New Zealand wares, there are 12 species of chimaeras belonging to five genera and three families. Below is a description of some of these species.

Family: Callorhinchidae.

Callorhinchus milii (Bory de Saint-Vincent 1823) (Figs. 6.6 and 6.7).

Fig. 6.7 The head of Callorhinchus milii (Bory de Saint-Vincent 1823). (Courtesy of Erik Tvskova)

Common name: ghost shark.

Etymology: *Callorhinchus:* Tautonymous with *Chimaera callorynchus* Linnaeus 1758 (but unnecessarily emended from *-rynchus* to *-rhinchus*: *callum* (L.), hard skin; *rhynchus* (L.), snout, referring to peculiar hoe-shaped proboscis; *milii:* In honor of Bory de Saint-Vincent's "old friend" (translation) Pierre Bernard Milius (1773–1829), a French naval officer, naturalist, and civil servant.

Diagnostic Features

This species of chimaeras has an elongated, smooth, and torpedo-shaped body with two widely separated, triangular dorsal fins (Roberts et al. 2015). They use their shoe-shaped snouts to dig the ocean bottom for invertebrates and small fishes (Bray 2014). The body of the ghost shark is silvery in color with shining reflections and dark, variable markings on the sides (Bray 2014). This species of ghost shark has three cone pigments for color vision (like humans).

Distribution

This species is known from southern Australian and New Zealand waters (Roberts et al. 2015). Nonetheless, it has been assumed with some confirming verification that the New Zealand population may differ from the population found in Australian waters (Roberts et al. 2015).

Habitat, Biology, and Ecology

The ghost shark is a marine species that enters brackish waters sometimes (Riede 2004). It lives at a depth range 0–227 m (Cox and Francis 1997). Individuals of this species reach a maximum total length of 1250 mm (Cox and Francis 1997), with a common total length of 750 mm TL (Armitage et al. 1994). The maximum age reported is 15 years (Bray 2014).

This species migrates into large estuaries and inshore bays in the spring to breed (Last and Stevens 1994). It feeds mainly on shellfish (Cox and Francis 1997). The females are oviparous (Dulvy and Reynolds 1997). The eggs are enclosed in horny shells (Breder and Rosen 1966) and are laid on sandy or muddy bottoms and take up to 8 months to hatch (Last and Stevens 1994). Embryos feed solely on yolk (Dulvy and Reynolds 1997). Young hatch at about 15 cm (Cox and Francis 1997). The males have a small, club-like protrusion on the head and long copulation organs near the pelvic fins (Wheeler 1977).

Economic Value

The meat of this species is known to have good quality for human consumption (Wheeler 1977).

Conservation Status

Walker et al. (2015) evaluated this species as Least Concern in the Red List of the IUCN based on the following reasons. This chimaera species is reasonably abundant and endemic to the continental shelf of each of southern Australia and New Zealand. The species has intermediate biological productivity; even though its maximum age is in the range of 9–20 years, it matures relatively early. In southern Australia, the species is utilized over its entire range. The present exploitation rates are regarded as maintainable.

Family: Chimaeridae.

Hydrolagus bemisi (Didier 2002) (Fig. 6.8).

Common name: pale ghost shark.

Etymology: *Hydrolagus: hydro-*, combining form of *hydor* (Gr.), water; *lagos* (Gr.), hare, i.e., "water rabbit," probably referring to three

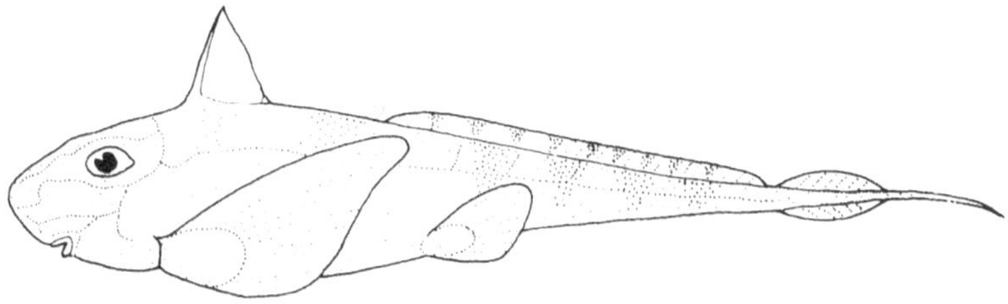

Fig. 6.8 Hydrolagus bemisi (Didier 2002). (Courtesy of Tony Ayling)

pairs of tooth plates, which tend to protrude from the mouth like a rabbit's incisors; *bemisi:* In honor of ichthyologist William E. Bemis, now at Cornell University, "longtime mentor and friend, and a leader in ichthyological research".

Diagnostic Features

Hydrolagus bemisi has a medium-sized body, with a tapering tail. The dorsal fin spine extends to the origin of the second dorsal fin when lowered. The second dorsal fin is elongated. The lateral line canals are horizontal and hyomandibular sharing a common branch. The caudal filament is long and stout (Didier 2002). The body is pale brown, fading to white ventrally. This species of chimaera can be separated from other chimaera species that live in New Zealand such as *H. novaezealandiae* and *H. homonycteris* as it has a pale silvery color with no pattern or spots (Roberts et al. 2015).

Distribution

This species is endemic to New Zealand waters.

Habitat, Biology, and Ecology

This species of chimaera is marine and lives at a depth range 400–1100 m (Didier 2002), but is usually found at 500–700 m (Didier 2002). Individuals of this species reach a maximum total length of 1120 mm (Didier 2002). This species can live between 15 and 22 years, although the lack of data still makes this unreliable (Finucci et al. 2021). The females are oviparous (Breder and Rosen 1966) and the eggs are encased in horny shells (Breder and Rosen 1966).

Economic Value

The flesh of this chimaera has a commercial value.

Conservation Status

Finucci and Kyne (2018) evaluated this species as Least Concern in the Red List of the IUCN based on the following reasons. The size is medium and inhabits deep water. It is confined to the New Zealand waters. The population structure has not been studied so far. The catch of this species has been declining recently. Individuals of this species are found in the bycatch of trawlers aiming for other species. Additionally, the species is extensively distributed and managed.

Hydrolagus homonycteris (Didier 2008) (Fig. 6.9).

Common name: black ghost shark.

Etymology: *Hydrolagus: hydro-*, combining form of *hydor* (Gr.), water; *lagos* (Gr.), hare, i.e., "water rabbit," probably referring to three pairs of tooth plates, which tend to protrude from the mouth like a rabbit's incisors; *homonycteris: homo-*, from *hominis* (L.), man; *nykteris* (Gr.), bat, referring to Thomas A. Griffiths, bat systematist known as "bat man," who introduced Didier to chimaeroid fishes.

Diagnostic Features

The ventral caudal fin is not indented at its origin to form a separate anal fin. The eyes are large, and the snout is slightly pointed. The dorsal spine is large. The claspers have areas of bristles on them and are rather short and thin. This species of chimaera can be separated from the rest of chimaera species by the following set of characters. The color of the body is an even dark black or blackish brown. The dorsal fin spine is longer than the height of first dorsal fin. The pelvic fins are conspicuously round. The males are with pelvic claspers that are dark at the base with pale tips (Didier 2008).

Fig. 6.9 Hydrolagus homonycteris (Didier 2008). (Courtesy of CSIRO National Fish Collection)

Distribution
The black ghost shark is reported from the southwestern Pacific Ocean off the coasts of southeastern Australia (Carrier et al. 2012) from Portland, Victoria, to Ulladulla, New South Wales, covering the states of New South Wales, Tasmania, and Victoria (Roberts et al. 2015). This species is also reported in New Zealand from the West Norfolk Ridge to the Campbell Plateau (Roberts et al. 2015). The species is found at the continental slope and seamounts (Reardon and Bray 2023).

Habitat, Biology, and Ecology
Hydrolagus homonycteris is a marine species that lives at a depth range of 866–1447 m (Didier 2008). It reaches a maximum total length of 1090 mm (Didier 2008). In Australia, it lives in waters 870–1450 m deep, while it lives at slightly shallower depths in New Zealand, at 500–1400 m (Reardon and Bray 2023).

Economic Value
The meat of this species is used for human consumption.

Conservation Status
Theiss et al. (2015) evaluated *Hydrolagus homonycteris* as Least Concern in the Red List of the IUCN based on the following reasons. This species is distributed on the continental slopes and seamounts at depths of 870–1450 m off southern Australia and at 500–1400 m off New Zealand. Because of the intersection in habitat, depth, and range, it has been a bycatch of benthic deepwater commercial trawl fisheries. The catch levels of this species are very low.

Hydrolagus trolli (Didier and Séret 2002) (Fig. 6.10).
Common name: pointy-nosed blue chimaera.
Etymology: *Hydrolagus: hydro-*, combining form of *hydor* (Gr.), water; *lagos* (Gr.), hare, i.e., “water rabbit,” probably referring to three pairs of tooth plates, which tend to protrude from the mouth like a rabbit’s incisors; *trolli:* In honor of American artist Ray Troll (b. 1954), Ketchikan, Alaska (USA), “one of the few true chimaeroid lovers of the world ... for his valiant efforts to increase ratfish awareness worldwide”.

Diagnostic Features
The orbit is surrounded by a dark line and dark shadowing beside edges of lateral-line canals. The snout is pointed. The preopercular and oral lateral-line canals share a common branch. The D1 lateral line is concave with pale margin. The males are with frontal tenaculum deeply curved and upturned distally with spines along the dorsal upturned edge. The prepelvic tenacula are deeply indented along the distal margin. The pelvic claspers are with a pale fleshy distal lobe, divided distally for one-third their length, where the tips are usually extending beyond the distal edge of pelvic fins. The ventral caudal fin is not deeply indented at its origin to form a separate anal fin. A venomous spine is present in front of the dorsal fin, which is used in defence.

Distribution
The pointy-nosed blue chimaera lives in the Pacific and Southern Oceans, with records from near New Caledonia, New Zealand, and southern Australia. The records from South Africa refer to

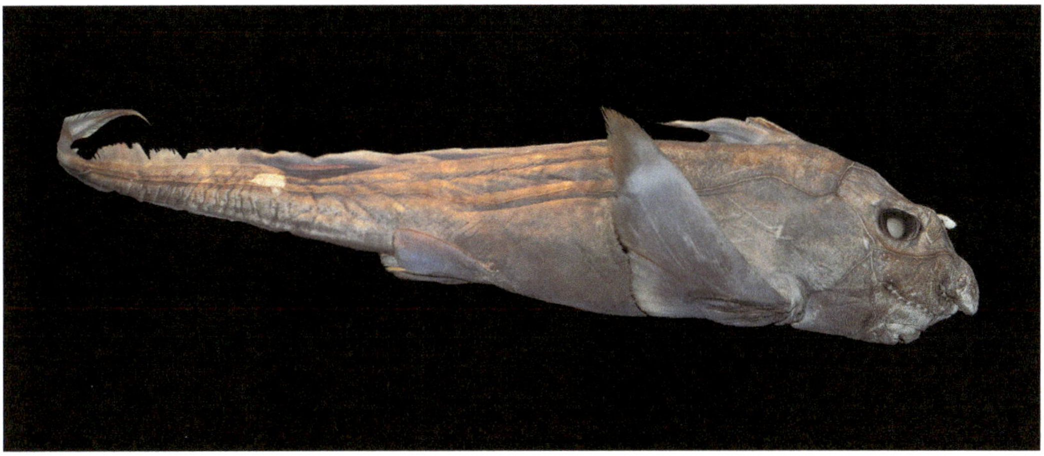

Fig. 6.10 Hydrolagus trolli (Didier and Séret 2002). (Courtesy of Citron Wikipedia)

other species. It is a deep-water species that has been recorded on deep continental and insular slopes at depths between 610 and 2000 m (2000 and 6560 ft), but more commonly below 1000 m (3300 ft) (Compagno and Dagit 2015).

Habitat, Biology, and Ecology

This chimaera is a marine species that inhabits at a depth range 612–1707 m (Didier and Séret 2002). The females are oviparous (Breder and Rosen 1966) and the eggs are encased in horny shells (Breder and Rosen 1966). Individuals of this species reach sexual maturity at 550–650 mm TL (Walovich et al. 2017).

Economic Value

The meat of this chimaera has a commercial value.

Conservation Status

Compagno and Dagit (2015) evaluated this chimaera as Least Concern in the Red List of the IUCN based on the following reasons. This species of chimaera has a wide geographical distribution in the Southern Ocean and could be relatively common in New Zealand waters, but uncommon in other regions. The living habitat of this chimaera is outside the trawling activity area, which makes this species being among the bycatches.

Family: Rhinochimaeridae

Harriotta raleighana (Goode and Bean 1895) (Fig. 6.11).

Common name: Pacific long-nosed chimaera.

Etymology: Harriotta: [an]a- (L.), belonging to; per Goode & Bean (1896), in honor of Thomas Harriott (ca. 1560–1621), English astronomer, mathematician, ethnographer, and translator, who published the first English work on American natural history (1588); raleighana: In honor of Sir Walter Raleigh (ca. 1554–1618), "philosopher and explorer, by whom the first English scientific expedition was sent to the New World" (Raleigh funded a 1585 expedition to Roanoke Island, North Carolina, with Thomas Harriott.

Diagnostic Features

The diagnostic features of *Harriotta raleighana* are the eyes are small; the interorbital space is broad; the snout tip is smooth; the first dorsal fin is low; the length of the dorsal spine anterior margin is 16% BDL; the pectoral and pelvic fins are broadly rounded; the interdorsal space is narrow; the length of the caudal dorsal margin is longer than the caudal ventral margin; the length of the caudal upper ray is longer than that of the lower ray; the infraorbital canal is strongly curved anterior to the eye, forming an S-shaped loop; the claspers are long and thick; and the tip of the

Fig. 6.11 Harriotta raleighana (Goode and Bean 1895). (Courtesy of Alexie Orlov, Russia)

clasper is extremely swelled so as to form a large, elongate, and fleshy knob, covered with 16–17 small, multifaceted, hemp-leaf-shaped spines, most of which were serially arranged along its dorsomesial margin (Nakayama et al. 2019).

Distribution

This species of chimaera has a cosmopolitan distribution.

Habitat, Biology, and Ecology

The Pacific long-nosed chimaera is a marine species living at a depth range of 200–3100 m (Coad and Reist 2018), but it is usually found at 500–2000 m (Angulo et al. 2014). Individuals of this species reach a maximum total length of 1200 mm for males and 10,250 mm for females (Cox and Francis 1997). The maximum reported weight of this species is 1.6 kg (Finucci 2017). This species feeds mostly on shellfish and crustaceans (Last and Stevens 1994). This species reproduces by laying eggs. These eggs are contained within a central chamber surrounded by a web-like structure. The female lays a pair of eggs several times per season (Didier 2004).

Economic Value

This species has no commercial value so far.

Conservation Status

Dagit et al. (2016) evaluated this species as Least Concern in the Red List of the IUCN based on the following reasons. This species of chimaera has extensive geographical distribution. It is common in the northern Atlantic, Northwest Pacific, and Southwest Pacific; nonetheless, few data are available about its biology. Individuals of this species get included in the deep-water trawler as a bycatch. The deep-water trawling might affect the population of this species if such activity is increased.

6.2.11 Venomous Sharks

Venomous sharks are restricted to those species acquiring dorsal fin spines, explicitly certain species of the families Heterodontidae, Squalidae, and Dalatiidae. These families have at least 11 genera known to acquire dorsal spines. Nonetheless, only the two species *Heterodontus francisci* and *Squalus acanthias* are known to be equipped with a venomous spine. The rest of the species, corresponding to the literature, have not been investigated by venomologists. Some of the dalatiid sharks have only a single elementary fin spine, or the spines are completely absent. Whether or not the dorsal fin spine of dalatiid sharks is a genuine venom apparatus is not known.

Oppian of Corycus in Cicilia (ca. AD 172–210) was the first to describe the dangers of spined sharks in his fishing poem, the *Halieutica*, in which he referred specifically to the dogfish, thus:

> *Cruel spines defend some fishes, as the Goby, fond of sands and rocks, the scorpion, swallows' fleet, dragons and dogfish, from their prickly mail, well names the spines. These in punctures sharp, a fatal poison from their spines inject.*

Awareness in the venomous competences of the spined sharks seemingly dropped until the latter part of the seventeenth century when William Dampier (1652–1715) (cited by Coutière 1899) stated that there was a dogfish living the waters about Guayaquil, Ecuador, which imposed fatal wounds. Bottard (1889) is said to have been the first to have studied the likelihoods of a venom apparatus in *Squalus acanthias*, but he presumed that none was present.

Ritter (1900) explored quite utterly the developmental anatomy of *Squalus* spines but did not comment upon their use as a venom organ. The importance of the dorsal spines of sharks as a true venom organ was fundamentally the discovery of the English surgeon H. Mir Evans (1920) who first recounted on a case of acute edema caused by a pierce from a dorsal spine of this fish. Later investigations by Evans (1921, 1943) showed a glandular structure lying in the posterior groove of the spine of the dogfish.

Roedel and Riply (1950) confirmed that the spines of *Squalus acanthias* were possibly mildly venomous and served as a defensive weapon. Halstead (1956, 1959) and Halstead and Mitchell (1963) have concisely discussed the venomous characteristics of spined sharks.

6.2.12 Mechanism of Intoxication

Wounds are imposed by the two dorsal spines that are situated closest to the anterior margins of each of the two dorsal fins. These spines are most often confronted by humans in trying to remove the fish from a hook and line or from a net in which the fish has become entrapped. Skin divers' lack of caution in holding spiny dogfish that have been speared.

6.2.13 Clinical Aspects

The symptoms are characterized of instant, extreme, stabbing pain that may last for a period of hours. The pain may be come with and followed by a generalized erythema and severe swelling of the affected part. Tenderness of the distressed part may stay for several days. According to Courtière (1899), dogfish stings may be fatal. Evans (1920, 1923, 1943) gave a clinical report on the attack of the dogfish.

6.2.14 Treatment

Wounds that resulted from spined sharks are normally of the puncture wound type. Since shark spines do not have an enveloping integumentary sheath and the bulk of the glandular tissue is located near the base of the spine, it would be an exceptional case for the glandular tissue to become inserted in the wound of the victim. In most cases, impacts resulting directly from the action of the venom are of a minor worry. Nonetheless, it is suitable to irrigate the wound with salt water and do whatever debridement may be necessary if the tissue bleeds, or if there is no bleeding, then a reasonable bleeding should be encouraged. The pain is usually gentle in comparison with most stingray stings, but care may be needed. The extremity should be submerged in hot water for a period of 30 min or more at as high a temperature as the victim can tolerate without doing further injury. The addition of sodium chloride or magnesium sulphate to the water is optional. Suturing infections from shark spines may sometimes occur, and antibiotic therapy may be needed. Raise of the injured limbs is recommended.

6.2.15 Human Interaction

In their list of the venomous creatures of New Zealand, Slaughter et al. (2009) included the dogfish *Squalus acanthias* as one of the venomous fish species in New Zealand. Haddad and Gadig (2005) reported on a case of sting caused by a member of the genus *Squalus* in Brazil. The incidence happened in Ubatuba (North Coast of São Paulo State, Brazil). The victim was a 54-year-old professional fisherman injured after handling a spiny dogfish of the genus *Squalus* (called the "cação-bagre" or "shark catfish" by the local fishermen). The victim injured near the little finger in his left hand by the spine anterior to the dorsal fin of the dogfish. An intense pain at the sting site (entirely disproportionate to the wound) was developed, which diminished over 6 hours, without treatment. The puncture site established erythema and edema, the latter lasting around 2 weeks, with local cutaneous thickening.

6.2.16 Species of the Venomous Sharks

Family: Squalidae.
Squalus acanthias (Linnaeus 1758) (Fig. 6.12).
Common name: spiny dogfish.
Etymology: *Squalus:* genus name from Latin *squalus*, meaning sh.ark (Hart 1973); *acanthias:* genus name from the Latin *squalus*, meaning shark; species name from the Greek *akanthias* referring to the spines (Hart 1973).

Diagnostic Features
The size of this dogfish is medium. The body is very slender, with a narrow head. The snout is moderately long. The anterior nasal flap is single-lobed. The first dorsal fin is small. The origin of the first dorsal fin is located just posterior to pectoral fin-free rear tips. The spine of the first dorsal fin is short. The caudal fin is pale in color, with poorly demarcated, whitish margin. A blackish caudal blotch is present at the apex of the upper lobe. The anterior margins of both lobes are whitish in juveniles. The dark caudal bar is absent. The dorsal and lateral surfaces of the body is bluish gray with an irregular array of moderately large white spots. The flank denticles are tricuspid (White et al. 2007).

Distribution
This species of dogfish is reported from the Western Atlantic Ocean, Indo-Pacific, Mediterranean, and Black Sea (Schilling et al. 2019). It is absent from the North Pacific (Mecklenburg et al. 2018). All records from

Fig. 6.12 Squalus acanthias (Linnaeus 1758). (Courtesy of Clinton Duffy, New Zealand)

North Pacific refer to *Squalus suckleyi* (Girard 1855) (Ref. Ebert et al. 2010).

Habitat, Biology, and Ecology

Squalus acanthias is a marine species that sometimes enters brackish water (Riede 2004). It lives at a depth range of 0–1460 m (Cox and Francis 1997), but is usually found at 50–300 m (Mecklenburg et al. 2002). Individuals of this species reach maturity at 809, with a maximum total length of 1600 mm (Mecklenburg et al. 2018). The maximum reported weight is 9.1 kg (Schilling et al. 2019) and the maximum reported age is 75 years (Cailliet et al. 2001).

This dogfish prefers a demersal, inshore, and offshore habitats (Compagno 1984). It is usually near the bottom, but also in midwater and at the surface (Cox and Francis 1997). It is reported to go in fresh water (McAllister 1990) but cannot survive there for more than a few hours (Compagno 1984). This species of dogfish is transoceanic migrations recorded, but rare (Pawson and Ellis 2005). Its longevity in the northern Atlantic is about 35–50 years (Fordham 2005), but most live only 20–24 years (Muus and Nielsen 1999). The gestation period is 2 years (Bianchi et al. 1999) and females are ovoviviparous (Breder and Rosen 1966). Individuals of this species feed on a diversity of prey, ranging from comb jellyfish, squid, mackerel, and herring to a wide array of benthic fishes, shrimps, crabs, and even sea cucumbers (Murdy and Musick 2013). Mating probably occurs in winter. The gestation period is 18–24 months, longest known of all chondrichthyans (Cox and Francis 1997). The size at birth is 18–30 cm (Cox and Francis 1997). In the northeast Atlantic, pups are born in winter, with a size at birth of about 26–28 cm (Pawson and Ellis 2005). The mother shark experiences a series of rhythmic contractions, just like in mammals, and the young are delivered head first (McEachran and Branstetter 1984).

Economic Value

The meat of this dogfish is utilized for human consumption, liver oil, vitamins, sandpaper, leather, and fertilizer (Compagno 1984). It is eaten fried, broiled, and baked (Frimodt 1995).

Conservation Status

Finucci et al. (2020) evaluated *Squalus acanthias* as Vulnerable in the Red List of the IUCN based on the following reasons. This dogfish is medium sized and has a global distribution, except for the North Pacific and Southern Ocean. It has been reported in both inshore and offshore marine habitats from the surface to 1978 m.

6.3 Venomous Stingrays

The notable stingray groups are Gymnurid (butterfly rays), Urolophid (round stingray), Myliobatid (bat or eagle rays), and Dasyatid (proper stingrays; Daly and Scharf 2012; Trickett et al. 2009). These fishes normally live and are encountered in the waters off coastal regions. They use concealment by partly dipping themselves in the sand (Daly and Scharf 2012; Auerbach 1991).

One eminent stingray attack was that of the Australian wildlife knowledgeable Stephen Robert "Steve" Irwin, "The Crocodile Hunter." The attack event that occurred to Steve is a characteristic stingray strike against humans and should be learnt. In this section of the book, I give a focus about the death of this great man and quote the death story of Steve Irwin as given by Wikipedia (2016):

> On 4 September 2006, Irwin was on location at Batt Reef, near Port Douglas, Queensland, taking part in the production of the documentary series Ocean's Deadliest. During a lull in filming caused by inclement weather, Irwin decided to snorkel in shallow waters while being filmed to provide footage for his daughter's television program.
>
> While swimming in chest-deep water, Irwin approached a stingray with an approximate span of 2 m (6.5 ft) from the rear, to film it swimming away.
>
> According to the incident's only witness, "All of a sudden [the stingray] propped on its front and started stabbing wildly with its tail. Hundreds of strikes in a few seconds." Irwin initially believed he only had a punctured lung; the stingray's barb pierced his heart, causing him to bleed to death (Selby 2014; Bond 2014). The stingray's behavior appeared to have been a defensive response to being boxed in. Crewmembers aboard Irwin's boat

administered CPR and rushed him to shore. Medical staff pronounced him dead at the scene (The Age 2006; Sydney Morning Herald 2006; Callinan 2006). Irwin's death is believed to be the only fatality from a stingray ever captured on video (CNN, Reuter 2006).

Queensland state police as part their mandatory investigations viewed footage of the incident. All copies of the footage were then destroyed at the request of Irwin's family (Gerard and Koch 2006). Production was completed on Ocean's Deadliest, which was broadcast in the US on the Discovery Channel on 21 January 2007. The documentary was completed with footage shot in the weeks following the accident, but without including any mention of Irwin's accidental death (The Daily Telegraph 2006; International Business Times 2007).

Background

Humans' knowledge of stingrays and their poisonous spines traced back to the time of Aristotle (324–322 BC), when he stated it in his book, *Historia Animalium* (Halstead 1970). During medieval times, not much information was presented about the poison of the stingray. All available data at that time was a duplication of that information given by Aristotle, Nicander, Dioscorides, and others (Halstead 1970). The plausible original illustration of a stingray was of the European common stingray, *Dasyatis pastinaca*, in Pierre Belon's *De Aquatilibus Libri Duo* published in Paris in 1553 (Schwartz 2005). Halstead (1970) wrote about the first fatal case by stingray described by Lycophoran (285–247 BC). It is about the death of Ulysses by a spear imagined having been tipped by a stingray sting. Other early European papers by Grevin (1568) and Eupharasen (1790) treated *D. pastinaca* and *Aetobatus narinari*, respectively.

Structure of the Venom Apparatus

In stingrays, the venomous system consists of (1) caudal appendages, (2) the spine enveloped in an integumentary sheath, (3) a venomous gland, and (4) the cuneiform area below the spine. The venomous constructions of different species of stingrays have diverse shapes. Such changes impact the capability of the fish to sting. Halstead and Bunker (1953) documented four types of venomous forms. It is feasible to locate more than one type within a single family of stingrays. The variations in shape of the four types of venomous apparatus are linked to the shape of the caudal fin and the thickness of its base (Fig. 6.13).

Halstead (1967) explained the typical gross anatomy of the venomous organ as follows (Fig. 6.14):

> The spine in the venomous structure is bilaterally retroserrate and covered with an integumentary sheath and composed of two layers, an inner vasodentine and a thinner outer layer of enamel. There are a number of shallow longitudinal furrows on the surface of the spine, where they are more clearly seen at its base. The dentations at the two margins of the spine are with medial grooves known as "ventrolateral glandular groove," which [is] covered with thin gray tissue supplied with blood vessels and connected with the integumentary sheath in the dermis. The epidermis will rapture under a pressure, when the spine enters the

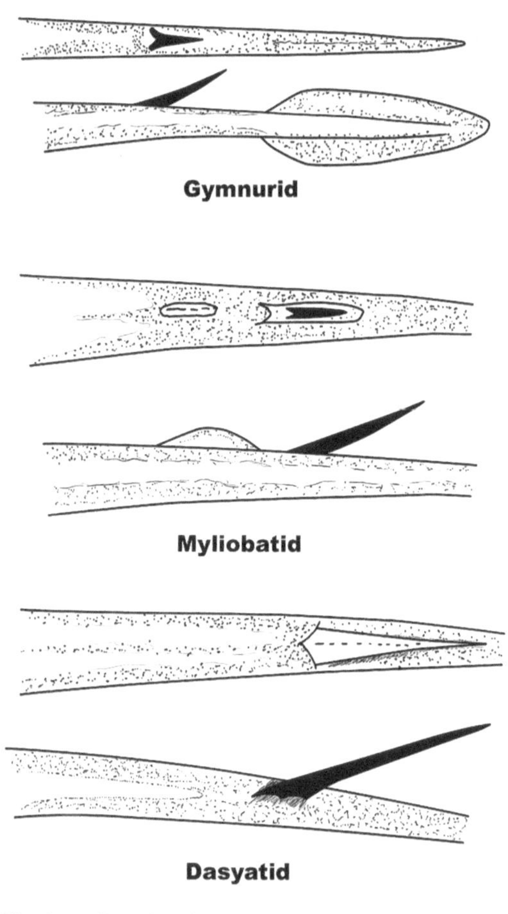

Fig. 6.13 Drawing showing the various types of stingray caudal appendages. (After Halstead 1967)

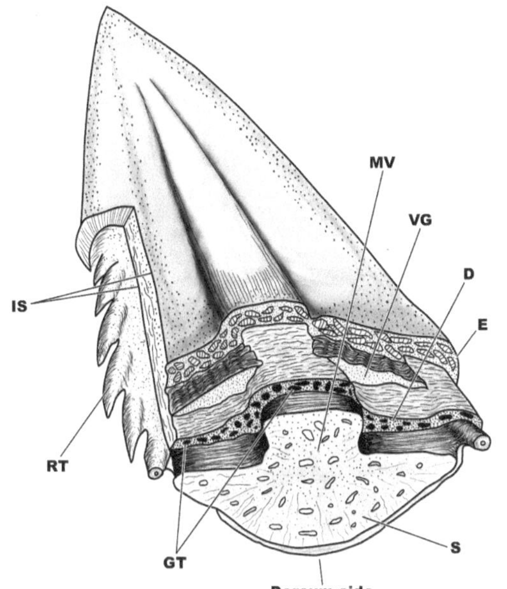

Fig. 6.14 Drawing showing the gross anatomy of a typical stingray sting. IS integumentary sheath; TR retrorse teeth; GT glandular triangle; E epidermis; VG venom gland; D dermis; MV median ventral ridge; S spine. (After Halstead 1967)

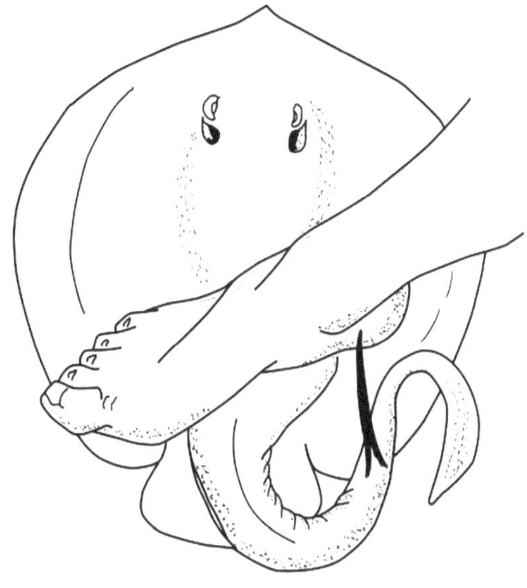

Fig. 6.15 Drawing showing the method by which a stingray usually inflicts its sting. (After Halstead 1967)

> body of the victim and resulted in damage to the stingray epidermis, and further traumatization of the recipient's wound and secretion and envenomation of poison inside the wound will occur. The integument of the cover which lies underneath the spine secretes venom. Therefore, the spine is fully covered with poison once it enters the body of the victim.

Schwartz (2005) discovered that the total number of serrations on the spine could reveal the environment of the stingray. For instance, total serrations above 100 indicate an open water ocean species (*Dasyatis centroura*, *Pteroplatytrygon violacea*, *Aetobatus narinari*); 70+ a midwater species (*Dasyatis pastinaca*, *Pteromylaeus bovinus*, *Himantura uarnak*, and *Taeniura meyeni*); 50+ near-substrate inhabitant (*Gymnura altavela*, *Myliobatis aquila*, *Dasyatis margarita*); 25–50 substrate species (*Taeniura grabata*, *Urogymnus ukpam*); and below 25 a fresh water species.

Causative Agent

Various investigators approve about the mechanism by which the stingray imposes its spines into the body of the victim. The primary points in this mechanism are that the stingray has no habit of striking humans, but when the ray is disturbed, it instinctively swings a barbed tail upward, which can impose deep puncture wounds (Gray et al. 1988). The spines have retroserrated teeth, making removal particularly not easy, which can cause a retained tail in the wound (Escoubas et al. 2000). The serrated spines are enclosed by an epithelial layer that has venom secretory cells located in the epithelium or in close contact with it (Fletcher et al. 1996; Saminathan et al. 2006) (Fig. 6.15). Serrated spines of rays may cause mechanical damage too (Wright 2009).

Injury is attained in a subcutaneous mass of granulomatous dermatitis and panniculitis with large zones of necrobiosis (Tartar et al. 2013). In the complex cases, the sting might break and provoke the retention of dentine pieces in the wound. Bacterial infections that are particularly caused by *Pseudomonas* sp. and *Staphylococcus* sp. are also regularly linked with these wounds (Haddad et al. 2004; Haris and Chapman 1995).

Investigations on toxicology and envenoming caused by elasmobranchs convey mostly cases connected with stingrays of suborder Myliobatoidei (Haddad et al. 2004), as they are the most clinically significant because their venom may result in increasing local pain which may be extended to imply the entire limb swelling and a characteristic bluish-white appearance of the wound.

Aquatic animals generate an enormous number of metabolic complex molecules, such as alkaloids, steroids, peptides, and proteins with chemical and pharmacological properties, distinct from those exhibited by the poisons of terrestrial animals (Ravi 2006). There seem to be various chemicals in the venom. There is some neurotoxicity (Ravi 2006), cardiotoxicity, and circulatory disturbances (Ravi 2006). Some studies exhibited that the venom of rays contains serotonin, 50-nucleotidase and phosphodiesterases (Fenner et al. 1989).

Symptoms

The spines and the venom gland may be broken during the attack and the spine may continue to be in the wound, which may be large and serrated; the victim then undergoes severe pain from the injected venom. Cuts made by a ray's stings can be escorted by intense local pain and can cause moderate to severe problems such as nausea, vomiting, salivation, sweating, respiratory depression, muscle fasciculations, convulsions, edema, and ischemic necrosis (Ravi 2007; Haris and Chapman 1995; Dehghani et al. 2009; Forrester 2005).

Treatment

Anticoagulants play a fundamental role as agents for the prevention and treatment of thromboembolic illness (Auerbach and Norris 2012). For more than five decades, anticoagulant drugs containing heparins, vitamin K-antagonists, and their byproducts have been the principal medicines in a clinical setting. Although their effectiveness stays undoubted, the harmful life-threatening side effects of these drugs have also been verified (Tartar et al. 2013).

Prevention

The natural position that the stingray takes is to bury its whole body in the sand or mud at the bottom of the area in which it is found. Therefore, it forms a threat to those who wade with bare feet in water occupied by them. To avoid stingray attacks, waders should shuffle the water along the bottom before stepping in.

6.3.1 Human Interaction

Vicki Waterhouse from *The New Zealand Herald* reported from Bay of Plenty on 24 April 2008 on case of stingray attack on human.

> *The report stated: A Tauranga man has told how he screamed when a stingray thrust its barb into his leg on a boat—and then had to endure an agonizing 2 1/2-hour trip back to shore. Dan Rawlinson, a commercial fisherman from Otumoetai, suffered paralysis after the attack on Tuesday just before 1 pm. Mr Rawlinson, 32, had been at sea for 4 days when he and his crew pulled in a net with a catch that included a stingray. But the stingray, about half the size of a car bonnet, attacked. 'It saw me and bent its tail around; it looked like it wanted revenge on me for catching it.' Mr Rawlinson said: 'I'm not usually out on the deck; I'm the captain so I'm usually inside. I went outside to do 5 minutes' worth of work—I should've stayed where I was.' The stingray jabbed him straight in the left calf. The barb went about 3 cm deep. The feeling was not one Mr Rawlinson wanted to remember. 'I screamed for a start ... I ripped it out and chucked the stingray back over the side,' he said. 'It was like 30 times worse than a bee sting.'*
> *The barbs of the ray's tail snagged on the inside of his leg as he tried to remove it, and with each jerk, his leg was torn further. 'It practically paralyzed my leg; I couldn't do anything with it,' he said.*
>
> *In agonizing pain, there was yet another hurdle for Mr Rawlinson to face—the boat was just off Motiti Island, 2 1/2 hours away from shore, and therefore just as far away from help. But he just got on with it, packed up, and drove the boat home. 'I just had my foot up on the rest and it was just throbbing,' he said. 'I'm obviously not allergic.' Mr Rawlinson eventually got back to dry land and was picked up by St John Ambulance as soon as he arrived. He was a little disappointed that he had gone through so much pain for something that looked quite minor, he said afterward. 'It just looks like a little puncture wound,' he said. 'I was just going to put a plaster on it.' Mr Rawlinson was given antibiotics and did not have to stay overnight in the hospital. 'There were no symptoms that the poison was running through my veins. It was just sort of an isolated spot on my leg, so I shouldn't die,' he laughed. But he cheerfully conceded it was a drama he could have done without. 'The life of a fisherman is*

> *not that easy.' The attack has made him think of crocodile hunter Steve Irwin, who died after a stingray struck him in the chest. 'What happened to Steve Irwin, it was where it went in'.*

Black et al. (2022) reported on a case of a stingray injury that happened to a 48-year-old man who was hiking toward the shore at Oreti Beach, near Invercargill in water about 30 cm deep, when he felt a severe pain in the right foot trailed unexpectedly by a stab wound to the left chest.

There was a sizable disruption in the water to the right of him, and a large marine creature briefly surfaced which he caught a glimpse of. He controlled to get out of the water without help, and an ambulance was called. The patient presented to a rural hospital with severe global chest pain, diaphoresis, dyspnea, and hypoxia (oxygen saturation of 90% while on 15 L of oxygen via Hudson mask). He showed to have a 4 cm wound in the left chest with a visible subcutaneous tissue; this was in the seventh intercostal space in the midaxillary line. He also had a 5 mm puncture site on the plantar aspect of the right foot just proximal to the fifth digit. On chest examination, he had bilateral air entry with no clear lateralizing signs. A small pneumothorax appeared on chest X-ray, and an urgent intercostal drain was placed based on severe symptoms with no large release of air and no significant improvement to his oxygenation.

He was further examined with extended Focused Assessment with Sonography in Trauma (e-FAST), which showed no free fluid in the abdomen, chest, or pericardium. A CT scan of the chest and abdomen was also performed which showed a small residual pneumothorax with no other significant injury seen in the chest or abdomen.

The patient was given a liberal IV opioids during his early presentation to ED; he was given 250 mcg of fentanyl, 8 mg of morphine, and 60 mg of ketamine with little effect on his pain. His chest wound was surveyed in the emergency room with no obvious intrusion into the thoracic cavity found. After a washout and exploration of the wound with no debris found, the incision was primarily closed with non-absorbable sutures. His pain was markedly improved the next day, and the intercostal drain was removed.

The New Zealand Herald on 16 July 2017 reported on a case of a stingray attack on human in the vicinity of Whangarei.

> The report stated: *A man had his arm sliced up by a stingray this morning near Whangarei. Coastguard were called around 10:20 am this morning to the 35-year-old who had 'severe laceration to his arm,' Coastguard Northern Region posted to Facebook. Volunteers on the* Circa *Rescue boat arrived around 10:45 am. After assessing his wound and giving first aid, the coastguard team transported the man back to Marsden Cove, and to a waiting ambulance." "I stood on what at the time felt like a really big kingfish, like a really, fat, solid fish. I stumbled on it and fell backward and then when I fell backward into the water, I got a massive jab in my side." "I initially thought I had been bitten by a shark. I checked down my side because it felt like I would have a hole there. I checked twice and then noticed I only had a small hole. It was confusing at the time." "The pain from it was totally insane. It's hard to describe because it's not like anything I have ever experienced before. There was no answer for the discomfort.*

6.3.2 Species of Stingrays

Family: Dasyatidae.

Bathytoshia brevicaudata (Hutton 1875) (Fig. 6.16).

Common name: short-tail stingray.

Diagnostic Features

The short-tail stingray is a large-sized fish, with a plain color. The disc is broad and rhombic. The belly is with transverse groove. The tail is short tapering strongly before caudal sting. There are spikes that are limited to the tail before sting. The pectoral fin is with a diagonal row of white spots dorsally and with narrowly rounded apex. The trunk is very thick. The short is snout and obtuse with a tip barely extended and anterior margins weakly convex. The eyes are small. The interorbital space is broad. The mouth is often with five to seven oral papillae and deep labial furrows. The lower jaw is weakly convex. The nasal curtain is very broadly skirt-shaped and its margin is

Fig. 6.16 Bathytoshia brevicaudata (Hutton 1875) (Courtesy of Alpha, Melbourne, Australia)

fringed. The nostrils are oblique oval. The skin is smooth and has no dermal denticles. The tail is very broadly depressed at the base, often with a shorter width disc and with one caudal sting, its ventral fold is short but prominent (extending to just beyond the tip of sting), and its dorsal fold is reduced to a hard ridge. The pelvic fins are reasonably small with narrowly rounded apices. The upper part of the body is homogeneously grayish brown and darkest on the tail tip and above the eye. Spiracles are white inside. The pores around the side of the head and the diagonal row of pores on each side of the anterior disc are white ventrally. The disc margin and undersurface of the tail are dusky (Last et al. 2016).

Distribution

This species of stingrays is distributed in the Indo-West Pacific region. It is reported from the southern Africa to New Zealand and off Japan and Peter the Great Bay (eastern Russia).

Habitat, Biology, and Ecology

Bathytoshia brevicaudata is a marine species that enters the brackish sometimes and lives at a depth range of 0–476 m (Compagno et al. 1989), but it is usually found at a range of 0–200 m (McMillan et al. 2011). Individuals of this species are reported to have a maximum total length of 4300 mm (Compagno et al. 1989), with a common length of 1250 mm (Armitage et al. 1994). This species inhabits offshore and on the outer shelf and uppermost slope and sometimes close inshore (Compagno et al. 1989). It is frequently in groups (Michael 1993). It feeds on fishes, bivalves, squid, and crustaceans (Michael 1993). Females are ovoviviparous (Dulvy and Reynolds 1997). The females exhibit ovoviparity (aplacental viviparity), with embryos feeding initially on yolk and then receiving additional nourishment from the mother by indirect absorption of uterine fluid enriched with mucus, fat, or protein through specialized structures (Dulvy and Reynolds 1997). Copulation lasts 3–5 min. Males have been observed to nudge the female's abdomen during parturition. Viviparous, young born at about 36 cm WD (Last and Stevens 1994). It is known as the largest stingray in the world weighing more than 350,000 g (Last and Stevens 1994). The barbed tail however can inflict a severe or potentially fatal wound (Last and Stevens 1994).

Economic Value

No economic value is reported for this species of stingray.

Conservation Status

Rigby et al. (2021) evaluated this species as Least Concern in the Red List of the IUCN based on the following reasons. This species is with an anti-tropical discontinued distribution in the Indo-Pacific. The species is a bycatch of several fisheries containing trawl, Danish seine, longline, and purse seine and is mainly discarded with likely high post-release survival, at least from line gear. Standardized catch-per-unit-effort data from southern Australia show a probable constant population that is at low risk of population decline. There is no verification of population decrease; the species is not expected to be close to achieving the population reduction threshold.

Pteroplatytrygon violacea (Bonaparte 1832) (Figs. 6.17 and 6.18).

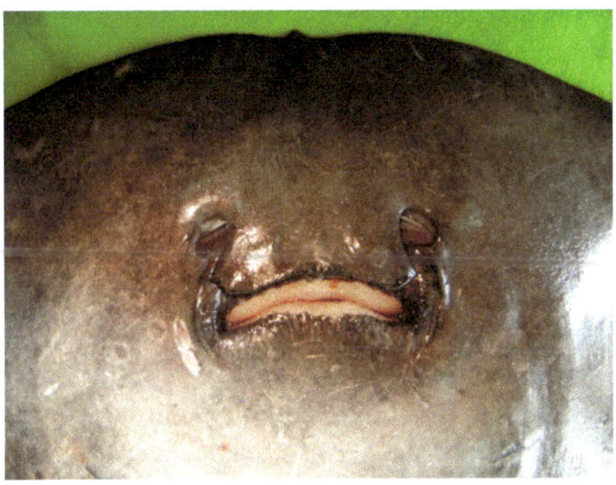

Fig. 6.17 Pteroplatytrygon violacea (Bonaparte 1832) (Courtesy of Hamid Badar Osmany, Pakistan)

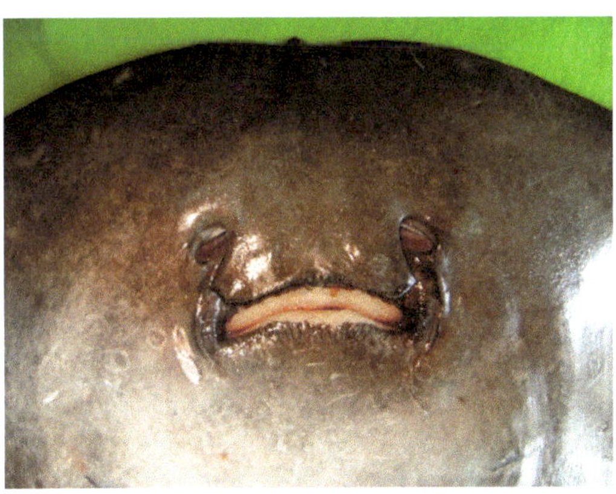

Fig. 6.18 Mouth of Pteroplatytrygon violacea (Bonaparte 1832) (Courtesy of Hamid Badar Osmany, Pakistan)

Common name: pelagic stingray.

Etymology: *Pteroplatytrygon:* Greek, *pteron* = wing, fin + Greek, platys = flat + Greek, *trygon* = a sting ray (Romero 2002); *violacea:* From the Latin *viola*, meaning violet.

Diagnostic Features

The body of this ray is thick, and the snout is wide and rounded. The pectoral disc is angular in shape. The tail is less than twice the body length and equipped with a long lower caudal finfold ending far in front of the tail tip. The upper finfold is absent. The disc is without thorns. There is usually one very long sting on the tail. The eyes are not projected (Compagno et al. 1989). The color of the body is consistently violet, purple, or dark blue-green dorsally and ventrally (Compagno 1986). Outstanding patterns are absent (Compagno 1986).

Distribution

This species of stingray has a cosmopolitan distribution mainly in the tropical and subtropical seas. It is reported from the Eastern Atlantic at the southeastern coasts of the Mediterranean and off Sicily. It is also recorded from the Cape Verde (Debelius 1998). This species is also found in the Eastern Pacific, where it is reported from California (USA), Baja California (Mexico), and the Galapagos Islands reported off Vancouver (McAllister 1990) and from Chile (Pequeño 1989). It is reported from the Western Atlantic too (Robins and Ray 1986.).

Habitats, Biology, and Ecology

The pelagic stingray is a marine species that lives at a depth range 1–381 m (Mundy 2005), but is usually found at a range of 1–100 m (Florida Museum of Natural History 2005). Individuals of this species reach a maximum disc width of 960 mm (Ebert 2003), with a common disc width of 800 mm (McEachran 1995). This species lives in open, tropical, and warm temperate waters usually in the first 100 m. It feeds on coelenterates (including medusae), squid, decapod crustaceans, and fish (Dulvy and Reynolds 1997). The venomous spine is located on the tail. Females are ovoviviparous (aplacental viviparity), with embryos feeding initially on yolk and then receiving additional nourishment from the mother by secondary absorption of uterine fluid supplemented with mucus, fat, or protein through specialized structures (Dulvy and Reynolds 1997). Female gives birth to two to nine young (Cox and Francis 1997).

Economic Value
This ray may be utilized for its meat and cartilage.

Conservation Status
Kyne et al. (2019) evaluated this species of ray as Least Concern in the Red List of the IUCN based on the following reasons. The size of this ray is medium and has a circumglobal distribution. It presents among the bycatch in pelagic longline fisheries around the world. It is mostly disposed of but is saved and used in some areas. The relative abundance indices across the world appear to fluctuate over time with no consistent substantial trend. Generally, abundance trends revealed to be constant or increasing in some regions.

Bathytoshia lata (Garman 1880) (Fig. 6.19).
Common name: brown stingray.

Fig. 6.19 Bathytoshia lata (Garman 1880) (Courtesy of Krajangdara, Tassapon)

Diagnostic Features
The brown stingray is a very large animal. The disc is broad and rhombic. It has a sharp spine over its disc and tail in adults. The shape of the body is reasonably elongate, with a gently tapering tail to caudal sting. The tail is slender and located ventrally as a fold, with a very small dorsal fold. The trunk is very thick. The pectoral fin top is narrowly rounded to angular in shape. The snout is short and broadly triangular with the tip stretched vaguely and anterior margins weakly undulate. The eyes are small. The interorbital space is broad. The mouth is also broad, with three to five oral papillae, weak labial furrows, and weakly convex lower jaw. The oval nostrils are diagonal. The body is with plain grayish brown or black on the upper surface. There is no distinct row of white pores on the disc, but frequently with uneven white flecks where the skin is damaged. The tail is dark dorsally before its sting, but usually white ventrally. The body surfaces are black past sting. The ventral surface is totally white, with a black ventral fold (Last et al. 2016).

Distribution
This species of rays is found in the Eastern Atlantic. It is reported from the southern France to Angola, including the Mediterranean Sea. It is also lives in the Indo-Pacific region, where it is common from southern Africa to Hawaii.

Habitat, Biology, and Ecology
The brown stingray reaches a maximum length of 2600 mm WD (Last et al. 2016), with a maximum reported weight of 290.0 kg (Last et al. 2016). Individuals of this species prefer the demersal or benthic habitats on the continental shelves (Last et al. 2016). They live on sand or mud bottoms and occasionally near coral reefs (Michael 1993). The crabs, prawns, and small bony fishes are their main food items (Last et al. 2016). The female is ovoviviparous (aplacental viviparity) (Dulvy and Reynolds 1997). The males mature at ca. 1000 mm WD and females at ca. 1100 mm WD, and pups born at ca. 350 mm WD (Last et al. 2016). It has a distinct pairing with embrace (Breder and Rosen 1966).

Economic Value
This species taken for its meat, cartilage, and thorny tail.

Conservation Status
Jabado et al. (2021) evaluated this stingray as Vulnerable in the Red List of the IUCN based on the following reasons. This species of ray is very large in size and common along its geographical distribution. Individuals of this species are usually included incidentally in the bycatch of different fishing activities. The current amendments to the taxonomy of this species and reporting of the species under different names (e.g., cow stingray, *Dasyatis ushiei*, or black stingray, *D. thetidis*) may have excluded its precise identity previously. A few data are existing to decide population changes, although the very large size of the brown stingray can be fundamentally liable to population reduction. There are essentially unmanaged fisheries that operate across the remaining regions where it has been recorded (e.g., West Africa, India, and Indonesia), and the extreme fishing stress is a reason of worry. Generally, it is doubted that the brown stingray has experienced a population reduction of 30–49% over the past three generation lengths (65 years) due to actual or probable levels of utilization.

Family: Myliobatidae
Myliobatis tenuicaudatus (Hector 1877) (Fig. 6.20).

Fig. 6.20 Myliobatis tenuicaudatus (Hector 1877). (Courtesy of Clinton Duffy, New Zealand)

Common name: Eagle ray.
Etymology: *Myliobatis:* Greek, *mylo* = mill + Greek, + Greek, *batis*, *-idos* = a ray (Raja sp.) (Romero 2002).

Diagnostic Features
The eagle ray is generally with a circular disc-like body, a projecting frog-like head, large fleshy pectoral fins, and a long tail armed with a spine that is capable of injecting a venom. The pectoral fins beat up and down so that the fish appears to "fly" through the water. The body has an olive-green, yellow, or dark brown color dorsally with pale blue or gray markings. The ventral surface of the body is white (Cox and Francis 1997). The caudal fin is absent (Cox and Francis 1997).

Distribution
This species of ray is distributed in the Southwest Pacific. It is reported from the Norfolk Island and New Zealand, including the Kermadec Islands.

Habitat, Biology, and Ecology
This ray is a marine species that enters brackish water sometimes and lives at a depth range of 0–160 m (Cox and Francis 1997). Individuals of this species reach a maximum length of 1500 mm WD (Michael 1993), with a common length of 1000 mm TL (Armitage et al. 1994). This ray lives in bays, estuaries, and near rocky reefs (Michael 1993). Individuals of this species are commonly solitary (Michael 1993). They feed on clams, oysters, worms, and crabs (Michael 1993). Females are ovoviviparous (Dulvy and Reynolds 1997). Although the spine on the tail can inflict a painful wound, the poison is quickly neutralized by immersion of the injured area in hot water and young are born at less than 500 mm (Cox and Francis 1997).

Economic Value
Meat of this species of ray may be used for human consumption.

Conservation Status
Kyne (2016) evaluated this species of ray as Least Concern in the Red List of the IUCN based on the following reasons. Thi species is common

in the coastal areas and across its geographical distribution. Even though it has been retained by several fishery activities, this species is frequently considered as unused in New Zealand.

6.4 Venomous Catfishes

Catfishes belong to the order Siluriformes that contains 34 valid families and over 400 genera with over 3000 known species (Ferraris 2007). Nevertheless, there is a deficiency of the actual number of species of catfishes despite the existing estimation about the number of species of this group. The venom glands in catfishes are positioned on the dorsal spine and on the spines of the two pectoral fins. They have axillary venom glands, and one dorsal, and two pectoral fin barbels to impose envenomation (Singletary et al. 2005). The fins are comprised of sharp retrose teeth that can lacerate the skin, increasing exposure and absorption of the venom (Baker 1997; Blomkalns and Otten 1999a, b). Soft tissue contaminations secondary to catfish envenomation are comparatively rare pathologic circumstances giving to emergency departments at hospitals. Difficulties connected to catfish envenomation include infection. The acuteness of the infection differs with the species of the catfish (Roth and Geller 2010).

6.4.1 The Venomous Catfish *Ameiurus Nebulosus* in New Zealand

The brown bullhead *Ameiurus nebulosus* was in New Zealand for the last 146 years. It is introduced in 1877 (Druett 1983). Since then, it has been distributed country-wise and became a pest.

Among the measures that city councils in New Zealand acted is to reduce the population of this species annually in an aim to remove it from the ecosystem. Among such activities is the development stages of the Western Springs Lakeside in Auckland. In this project, a massive removal of the pest fish such as koi carp, goldfish, and catfish was performed in 2018 (Auckland City Council 2018). Another similar project is that on the Tauriko Lakes and Papamoa Stormwater System management that took place in 2019 (Tauranga City Council 2019). The other project is the first pest fish survey of the Motukaraka Drainage System, Mercer, that took place on 29 October–1 November 2019 (New Zealand Waterways Restoration 2019).

MacDonald and Anaru (2020), in their work on management of fish pest in New Zealand, mentioned that the Māori who are the indigenous people of New Zealand and throughout the country and local government relationships with hapū (a group or family from a common ancestor) and iwi (a postcolonial term for groups of hapū who have common ancestral links or share a geographical location) are keen to the sustainable management of natural resources. MacDonald and Anaru (2020) describe an inventiveness formed in response to an incursion of brown bullhead catfish (*Ameiurus nebulosus*) in Lake Rotoiti and (as of December 2018) Lake Rotorua, two of the ancestral lakes under the mana whakahaere (a self-determination and decision-making role) of Te Arawa Lakes Trust. Their program includes a full-time staff member from Te Arawa Lakes Trust working directly alongside the Bay of Plenty Regional Council to progressively include the catfish population in Lakes Rotoiti and Rotorua. This project has removed over 80,000 catfish from the lakes, generated significant research findings, engaged over 500 volunteers, and delivered an educational program to over 1500 school students across the district. This partnership approach to a biosecurity project has reinforced relationships between the local government, iwi, hapū, and the wider community while offering positive environmental consequences.

Background

The capability of catfishes to instigate painful wounds has been well-known for a long time, but it is not very well recognized. Autenrieth (1833) stated that Richter (1764) cited in Autenrieth (1833) wrote about a deadly event that occurred to a Spanish angler after a strike by a sea catfish. Günther (1864) was the first to assemble all the information about the occurrence of catfish and confirmed that there is a deficiency of scientific information concerning the venom organs of these fishes. Subsequently, Bottard (1899) illustrated for the first time the venom device of the oriental catfish *Plotosus lineatus*. In the early twentieth century, the works of Reed (1900a, b, 1906, 1907) and Reed and Lloyd (1916) mark the start of an inclusive account of the venom organs of a catfish. The toxicological and immunological characteristics of catfish venom were dealt with by Toyoshima (1918). Mansueti (1951) stated on the stinging skill of a catfish. Numerous investigations on the venom construction of the catfish have endured to appear since then (Halstead et al. 1953; Halstead 1959; Halstead and Mitchell 1963).

6.4.2 The Venom Apparatus

Cocan et al. (2020) in their study on the venomous glands of the brown bullhead *Ameiurus nebulosus* analyzed the shape and topography of the venomous glands in this species using MRI techniques. The main aim of their study was to offer an excellent insight of the defence process of this species. The existent features and structure of venomous glands made brown bullhead one of the top invasive species spreading its habitat worldwide. They found that in the studied species, the venomous glands, together with the first radius of the pectoral and dorsal fins, as well as the muscular fascicles attached to the radius, are establishing the venomous system. The triangular location of pectoral and dorsal spines and of their afferent venomous glands characterizes an effective protecting method versus aquatic, terrestrial, and aerial predators. Using MRI techniques for anatomical research of venomous glands in the brown bullhead demonstrated to be very accurate in detecting both the shape and the topography of the venomous glands. Also, by using MRI techniques, Cocan et al. (2020) emphasized the complicated defence method that the brown bullhead has grown versus its natural predators.

Rojo (2013) reported on the pectoral spines of the *Ameiurus nebulosus*. He illustrated these structures as the result of the calcification of the first pectoral radius. The MRI image obtained by Cocan et al. (2020) on the sagittal section permits the recognition of three processes placed in three diverse tracks (dorsal, ventral, and anterior) at the base of the pectoral spine. The occurrence of these processes has also been stated in the literature for the species *Ictalurus lambda* (Hubbs and Hibbard 1951), *Clarias gariepinus*, and *Heteropneustes fossilis* (Satora et al. 2008). Muscle bundles are included at the level of the three bone processes stated above. The contraction of these muscles has the task of putting the pectoral spines in a perpendicular angle on the body (Miano et al. 2013; Moldowan et al. 2015). In addition to the imperative protecting task it exhibits, it appears that the pectoral spines in many species are also involved in a role in interspecific communication by making high-frequency sounds (Fine et al. 1997; Fine and Ladich 2003; Kaatz et al. 2010).

As in the case of pectoral fins, in the dorsal fin, the first ray experiences extreme calcification developments. The venomous glands of *Ameiurus nebulosus*, together with the first hard ray of the pectoral and dorsal fins and the muscular fascicles attached to the radius, form the venomous apparatus. Comparable results have been recorded in the literature for *Noturus gladiator* (Egge and Simons 2011) and *Noturus stigmosus* (Wright 2012).

When in an erect position, in the brown bullhead, as in other catfish species, the dorsal spine together with the two pectoral spines becomes a device of defence, boosting virtually the dimensions of the body over the ingestion ability of predators (Halstead et al. 1953; Endler 1986; Haddad 2000; Fine and Ladich 2003; Bosher et al. 2006; Rojo 2013). Likewise, the same guard mechanism is explained in the literature in the case of individuals of the species Diodontidae (Leis 2006; Byeon et al. 2011).

If the predator implements the attack, due to the pressure and rupture of the spines that have penetrated the tissues of the predator, the venomous apparatus discharges a certain quantity of venom (Blomkalns and Otten 1999a, b; Satora et al. 2008).

Unlike the brown bullhead (*Ameiurus nebulosus*), which reports a dorsal spine and two pectoral spines, the data only expresses the occurrence of the pectoral spines in African catfish (*Clarias gariepinus*) (Satora et al. 2008). Haddad and Gadig (2005) noted only the existence of the dorsal spine in case of sharks from the Heterodontidae, Squalidae, and Dalatiidae families. From these three shark families listed, the presence of a venomous gland at the base of the dorsal spine (Halstead 1988a, b) is recorded only for individuals belonging to Heterodontidae and Squalidae.

The common length of the brown bullhead is usually between 20 and 25 cm (Muus and Dahlström 1968), having almost the same size as other prey fish (rudd, bream, Prussian carp). Although the brown bullhead and the species mentioned above are assigned to the same ecological habitat and are companionable, they have unique guard method. The presence of scales on rudd, bream, and Prussian carp is in fact a simple defence mechanism, but the pectoral and dorsal spines and venomous apparatus of the brown bullhead are a complex defence mechanism because of muscular contraction when the species is in danger. In this case, the dorsal and pectoral spines are positioned in a triangular formation (Rojo 2013), making it very hard to be caught by predators. The effectiveness of this triangular arrangement and venomous apparatus is enhanced when the brown bullhead is attacked by predators that used their mouths to catch prey (cormorant, pelican, pike, zander, turtle, otter). In case of predators that use their claws to catch the prey (Canadian bald eagle, white-tailed eagle), the defence mechanism is less effective (Jackman et al. 1999).

Causative Agent

The typical sting method of catfishes is shown in Fig. 6.21. When a spine inserts a potential predator, the skin neighboring the venom gland cells is smashed, freeing the venom into the wound. Catfish venom is known to demonstrate neurotoxic and hemolytic features and can yield a diversity of further consequences such as severe pain, ischemia, muscle spasm, and respiratory distress, although any single species' venom may not display all these properties (Halstead 1978). The neurotoxic and hemolytic features could be due to a 15-kDa protein, termed toxin-PC (Auddy and Gomes 1996). In certain catfish species, the venom comprises from two to eight toxic proteins with rough molecular weights of 10 kDa (Calton and Burnett 1975). Both the mechanism by which these toxins act and their physiological goals are very inadequately comprehended. It is deemed that cytolytic activity produced by pore formation in cell membranes is a probable justification, as this activity is present in other "pain-producing" venoms, such as those produced by bees (Pawlak et al. 1991) and platypus (Kourie 1999), and reactions steady with this mechanism have been noted regarding piscine venoms (Church and Hodgson 2002).

Fig. 6.21 Drawing showing a catfish in the act of stinging. (After Halstead 1967)

Symptoms

Catfish envenomation shows as local intense pain, edema, erythema, and paleness and sometimes as cutaneous necrosis (Baker 1997; Blomkalns and Otten 1999a, b; Haddad 2000, Haddad et al. 2003a, b; Haddad and Lastoria 2005). Systemic appearances, such as cardiac arrhythmias, dyspnea, or neurological symptoms, do not appear during the acute envenomation stage (Haddad 2000; Haddad and Lastoria 2005). The cuts caused by catfish are usually painful, but when the fish have venomous stings, the pain is intense and persists longer. The pain can extend to the adjacent organs. There is local inflammation, but signs subside after about 6 h, without other problems. Worries occur when stings break in the wound, representing as foreign bodies and assisting derived bacterial and fungal infections, which are a major problem.

Treatment

The treatment of catfish envenomation is comparable to that of stingray stings.

Prevention

Envenomation by catfish can be stopped by avoiding direct contact with the dorsal and pectoral spines of the fish. To handle the catfish, one should hold the fish firmly behind the dorsal and pectoral fins.

6.4.3 Venomous Catfish Species

Family: Ictaluridae.

Ameiurus nebulosus (Lesueur 1819) (Fig. 6.22).

Common name: brown bullhead.

Etymology: *Ameiurus:* Greek, a = without + Greek, *meiouros*, -os, -on – without tail (Romero 2002); *nebulosus: nebulosus* meaning clouded, in relation to mottled and gray coloring (Scott and Crossman 1973).

Diagnostic Features

The body is triangular in cross section. A dorsal fin that bears a spine is present (Missouri Department of Conservation (2023). There is a single adipose fin posterior to the dorsal fin. The tail is only slightly notched, with the dorsal and ventral lobes pointing inward. The pectoral fins have spines that bear five to eight serrated teeth on their posterior edge (Florida Museum 2017). There are eight barbels around the mouth (Guth 2011). The barbels on the chin are black to yellowish brown (Michigan Department of Natural Resources 2017). There is no scales on the body (University of Wisconsin Sea Grant Institute 2023). The mouth is terminal (University of Wisconsin Sea Grant Institute 2023; Guth 2011) or slightly subterminal (University of Wisconsin Sea Grant Institute 2023), with the upper jaw projecting slightly past the lower jaw (Missouri Department of Conservation (2023). This position enables bottom feeding. The color of the body is darker brown, green dorsally, and growing lighter green and yellow to the ventral surface. The belly is off-white or cream. Furthermore, there are darker, brown-black speckles along the whole surface of the fish. This species can be recognized from black and yellow bullheads by their yellow-black chin barbels, the lack of a hypopigmented bar at the base of the tail (which is present in black bullheads), and their 21–24 anal fin rays (University of Wisconsin Sea Grant Institute 2023).

Fig. 6.22 Ameiurus nebulosus (Lesueur 1819). (Courtesy of Wikipedia)

Distribution
This species of catfish is originally native to North America. It is introduced into several countries.

Habitat, Biology, and Ecology
The brown bullhead is a fresh water species that lives at a depth range 0–12 m (Scott and Crossman 1973). Individuals of this species reach a maximum total length of 550 mm (IGFA 2001), with a common total length of 250 mm TL (Muus and Dahlström 1968). The maximum recorded weight is 2.7 kg (IGFA 2001) and the maximum reported age is 9 years (Kottelat and Freyhof 2007). This species occurs in pools and sluggish runs over soft substrata in creeks and small to large rivers. Individuals of this species also live in marshes, lakes, and ponds (Page and Burr 2011). It is seldom found in brackish waters (Scott and Crossman 1973). It is a nocturnal feeder that feeds on mollusks, insects, leeches, crayfish and plankton, worms, algae, plant material, and fishes and has been reported to feed on eggs of least cisco, herring, and lake trout (Scott and Crossman 1973). The young with a length of 30–60 mm feed mainly on chironomid larvae, cladocerans, ostracods, amphipods, bugs, and mayflies (Scott and Crossman 1973). Thi species has a high tolerance for high carbon dioxide and low oxygen concentrations and temperatures up to 31.6 °C although experiments show the upper lethal temperature to be 37.5 °C and is resistant to domestic and industrial pollution (Scott and Crossman 1973).

Economic Value
The meat of this species of catfish is used for human consumption. It is usually prepared hot-smoked and cooked in various ways (Scott and Crossman 1973).

Conservation Status
NatureServe (2013) evaluated this species of catfish as Least Concern in the Red List of the IUCN based on the following reasons. The distribution range of this species is very wide, with large population size and stable or increasing trend, and lacks major threats.

6.5 Venomous Scorpaenid Fishes

The scorpaenid fishes (family: Scorpaenidae) are considered among the most dangerous venomous fishes. They are slow-moving animals and widely distributed in tropical and temperate shallow waters. These fishes favor hiding near rocks, in reefs, or under plants, which are increasing factors to human accidents (Russell 1965a, b; Haddad 2000). Human events such as careless fish handling or the fish being stepped on will lead to the release of the venom from the poisonous glands into the wound (Russell 1965a, b). Pain and edema are the most apparent symptoms of such poisoning.

Background
The works of Aristotle (384–322 BC; Coutière 1899) about the sea hog (*Scorpaena porcus*) could be considered the first announcement in the history of scorpaenid fish poisoning. No known works appeared on the venom of the scorpion fish from the time of Aristotle until the middle of the sixteenth century when Rondelet (1507–1557), Salviani (1514–1572), and Belon (1517–1564; cited by Halstead 1970) issued their work on scorpionfishes and their venom. Autenrieth (1833) wrote on the stings of numerous scorpionfish species. Bottard (1899, cited in Coutière 1899) depicted for the first time the morphology of the venom organs of scorpionfishes in his *Les Poissons Venimeux*. In the early twentieth century, the nearly thorough anatomical description on the venom organs of scorpaenid was that of Pawlowsky (1906, 1913, 1929), where he studied a large number of species of this group. As to the toxicological findings, the works of Duhig (1929) and Duhig and Jones (1928a, b) are deemed establishing in this field. Since then, an augmented attention in the study of the cases of scorpionfish stings and venom was detected and issued in several publications (Tang 1953, 1954; Halstead et al. 1956; Endean 1961; Kizer et al. 1985; Patel and Wells 1993; Haddad et al. 2003a, b).

6.5.1 Morphology of the Venom Apparatus of Scorpaenid Fishes

There are three kinds of venom organs in scorpionfishes which vary in their structure and are found in the three genera *Pterois*, *Scorpaena*, and *Synanceia* (Fig. 6.23). Generally, the venom organs of scorpionfishes are more complex than other poisonous fishes. The following is a summary of the description of venom organs of *Pterois* and *Scorpaena* which is based on Halstead (1970).

Venomous Organs of the Members of the Genus Pterois (Fig. 6.24)

There are 13 dorsal spines, 3 anal spines, and 2 pelvic spines included in the venom organs of this grouping of scorpionfishes. All these spines are associated with a suitable poisonous gland and enveloped by integumental coverings. This cover is part of the poisonous assembly, where the venom is produced within its inner surface. The spine of the dorsal fin is elongated and straight excluding for its terminal end, where it is slightly curved. The outline of the tip of the spine is triangular in cross section. A groove known as the "anterolateral-glandular groove" flows along the dorsal side of the spine and is linked with the lateral condyle at the base of the spine, with a deep channel running along its length. The anal spines are short, firm, and slightly curved at the tip, which is triangular. The anterolateral-glandular groove is analogous in shape to that found in the dorsal spine, but shallow. There are some variations in the structure of the spine among the anal fin spines as some of the spine's shortage of the anterolateral-glandular groove. The spines of the ventral fin are unlike those of dorsal and anal fins. In these spines, the anterolateral-glandular groove is missing, and as an alternative, there are superior and inferior grooves initiating from condyles near each of them.

Venomous Organs of the Members of the Genus Scorpaena (Fig. 6.25)

There are 12, 3, and 2 dorsal, anal, and pelvic spines, respectively, in the venom organ of the member of the genus *Scorpaena*. Like those spines of the genus *Pterois*, they are wrapped with a glandular integumental covering. The spine itself is slender with a fusiform shape. The spines of the dorsal fin are elongated, straight, and slightly curved at the tip, which is trigonal in outline. The anterolateral-glandular groove begins in

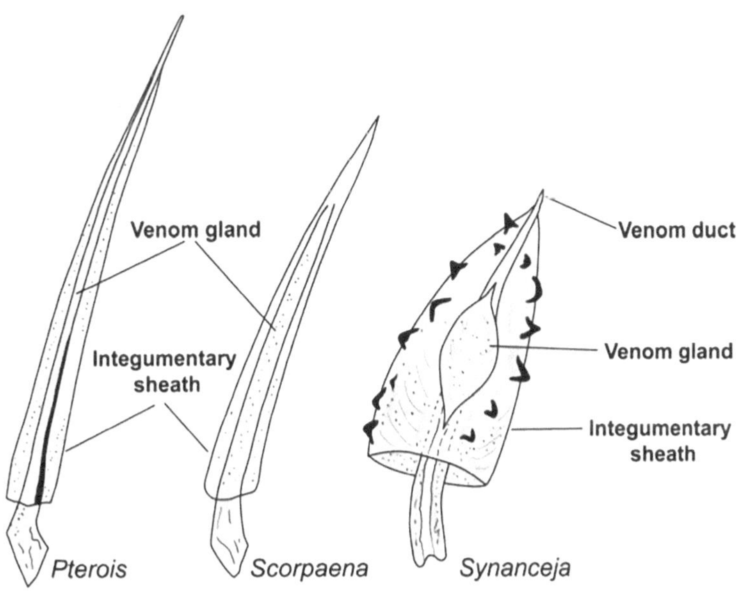

Fig. 6.23 Drawing showing morphological differences in the stings of *Pterois*, *Scorpaena*, and *Synanceia*. (After Halstead 1967)

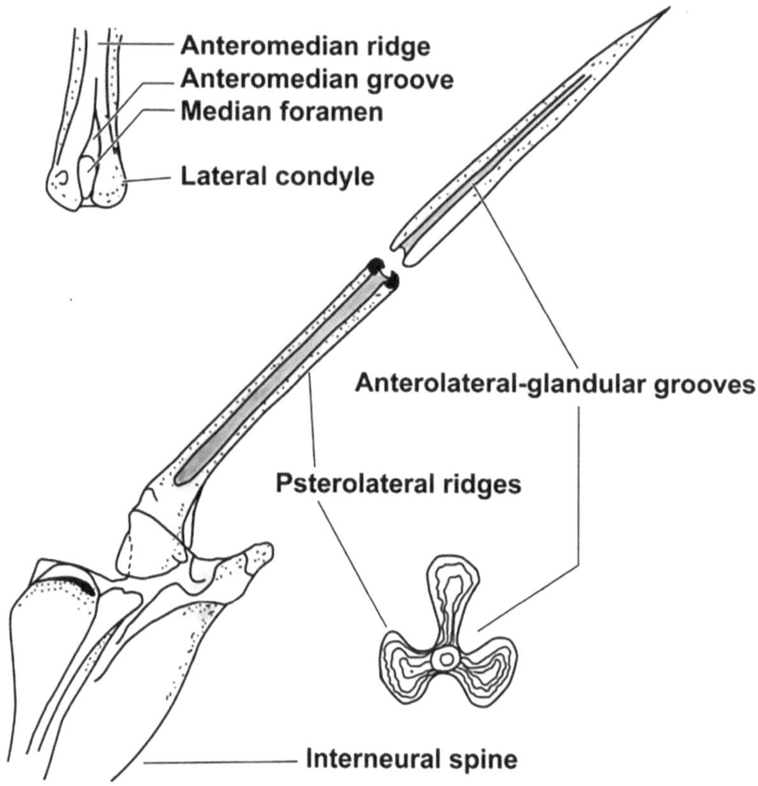

Fig. 6.24 Drawing of a typical dorsal spine of *Pterois volitans*. (After Halstead 1967)

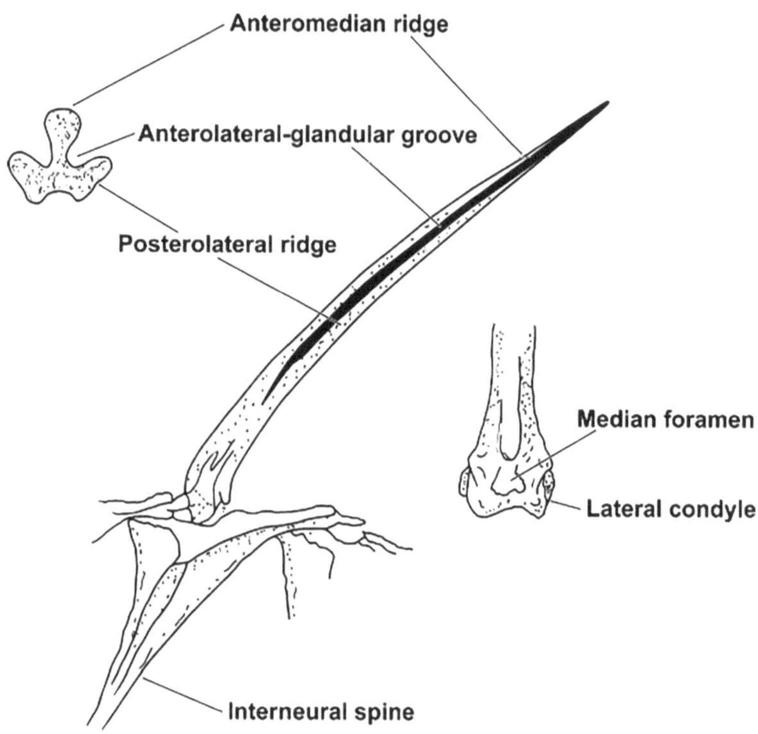

Fig. 6.25 Drawing of a typical dorsal spine of *Scorpaena guttata*. (After Halstead 1967)

the middle third of the spine. It is shallow along the length of the spine and gets deeper near the tip. The anal spines are comparable in shape to those of the dorsal spines. The grooves are narrow and shallow in the basal part of the spine and turn wider and deeper toward the tip of the spine. The pelvic spines are also like those of the dorsal and anal spines in shape and formation.

The microscopic structure of the venom organs of all scorpionfishes is the same. The spine is comprised of a dentine structure and is inserted in the layer of the skin. The peripheral sides of the spines are compiled of viterodentine. Within the shaft of the spine, the cross section shows the existence of a broad groove of varying dimensions, which gets smaller in size near the tip of the spine. The integumental cover that wraps the spine is a thin layer of a dense fibrous connective tissue. In this epithelial tissue, large mucous cells are abundant, with a clear cytoplasmic area. The epidermis and dermis of the sheath are disconnected by a thick pigmented layer neighboring the basement membrane of the epidermis. Contained by the anterolateral-glandular groove, there is a cluster of large polygonal glandular cells packed with granular cytoplasm. In the longitudinal section of the spine, the dentine-like substance is displayed to be thick near the base, and narrow cavities appear in the structure of the spine near the tip.

Causative Agent

Two types of envenomation methods are used by the different species of scorpionfishes. The victim gets a stab from the spines either by mishandling of a fish through removing them from a net or a hook or by careless stepping on the dorsal stings of a stonefish. In both cases, the venom will be inserted into the body of the victim with the aid of the spines. In the process of envenomation and due to the pressure resulting from the stab of the spines into the body of the victim, the glandular sheath covering the spines breaks and the venom is injected in the wound by the anterolateral-glandular groove.

The scorpaenid venom is an unstable protein, with a molecular weight 150,000. It can be denaturized by heat, a characteristic used in its treatment. Saunders and Taylor (1959) and Choromanski et al. (1984) suggested that the tissue from the lionfish spine contains one or more soluble lethal toxins. As with other soluble venom preparations from Scorpaenidae, the toxic constituent in soluble extracts is not stable under conventional storage conditions. The scorpaenid venom is comprised of several biologically active factors: (1) hyaluronidase fraction, (2) capillary permeability factor, (3) a toxic or lethal fraction (Wiener 1959), and (4) a pain-producing factor (Austin et al. 1965).

Symptoms

Halstead (1970) gave the following general symptoms when the venom of a scorpaenid fish species enters the body of a victim. The venom can lead to sharp throbbing pain, which can remain for several hours. The skin around the wound may become swollen, red, and hot. Gangrene will follow subsequently because of secondary bacterial infection. Systemic stoppage contains cardiac failure, delirium, convulsions, and nervous disturbances. Victims may have a primary shock accompanied with faintness, weakness, nausea, loss of consciousness, rapid weak pulse, low blood pressure, and respiratory distress. There are some changes in the symptoms initiated by introducing the venom of different species of scorpaenid fishes.

Scorpaenid venom involves high concentrations of acetylcholine that influences the physiology of isolated heart and skeletal muscle preparations. The part of acetylcholine in the spine tissue could improve the act of the toxic constituent, maybe by provoking local vasodilation at the stabbing site or by producing pain by direct or indirect action on sensory neurons (Saunders and Taylor 1959; Choromanski et al. 1984). The toxin makes a period of muscle fibrillation followed by neuromuscular obstruction, which could account for the toxic symptoms, that is, respiratory distress, uncoordinated movement, muscular weakness, and paralysis, followed by death (perhaps because of respiratory arrest; Cohen and Olek 1989).

The capillary permeability factor of the venom precedes to enormous edema after envenomation

(Poh et al. 1991) and may account for hemorrhagic pulmonary edema. The stonustoxin is the lethal part of the toxin and could give significant hypotension, which is the main reason of death (Low et al. 1994).

Treatment

The following treatment phases were concise from Halstead (1970). Relief of the pain, stopping the impact of the venom, and preventing secondary infection are the main aims for scorpaenid fish poisoning. The wound should be completely cleaned to remove the venom as much as possible. In the scorpionfish stings, bleeding is encouraged, and a bandage should be applied above the wound and released every few minutes. Hot water as much as the victim can tolerate should be applied to the wound as soon as possible, and the process of cleaning the wound should last for more than an hour. No surgical closure of the wound is suggested and antitetanus agents should be applied. Envenomation of the stonefish is more dangerous and riskier than scorpionfishes and needs urgent thorough care. For such cases, antivenin is used, which involves refined and concentrated hyperimmune horse serum, which neutralizes the stonefish venom.

6.5.2 Species of the Scorpaenid Fishes

Family: Scorpaenidae.

Pterois volitans (Linnaeus 1758) (Fig. 6.26).

Common name: red lionfish.

Etymology: *Pterois:* Greek, *pteron* = wing, fin (Romero 2002).

Diagnostic Features

This scorpaenid species has varying body colors, and this is regularly linked to the environment. The coastal species are generally darker, sometimes almost black in estuaries, often with large tentacles above the eyes (Kuiter and Tonozuka 2001a, b). This species has greatly elongated dorsal fin spines. There is a connecting tissue in between the spines of the pectoral fin rays. The pectoral fin is long, and the membranes of all fins are spotted.

Fig. 6.26 Pterois volitans (Linnaeus 1758). (Courtesy of Alexander Vasenin)

Distribution

This scorpaenid species is distributed in the Pacific Ocean. It is reported from the Cocos-Keeling Islands and Western Australia (Randall et al. 1997). It is also found in the eastern Indian Ocean to the Marquesas and Oeno (Pitcairn group), north to southern Japan and southern Korea, south to Lord Howe Island, northern New Zealand, and the Austral Islands.

Habitat, Biology, and Ecology

The red lionfish is a marine species that lives in the reef areas at a depth range 2–55 m (Allen and Steene 1988). Individuals of this species reach a maximum total length of 457 mm (Ehemann 2017). The maximum recorded weight of this species is 1.4 kg (Ehemann 2017) and the maximum known age is 10 years (Hinton 1962). It lives in lagoon and seaward reefs from turbid inshore areas to depths of 50 m. Individuals are frequently isolated; they hide in undisplay locations at daytime regularly with head down and almost immobile. The young are pelagic, exploring over great distances and the reason for their broad geographical range (Kuiter and Tonozuka 2001a, b). This species feeds on small fishes, shrimps, and crabs at night, catching its preys by its widespread pectorals, stunning it, and then swallowing it in one sweep. The dorsal spines are venomous; the sting can be treated by heating the afflicted part and application of corticoids (Cornic 1987).

Economic Value

The meat of this species has a good economic market.

Conservation Status

Motomura and Matsuura (2016c) evaluated this species as Least Concern in the Red List of the IUCN based on the following reasons. It has a wide geographical distribution and greatly available in the areas where it is found. Thi species has been introduced in several geographical regions, where its population swiftly increased.

Scorpaena cardinalis (Solander and Richardson 1842) (Fig. 6.27).

Common name: eastern red scorpionfish.

Etymology: *Scorpaena:* Latin, *scorpaena* = a kind of fish, 1706 (Romero 2002).

Diagnostic Features

The uncovered cycloid scales are covering the pectoral fin base and the anteroventral surface of the body (some scales covered by a thin skin, especially in large adults). There is no longitudinal ridge on the lateral surface of the maxilla. The lateral lacrimal spine has two spinous points. The anterior lacrimal spine has one to three small spinous points on its posterior margin. The posterior lacrymal spine has one to three spinous points, directed posteroventral throughout life; the median interorbital ridge is present. There is no coonal spines. The occipital pit and extra preopercular spine are present. The pterotic spine is simple and smooth distally throughout life. The space between the upper and the lower opercular spines is covered by a thin skin without sensory pores or canals. There are few skin flaps or tentacles on the anterodorsal surface of the body. The large black blotch on the spinous portion of the dorsal fin in both sexes is absent. Two large white blotches on the caudal fin base are present (Motomura et al. 2011).

Fig. 6.27 Scorpaena cardinalis (Solander and Richardson 1842). Bernard Yau, Australia

Distribution

The eastern red scorpionfish is distributed in the Southwest Pacific. It is reported from the northern New Zealand and offshore islands of the Tasman Sea (Norfolk and Lord Howe Islands, Middleton and Elizabeth Reefs, New South Wales, Australia; not recorded from the coastal waters of Australian mainland).

Habitat, Biology, and Ecology

This scorpaenid fish is a marine species that lives in the reef areas at a depth range of 0–154 m (Motomura et al. 2011). Individuals of this species reach a maximum total length of 405 mm (Stewart and Hughes 2010). The maximum recorded age of this species is 33 years (Stewart and Hughes 2010). It lives in rocks and in crevices and caverns at all depths. It is sporadically located in large rock pools in the lower littoral zone. The eastern red scorpionfish is nocturnal in its habit. The main food items of this species are the small mobile benthic animals (Russell 1983).

Economic Value

The economic value of this species is not known.

Conservation Status

Motomura and Matsuura (2016a) evaluated this species as Least Concern in the Red list of the IUCN based on the following reasons. The species is confined to the northern New Zealand, Lord Howe, and Norfolk Islands. It is reasonably common.

Scorpaena onaria (Jordan and Snyder 1900) (Fig. 6.28).

Common name: western scorpionfish.

Etymology: *Scorpaena:* Latin, *scorpaena* = a kind of fish, 1706 (Romero 2002).

Diagnostic Features

The body of this species is moderately compressed near the head and much compressed near the tail. The nape is slightly curve. Several small papillae are present on the head, particularly on the upper head and snout. There are a few small tentacles along the lateral line, which are considered the only tentacles on the flanks. The mouth is large and oblique. Teeth on jaws, vomer, and palatine are present. The color of the body is reddish. There are scattered irregular pale blotches and some black dots on the body. A large black patch at the posterior part of the dorsal fin is present (Shao and Chen 1988).

Distribution

This scorpaenid species is found in the Indo-West Pacific region. It is reported from Japan and South Korea. It is also recorded from Taiwan (Shao 1997) and New Caledonia (Fricke et al. 2011).

Habitat, Biology, and Ecology

This scorpaenid fish is a marine species that lives at a depth range 223–500 m (Fricke et al. 2011). It attains a maximum total length of 300 mm TL (Masuda et al. 1984). Individuals of this species are sexually dimorphic; only the males have a black blotch on the spiny part of the dorsal fin (Motomura and Matsuura 2016b). The species is an ambush predator feeding on small fish and crustaceans. The female is an oviparous species. There are venom glands at the base of the dorsal fin spines (Taiwan Data Base (2023).

Fig. 6.28 Scorpaena onaria (Jordan and Snyder 1900). (Courtesy of Izuzuki Japan)

Economic Value

This species has no economic value except for its venom that may be used in pharmaceutical industries.

Conservation Status

Motomura and Matsuura (2016b) evaluated this scorpaenid species as Least Concern in the Red List of the IUCN based on the following reasons. The species is a deep-water dweller and has wide geographical distribution, where it is common. Despite being aimed by fishery activities, the population of this species is significantly not affected.

Scorpaena papillosa (Schneider and Forster 1801) (Fig. 6.29).

Common name: red rock cod.

Etymology: *Scorpaena:* Latin, *scorpaena* = a kind of fish, 1706 (Romero 2002).

Diagnostic Features

The mouth of this scorpaenid species is large encompassing small, thin teeth that form velvety bands (villiform). There is a small row of spines

Fig. 6.29 Scorpaena papillosa (Schneider and Forster 1801). (Courtesy of Brian Gratwicke)

beneath the eyes and a gill cover margin containing three spines (Paulin and Roberts 1992).

Distribution

The red rock cod is distributed in the Indo-West Pacific region from the southern Australia and New Zealand (Paulin et al. 1989).

Habitat, Biology, and Ecology

This scorpaenid fish is a marine species that enters brackish water sometimes and lives at a depth range of 5–50 m (May and Maxwell 1986). Individuals of this species reach a maximum total length of 300 mm (May and Maxwell 1986). They live on rocky bottoms (May and Maxwell 1986). Also, they can be located in shallow estuaries from muddy environments to offshore reefs in kelp reef (Kuiter 1993). Thi species has a nocturnal habit. The juveniles are sometimes noticed in large rock pools. Individuals of this species feed on varied invertebrate animals, but essentially on crustaceans and fishes (Paulin and Roberts 1992).

Economic Value

No economic value is known for this species.

Conservation Status

This species is not evaluated for conservation status.

References

Allen GR, Erdmann MV. Reef fishes of the east indies. Perth: University of Hawai'i Press, Volumes I-III. Tropical Reef Research; 2012.

Allen GR, Steene RC. Fishes of Christmas Island Indian Ocean. Christmas Island Natural History Association, Christmas Island, Indian Ocean, 6798, Australia; 1988. p. 197.

Angulo A, López MI, Bussing WA, Murase A. Records of chimaeroid fishes (Holocephali: Chimaeriformes) from the Pacific coast of Costa Rica, with the description of a new species of *chimera* (Chimaeridae) from the eastern Pacific Ocean. Zootaxa. 2014;3861(6):554–74.

Armitage RO, Payne DA, Lockley GJ, Currie HM, Colban RL, Lamb BG, Paul LJ, editors. Guidebook to New Zealand commercial fish species. Revised edition. New Zealand Fishing Industry Board, Wellington, New Zealand; 1994. p. 216.

Arratia G. The skin of catfishes – a review. In: Catfishes, 1. Museum für Naturkunde der Humboldt Universität, Invalidenstr, Berlin; 2003. p. 177–199.

Atkinson PRT, Boyle A, Hartin D, McAuley D. Is hot water immersion an effective treatment for marine envenomation? Emerg Med J. 2006;23(7):503–8.

Auckland City Council. Western Springs Lakeside Park development plan. Draft for consultation. Waitamata Local Board, Auckland City Council; 2018. p. 115.

Auddy B, Gomes A. Indian catfish (Plotosus canius, Hamilton) venom. Occurrence of lethal protein toxin (toxin-PC). Adv Exp Med Biol. 1996;391:225–9.

Auerbach PS. Marine envenomations. N Engl J Med. 1991;325:486–93.

Auerbach PS, Norris RL. Disorders caused by venomous snakebites and marine animal exposures. In: Longo DL, Fauci AS, Kasper DL, Hauser SL, Larry Jameson J, Loscalzo J, editors. Harrison's principles of internal medicine. 18th ed. New York: McGraw-Hill; 2012.

Austin L, Gillis RG, Youatt G. Stonefish venom: some biochemical and chemical observations. Aust J Exp Biol Med Sci. 1965;43:79.

Autenrieth HF. Ueber das gift der fische. Tübingen: C. F. Osiander; 1833. p. 287.

Baker DH. An unusual foreign body: catfish spine. Pediatr Radiol. 1997;27:585.

Bianchi G, Carpenter KE, Roux JP, Molloy FJ, Boyer D, Boyer HJ. FAO species identification guide for fishery purposes. Field guide to the living marine resources of Namibia. FAO, Rome; 1999. p. 265.

Black BJ, Londahl MJ, Richter KK. Challenges in the treatment of a stingray injury: a rare case report in a rural New Zealand hospital. The New Zealand Medical Journal (Online). 2022;135(1558):96–102.

Blomkalns AL, Otten EJ. Catfish spine envenomation: a case report and literature review. Wilderness Environ Med. 1999a;10:242.

Blomkalns AL, Otten EJ. Catfish spine envenomation: a case report and literature review. Wilderness Environ Med. 1999b;10:242–6.

Bond N. Cameraman Justin Lyons reveals Steve Irwin's final words: 'I'm dying'. The Advertiser, South Australia; 10 March 2014. Retrieved 31 Jul 2016.

Bosher BT, Newton SH, Fine ML. The spines of the channel catfish, Ictalurus punctatus, as an anti-predator adaptation: an experimental study. Ethology. 2006;112:188–95. https://doi.org/10.1111/j.1439-0310.2006.01146.x.

Bottard A. Les poissons verimeux. Paris: Octave Doin; 1889. p. 198.

Bottard A. Les poissons venimeux. The'se Paris, Octave Doin e'd; 1899. p. 198.

Bray D. Elephant fish, *Callorhinchus milii. Fishes of Australia*. Archived from the original on 21 February 2019; 2014. Retrieved 26 July 2023.

Breder CM, Rosen DE. Modes of reproduction in fishes. Neptune City: T.F.H. Publications; 1966. p. 941.

Byeon MS, Park JY, Yoon SW, Kang HW. Structure and development of spines over the skin surface of the river puffer Takifugu obscurus (Tetraodontidae, Teleostei) during larval growth. J Appl Ichthyol. 2011;27:67–72. https://doi.org/10.1111/j.1439-0426.2010.01606.x.

Cailliet GM, Andrews AH, Burton EJ, Watters DL, Kline DE, Ferry-Graham LA. Age determination and validation studies of marine fishes: do deep-dwellers live longer? Exp Geront. 2001;36:739–64.

Callinan R. Death of a crocodile hunter. Time; 4 September 2006. Retrieved 31 Jul 2023.

Calton GJ, Burnett JW. Catfish (Ictalurus catus) fin venom. Toxicon. 1975;13:399–403.

Carrier JC, Musick JA, Heithaus MR. Biology of sharks and their relatives, vol. 101. CRC Press; 2012. p. 122.

Cavazzani E. L'ichtyotoxique chez le "Petromuzon marinus". Arch Ital Biol. 1892;18:182–6.

Choromanski JM, Murray TF, Webber LJ. Responses of the isolated buffalo sculpin heart to stabilized venom of the lionfish (Pterois volitans). Proc West Pharmacol Soc. 1984;27:229–32.

Church JE, Hodgson WC. The pharmacological activity of fish venoms. Toxicon. 2002;40:1083–93.

Clarke TL. Some observations of fish poisoning in the Virgin Islands. West Indian Bull. 1918;17:56–67.

CNN. Reuters. Stingray deaths rare and agonizing. CNN. Reuters; 4 September 2006. Archived from the original on 21 September 2006. Retrieved 31 Jul 2023.

Coad BW, Reist JD. Marine fishes of Arctic Canada. Toronto: University of Toronto Press; 2018. p. 618.

Cocan D, Mireşan V, Popescu F, Constantinescu R, Coroian A, Laţiu C, Flaviu Turcu RV, Fărcăşanu AŞ, Martonos C. MRI investigations on venomous glands of Brown bullhead, Ameiurus nebulosus (Lesueur, 1819) (Actinopterygii: Ictaluridae). Pak J Zool. 2020;52(4)

Cohen AS, Olek AJ. An extract of lionfish (Pterois volitans) spine tissue contains acetylcholine and a toxin that affects neuromuscular transmission. Toxicon. 1989;27(12):1367–76.

Collette BB. A review of the venomous toadfishes, subfamily Thalassophryninae. Copeia. 1966;4:846–64.

Compagno LJV. FAO Species Catalogue. Vol. 4. Sharks of the world. An annotated and illustrated catalogue of shark species known to date. Part 1 - Hexanchiformes to Lamniformes. FAO Fish. Synop. 125(4/1):1–249. Rome, FAO; 1984.

Compagno LJV. Dasyatidae. In: Smith MM, Heemstra PC, editors. Smiths' sea fishes. Berlin: Springer-Verlag; 1986. p. 135–42.

Compagno LJV, Dagit DD. *Hydrolagus trolli*. The IUCN Red List of Threatened Species; 2015. https://doi.org/10.2305/IUCN.UK.2015-4.RLTS.T60197A70709551.en

Compagno LJV, Ebert DA, Smale MJ. Guide to the sharks and rays of southern Africa. London: New Holland (Publication) Ltd; 1989. p. 158.

Cornic A. Poissons de l'Ile Maurice. Editions de l'Océan Indien, Stanley Rose Hill, Ile Maurice; 1987. p. 335.

Coutière H. Poissons venimaux et posissons vènèneux. Paris: The'se Agrèg. Carrè et Naud; 1899. p. 221.

Cox G, Francis M. Sharks and rays of New Zealand. Canterbury University Press; 1997. p. 68.

Dagit DD, Walls RHL, Buscher E. *Harriotta raleighana*. The IUCN Red List of Threatened Species; 2016.: https://doi.org/10.2305/IUCN.UK.2016-1.RLTS.T60140A3088899.en

Daly JS, Scharf MJ. Injuries caused by venomous fish spines. In: Goldsmith L, Katz S, Gilchrest B, Paller A, Leffell D, Wolff K, editors. Fitzpatrick's dermatology in general medicine. 8th ed. New York: McGraw-Hill; 2012.

De la Cruz Agüero J, Arellano Martínez M, Cota Gómez VM, de la Cruz-Agüero G. Catalogo de los peces marinos de Baja California Sur. IPN-CICIMAR, La Paz, Mexico; 1997. p. 346.

Debelius H. Fischführer Mittelmeer und Atlantik. Hamburg: Jahr Verlag GmbH & Co.; 1998. p. 305.

Dehghani H, Sajjadi MM, Rajaian H, Sajedianfard J, Parto P. Study of patient's injuries by stingrays, lethal activity determination and cardiac effects induced by Himantura gerrardi venom. Toxicon. 2009;54:881–6.

Didier DA. Two new species of chimaeroid fishes from the southwestern Pacific Ocean (Holocephali, Chimaeridae). Ichthyol Res. 2002;49(4):299–306.

Didier DA. Phylogeny and classification of extant Holocephali. In: Biology of sharks and their relatives, vol. 4; 2004. p. 115–38.

Didier DA. Two new species of the genus *Hydrolagus* Gill (Holocephali: Chimaeridae) from Australia. In: Last PR, White WT, Pogonoski JJ, editors. Descriptions of New Australian Chondrichthyans. CSIRO Marine and Atmospheric Research Paper no. 22; 2008.

Didier DA, Séret B. Chimaeroid fishes of New Caledonia with description of a new species of *Hydrolagus* (Chondrichthyes, Holocephali). Cybium. 2002;26(3):225–33.

Druett J. Exotic intruders: the introduction of plants and animals in New Zealand. Auckland: Heinemann Publisher; 1983. p. 291.

Duhig JV. The nature of the venom of "Synanceia horrida" (the stonefish). Z Immunita¨tsforsch. 1929;62:185–9.

Duhig JV, Jones G. The venom apparatus of the stone fish (Synanceia horrida). Mem Queensland Mus. 1928a;9:136–50.

Duhig JV, Jones G. Haemotoxin of the venom of Synanceia horrida. Aust J Exp Biol. 1928b;5:173–9.

Dulvy NK, Reynolds JD. Evolutionary transitions among egg-laying, live-bearing and maternal inputs in sharks and rays. Proc R Soc Lond Ser B Biol Sci. 1997;264:1309–15.

Duron M, Queró JC. Diodontidae. In: Quero JC, Hureau JC, Karrer C, Post A, Saldanha L, editors. Check-list of the fishes of the eastern tropical Atlantic (CLOFETA). JNICT, Lisbon; SEI, Paris; and UNESCO, Paris. Vol. 2; 1990. p. 1073–1076.

Ebert DA. Sharks, rays and chimaeras of California. California Natural History Guides No. 71. University of California Press; 2003. p. 284.

Ebert DA, White WT, Goldman KJ, Compagno LJV, Daly-Engel TS, Ward RD. Resurrection and redescription of *Squalus suckleyi* (Girard, 1854) from the North Pacific, with comments on the *Squalus acan-*

thias subgroup (Squaliformes: Squalidae). Zootaxa. 2010;2612:22–40.

Eger WH. An exotoxin produced by the puffer, Arthron hispidus, with notes on the toxicity of other plectognath fishes. M.S. Thesis. University of Hawaii; 1963. p. 88.

Egge JJD, Simons AM. Evolution of venom delivery structures in madtom catfishes (Siluriformes: Ictaluridae). Biol J Linn Soc. 2011;102:115–29. https://doi.org/10.1111/j.1095-8312.2010.01578.x.

Ehemann NR. A new western Atlantic size record for *Pterois volitans* (Actinopterygii: Scorpaeniformes: Scorpaenidae) with new morphological data on head spinules. Acta Ichthyol Piscat. 2017;47(3):297–301.

Endean R. A study of the distribution, habitat, behavior, venom apparatus and venom of the stonefish. Aust J Mar Freshw Res. 1961;12:177–90.

Endler JA. Defense against predators. In: Feder ME, Lauder GV, editors. Predator-prey relationships: perspectives and approaches from the study of lower vertebrates. Chicago: University of Chicago Press; 1986. p. 109–34. https://doi.org/10.2307/1445310.

Englesen H. Om giftfiskog giftige fisk. Nord Hyg Tidskrift. 1922;3:316–25.

Eschmeyer WN, Herald ES, Hammann H. A field guide to Pacific coast fishes of North America. Boston: Houghton Mifflin Company; 1983. p. 336.

Escoubas P, Diochot S, Corzo G. Structure and pharmacology of spider venom neurotoxins. Biochimie. 2000;82:893–907.

Eupharasen BA. Raja (narinari). K Vetinsk Akad Handl. 1790;11:217–9.

Evans HM. The poison of the spiny dogfish. Br Med J. 1920;1(3087):287–8.

Evans HM. The poison organs and venoms of venomous fish. Br Med J. 1921;2(3174):690–2.

Evans HM. The defensive spines of fishes, living and fossil, and the glandular structure in connection therewith, with observations on the nature of fish venoms. Philosophical Transaction of the Royal Society of London, B. 1923;212:1–33.

Evans HM. Sting fish and seafarer. London: Faber and Faber Ltd; 1943. p. 180.

Evans HM. Sharks: vicious and venomous. Lancet. 1944;247(6331):859–60.

Fenner PJ, Williamson JA, Skinner RA. Fatal and non-fatal stingray envenomation. Med J Aust. 1989;151:621–5.

Ferraris CJ. Checklist of catfishes, recent and fossil (Osteichthyes: Siluriformes), and catalogue of siluriform primary types. Zootaxa. 2007;1418:1–628.

Fine ML, Ladich F. Sound production, spine locking and related adaptations. In: Kapoor BG, Arratia G, Chardon M, Diogo M, editors. Catfishes. Enfield: Science Publishers, Inc.; 2003. p. 248–90.

Fine ML, Friel JP, McElroy D, King CB, Loesser KE, Newton S. Pectoral spine locking and sound production in the channel catfish (Ictalurus punctatus). Copeia. 1997;1997:777–90. https://doi.org/10.2307/1447295.

Finucci B. Ecology of New Zealand deep-sea Chondrichthyans. Victoria University of Wellington: Doctor of Philosophy in Marine Biology; 2017. p. 297.

Finucci B, Kyne PM. *Hydrolagus bemisi*. The IUCN Red List of Threatened Species; 2018. https://doi.org/10.2305/IUCN.UK.2018-2.RLTS.T41826A116737232.en

Finucci B, Cheok J, Chiaramonte GE, Cotton CF, Dulvy NK, Kulka DW, Neat FC, Pacoureau N, Rigby CL, Tanaka S, Walker TI. *Squalus acanthias*. The IUCN Red List of Threatened Species; 2020. https://doi.org/10.2305/IUCN.UK.2020-3.RLTS.T91209505A124551959.en

Finucci B, Cheok J, Ebert DA, Herman K, Kyne PM, Dulvy NK. Ghosts of the deep- biodiversity, fisheries, and extinction risk of ghost sharks. Fish Fish. 2021;22:391–412.

Fitzgerald FT. Animal bites and stings. In: Humes HD, editor. Kelly's textbook of internal medicine. Philadelphia: Lippincott Williams & Wilkins; 2000.

Flaschentrager B, Abdallah MM. Some toxic fishes of the Red Sea. Alexandria Med J. 1957;3:177–88.

Fletcher JE, Hubert M, Wieland SJ, Gong Q, Jiang M. Similarities, and differences in mechanisms of cardiotoxins, melittin and other myotoxins. Toxicon. 1996;34:1301–11.

Florida Museum. The brown bullhead. University of Florida. October 23, 2017. Retrieved July 23, 2023.

Florida Museum of Natural History. Biological profiles: pelagic stingray; 2005. Retrieved on 26 August 2005, from https://www.flmnh.ufl.edu/fish/Gallery/Descript/PelagicStingray/PelagicStingray.html. Ichthyology at the Florida Museum of Natural History: Education-Biological Profiles. FLMNH, University of Florida.

Fordham S. Spiny dogfish. In: Fowler SL, Camhi M, Burgess GH, Cailliet GM, Fordham SV, Cavanagh RD, Simpfendorfer CA, Musick JA, editors. Sharks, rays and chimaeras: the status of the chondrichthyan fishes. Cambridge: IUCN SSC Shark Specialist Group. IUCN; 2005.

Forrester BM. Pattern of stingray injuries reported to Texas poison centres from 1998 to 2004. Hum Exp Toxicol. 2005;24:639–42.

Francis MP. Checklist of the coastal fishes of Lord Howe, Norfolk, and Kermadec Islands, Southwest Pacific Ocean. Pac Sci. 1993;47(2):136–70.

Fredericq L. Die secretion von schutz- und nutzstoffen. In: Winterstein H, editor. Handbuch der vergleichenden physiologie, vol. II. Jena: G. Fischer; 1924. p. 1–87.

Fricke R. Fishes of the Mascarene Islands (Réunion, Mauritius, Rodriguez): an annotated checklist, with descriptions of new species. Köeltz Scientific Books, Koenigstein, Theses Zoologicae, Vol. 31; 1999. p. 759.

Fricke R, Kulbicki M, Wantiez L. Checklist of the fishes of New Caledonia, and their distribution in the Southwest Pacific Ocean (Pisces). Stuttgarter Beiträge zur Naturkunde A, Neue Serie. 2011;4:341–463.

Frimodt C. Multilingual illustrated guide to the world's commercial Coldwater fish. Oxford: Fishing News Books, Osney Mead; 1995. p. 215.

Fujita T. Gustatory cells as paraneurons. In: Kurihata K, Suzuki N, Ogawa H, editors. Olfaction and Tase XI. Tokyou: Springer; 1994. p. 2–4.

Gasparini JL, Floeter SR. The shore fishes of Trindade Island, western South Atlantic. J Nat Hist. 2001;35:1639–56.

Gerard I, Koch T. Steve Irwin's freak death filmed. The Australian; 4 September 2006. Retrieved 31 Jul 2023.

Goodrich ES. Vertebrata craniata. 518p. In: Lancaster R, editor. A treatise on zoology. London: Part IX. A. and C. Black; 1909.

Gray WR, Olivera BM, Cruz LJ. Peptide toxins from venomous conus snails. Annu Rev Biochem. 1988;57:665–700.

Grevin J. Deux livres des venesis. Paris: C. Plantin; 1568. p. 423.

Günther A. On a poison organ in a genus of batrachoid fishes. Proc Zool Soc Lond. 1864;1:155–8.

Guth R. Sterling R, Leonard J, Dewey T, editors. Brown catfish *Ameiurus nebulosus*. *BioKIDS*. Animal Diversity Web; 2011. Retrieved July 23, 2022.

Haddad V Jr. Atlas of dangerous aquatic animals of Brazil: a guide to identification and treatment. Sao Paulo: Ed. Roca; 2000.

Haddad V Jr, Gadig OBF. The spiny dogfish (*Squalus cubensis*/megalops group): the envenoming of a fisherman, with taxonomic and toxicological comments on the Squalus genus. Toxicon. 2005;46(7):828–30.

Haddad V Jr, Lastoria JC. Injuries caused by Pimelodidade freshwater catfishes (mandijubas): clinical and therapeutic aspects. Diagno'stico Tratamento. 2005;10(3):132–3.

Haddad V Jr, Garrone Neto D, de Paula NJB, Marques FPL, Barbaro KC. Freshwater stingrays: study of epidemiologic, clinic and therapeutic aspects based on 84 envenomings in humans and some enzymatic activities of the venom. Toxicon. 2004;43:287–94.

Haddad V, Pardal PP, Cardoso JL, Martins IA. The venomous toadfish Thalassophryne nattereri (niquim or miquim): report of 43 injuries provoked in fishermen of Salinopolis (Para state) and Aracaju (Sergipe state). Brazil Rev Inst Med Trop. 2003a;45:221–3.

Haddad V, Martins IA, Makyama HM. Injuries caused by scorpionfishes (Scorpaena plumieri Bloch, 1789 and Scorpaena brasiliensis Cuvier, 1829) in the Southwestern Atlantic Ocean (Brazilian coast): epidemiologic, clinic and therapeutic aspects of 23 stings in humans. Toxicon. 2003b;42(1):79–83.

Halstead BW. Animal phyla known to contain poisonous marine animals. In: Buckly EE, Porges N, editors. Venoms. Washington: American Association of Advancement of Science; 1956. p. 9–27.

Halstead BW. Dangerous marine animals. Cambridge: Cornell Maritime Press; 1959. p. 146.

Halstead BW. Fish poisonings-their diagnosis, pharmacology, and treatment. Clin Pharmacol Ther. 1964;5(5):615–27.

Halstead BW. Poisonous and venomous marine animals of the world, vol. II. Washington: Government Printing Office; 1967.

Halstead WB. Poisonous and venomous marine animals of the world, vertebrates continued, vol. 3. Washington: US Government Printing Office; 1970.

Halstead BW. Poisonous and venomous marine animals of the world. Princeton: The Darwin Press; 1978.

Halstead BW. Poisonous and venomous marine animals of the world. 2nd ed. Princeton: The Darwin Press; 1988a.

Halstead BW. Poisonous and venomous marine animals of the world. 2nd ed. New York: Darwin Press Inc; 1988b. p. 701–11.

Halstead BW, Bunker NC. The venom apparatus of the ratfish *Hydrolagus colliei*. Copeia. 1952;3:128–38.

Halstead BW, Bunker NC. Stingray attacks and their treatment. Am J Trop Med Hyg. 1953;2(1):115–28.

Halstead BW, Mitchell LR. A review of the venomous fishes of the Pacific area. In: Keegan HL, MacFarlane WV, editors. Venomous and poisonous animals and noxious plants of the Pacific region. New York: Pergamon Press; 1963. p. 173–202.

Halstead BW, Kuninobu LS, Hebard HG. Catfish stings and the venom apparatus of the Mexican catfish, *Galeichthys felis* (Linnaeus). Trans Am Microsc Soc. 1953;72:297–314.

Halstead BW, Chitwood MJ, Modglin FR. Stonefish stings, and the venom apparatus of Synanceja horrida (Linnaeus). Trans Am Microsc Soc. 1956;75:381–97.

Halstead BW, Aucrbach PS, Campbell DR. A colour atlas of dangerous marine animals. Ipswich: Wolfe Medical Publications Ltd, W.S. Cowell Ltd; 1990. p. 192.

Hardy AC. The open sea – its natural history: the world of plankton. Boston: Houghton Mufflin; 1956.

Hardy R, Mackie PR. Observations on the chemical composition and toxicity of ratfish (*Chimaera monstrosa*). J Sci Food Agric. 1971;22(7):382–8.

Haris PI, Chapman D. The conformational analysis of peptides using Fourier transform IR spectroscopy. Biopolymers. 1995;37:251.

Hart JL. Pacific fishes of Canada. Bull. Fish. Res. Board Can. 1973;180:740.

Hashimoto Y. On the toxicity of a puffer, "Nashifugu" (*Spheroides vermicularis radiates*) (in Japanese). Bull Jpn Soc Sci Fish. 1950;16:43–6.

Hayes AJ, Sim AJW. Ratfish (*chimaera*) spine injuries in fishermen. Scott Med J. 2011;56(3):161–3.

Heemstra PC. Additions and corrections for the 1995 impression. In: Smith MM, Heemstra PC, editors. Smiths' sea fishes. Berlin: Springer-Verlag; 1995. p. v–xv.

Hinton. Horned shark, gar, mormyriad, characin, carp, armored catfish, arowana, upside down catfish. Not given; 1962.

Hubbs CL, Hibbard CW. Ictalurus lambda, a new catfish, based on a pectoral spine from the lower Pleistocene of Kansas. Copeia. 1951;1951:9–14. https://doi.org/10.2307/1438044.

IGFA. Database of IGFA angling records until 2001. Fort Lauderdale: IGFA; 2001.

International Business Times. Crocodile hunter's last show completed. International Business Times; 6 January 2007. Archived from the original on 7 Jan 2007. Retrieved 31 Jul 2023.

Jabado RW, Chartrain E, De Bruyne G, Derrick D, Dia M, Diop M, Doherty P, Finucci B, Leurs GHL, Metcalfe K, Pires JD, Seidu I, Soares A-L, Tamo A, Vander Wright WJ, Williams AB. *Bathytaton lata*. The IUCN Red List of Threatened Species; 2021. https://doi.org/10.2305/IUCN.UK.2021-2.RLTS.T104071039A104072486.en

Jackman RE, Hunt WG, Jenkins JM, Detrich PJ. Prey of nesting bald eagles in Northen California. J Raptor Res. 1999;33:87–96.

Jawad LA, Al-Saboonchi A, AL-Dirawi AMH, Maki A. The spot-fin porcupine fish Diodon hystrix (Diodontidae: Pisces) in Iraq: a new record for the Arabian Gulf region. Cah Biol Mar. 2019;60(5):403–7.

Kaatz IM, Stewart DJ, Rice AN, Lobel PS. Differences in pectoral fin spine morphology between vocal and silent clades of catfishes (order Siluriformes): ecomorphological implications. Curr Zool. 2010;56:73–89.

Kizer KW, McKinney HE, Auerbach PS. Scorpaenidae envenomation. A five-year poison Centre experience. JAMA. 1985;253:807–10.

Kobert R. Compendium der praktischen toxikologie. Stuttgart: F. Rnke; 1894. p. 90–3.

Kobert R. Ueber giftfische und fischgifte. Med Woche. 1902; 19:199–201; 20:209–12; 21:21–225.

Kottelat M, Freyhof J. Handbook of European freshwater fishes. Berlin: Publications Kottelat, Cornol and Freyhof; 2007. p. 646.

Kourie JI. A component of platypus (*Ornithorhynchus anatinus*) venom forms slow-kinetic cation channels. J Membr Biol. 1999;172:37–45.

Kuiter RH. Coastal fishes of South-Eastern Australia. Honolulu: University of Hawaii Press; 1993. p. 437.

Kuiter RH, Tonozuka T. Pictorial guide to Indonesian reef fishes. Part 1. Eels- Snappers, Muraenidae - Lutjanidae. Zoonetics, Australia. 1–302; 2001a.

Kuiter RH, Tonozuka T. Pictorial guide to Indonesian reef fishes. Part 3. Jawfishes - Sunfishes, Opistognathidae - Molidae. Zoonetics, Australia; 2001b. p. 623 893.

Kyne PM. *Myliobatis tenuicaudatus*. The IUCN Red List of Threatened Species; 2016. https://doi.org/10.2305/IUCN.UK.2016-1.RLTS.T70686656A70687032.en

Kyne PM, Barreto R, Carlson J, Fernando D, Francis MP, Fordham S, Jabado RW, Liu KM, Marshall A, Pacoureau N, Romanov E, Sherley RB, Winker H. *Pteroplatytrygon violacea*. The IUCN Red List of Threatened Species; 2019. https://doi.org/10.2305/IUCN.UK.2019-1.RLTS.T161731A896169.en

Larson E, Lalone RG, Rivas LR. Comparative toxicity of the Atlantic puffer fishes of the genera *Spheroides*, *Lactophrys*, *Lagocephalus* and *Chilomycterus*. Fed Proc. 1960;19:388.

Last PR, Stevens JD. Sharks and rays of Australia. Australia: CSIRO; 1994. p. 513.

Last PR, White WT, de Carvalho MR, Séret B, Stehmann MFW, Naylor GJP. Rays of the world. CSIRO Publishing, Comstock Publishing Associates; 2016. p. i-ix+1-790.

Leis JM. Diodontidae. Porcupine fishes (burrfishes). In: Carpenter KE, Niem V, editors. FAO species identification guide for fishery purposes. The living marine resources of the Western Central Pacific. Vol. 6. Bony fishes part 4 (Labridae to Latimeriidae), estuarine crocodiles. FAO, Rome; 2001. p. 3958–3965.

Leis JM. Nomenclature and distribution of the species of the porcupinefish family Diodontidae (Pisces, Teleostei). Mem Mus Victoria. 2006;63:77–90. https://doi.org/10.24199/j.mmv.2006.63.10.

Lieske E, Myers R. Collins pocket guide. Coral reef fishes. Indo-Pacific & Caribbean including the Red Sea. Haper Collins Publishers; 1994. p. 400.

Liu J, Zapfe G, Shao K-T, Leis JL, Matsuura K, Hardy G, Liu M, Robertson R, Tyler J. *Diodon hystrix* (errata version published in 2016). The IUCN Red List of Threatened Species; 2015: e.T193668A97664783. Accessed on 29 June 2023.

Low KSY, Gwee MCE, Yusen R, Gopalakrishnakone P, Khoo HE. Stonustoxin: effects on neuromuscular function in vitro and in vivo. Toxicon. 1994;32:573–81.

Maass TA. Gift-tiere. In: Junk W, editor. Tabulae biologicae, vol. 13. Zaltbommel: N.V. van de Garde and Co's Drujjeriji; 1937. p. 272.

MacDonald L, Anaru W. Iwi and local government partnerships in biosecurity: a case study of Te Arawa catfish Killas in response to a pest fish incursion in the Bay of Plenty, New Zealand. Perspectives in Biosecurity. 2020;5:17–22.

Macober RD. An observation on pufferfish toxin. J Wash Acad Sci. 1956;46:85.

Mansueti R. Madtoms pack a punch. Nat Mag. 1951;44:314–6.

Masuda H, Amaoka K, Araga C, Uyeno T, Yoshino T. The fishes of the Japanese archipelago, vol. 1. Tokyo: Tokai University Press; 1984. p. 437.

Matsuura K. Ostraciidae. Boxfishes. In: Carpenter KE, Niem V, editors. FAO species identification guide for fishery purposes. The living marine resources of the Western Central Pacific. Vol. 6. Bony fishes part 4 (Labridae to Latimeriidae), estuarine crocodiles. FAO, Rome; 2001. p. 3948–3951.

May JL, Maxwell JGH. Trawl fish from temperate waters of Australia. Tasmania: CSIRO Division of Fisheries Research; 1986. p. 492.

McAllister DE. A list of the fishes of Canada. Syllogeus No. 64. Nat. Mus. Nat. Sci., Ottawa, Canada; 1990. p. 310.

McEachran JD. Dasyatidae. Rayas-látigo. In: Fischer W, Krupp F, Schneider W, Sommer C, Carpenter KE, Niem V, editors. Guia FAO para Identification de Especies para los Fines de la Pesca. Pacifico Centro-Oriental. 3 Vols. FAO, Rome; 1995. p. 752–755.

McEachran JD, Branstetter S. Squalidae. In: Whitehead PJP, Bauchot M-L, Hureau J-C, Nielsen J, Tortonese E, editors. Fishes of the North-Eastern Atlantic and Mediterranean, vol. 1. Paris: UNESCO; 1984. p. 128–47.

McMillan PJ, Francis MP, James GD, Paul LJ, Marriott PJ, Mackay E, Wood BA, Griggs LH, Sui H, Wei F. New Zealand fishes. Volume 1: A field guide to common species caught by bottom and midwater fishing. New Zealand Aquatic Environment and Biodiversity Report No. 68; 2011, p. 329.

Mecklenburg CW, Mecklenburg TA, Thorsteinson LK. Fishes of Alaska. Bethesda: American Fisheries Society; 2002. p. xxxvii+1037.

Mecklenburg CW, Lynghammar A, Johannesen E, Byrkjedal I, Christiansen JS, Dolgov AV, Karamushko OV, Mecklenburg TA, Møller PR, Steinke D, Wienerroither PR. Marine fishes of the Arctic Region. Volume I. Conservation of Arctic Flora and Fauna, Akureyri, Iceland; 2018. [CAFF Monitoring Series Report 28:1–454.]

Miano JP, Loesser-Casey KE, Fine ML. Description and scaling of pectoral muscles in ictalurid catfishes. J Morphol. 2013;274:467–77. https://doi.org/10.1002/jmor.20108.

Michael SW. Reef sharks and rays of the world. A guide to their identification, behavior, and ecology. Monterey: Sea Challengers; 1993. p. 107.

Michigan Department of Natural Resources. Brown Bullhead, Ameiurus nebulosus. Michigan Department of Natural Resources; 2017. Archived from the original on February 20, 2018. Retrieved July 23, 2022.

Missouri Department of Conservation. The brown bullhead; 2023. Retrieved July 23, 2023.

Moldowan PD, Keevil MG, Kell S, Brooks RJ, Litzgus DJ. Anti-predator defense of brown bullheads (*Ameiurus nebulosus*) and interactions with snapping turtles (Chelydra serpentina). Can Field Natural. 2015;129:189–93. https://doi.org/10.22621/cfn.v129i2.1701.

Motomura H, Matsuura K. *Scorpaena cardinalis*. The IUCN Red List of Threatened Species 2016; 2016a. https://doi.org/10.2305/IUCN.UK.2016-3.RLTS.T69917385A70009717.e

Motomura H, Matsuura K. *Scorpaena onaria*. *IUCN Red List of Threatened Species*; 2016b: e.T79800168A79800171.

Motomura H, Matsuura K. *Pterois volitans*. The IUCN Red List of Threatened Species; 2016c. https://doi.org/10.2305/IUCN.UK.2016-3.RLTS.T190468A78940195.en

Motomura H, Struthers CD, McGrouther MA, Stewart AL. Validity of *Scorpaena jacksoniensis* and a redescription of *S. Cardinalis*, a senior synonym of *S. Cookii* (Scorpaeniformes: Scorpaenidae). Ichthyol Res. 2011;58(4):315–32.

Mundy BC. Checklist of the fishes of the Hawaiian archipelago. Bishop Mus Bull Zool. 2005;6:1–704.

Muramoto K, Kamiya H. The amino-acid sequence of a lectin from conger eel, *Conger myriaster*, skin mucus. Biochim Biophys Acta. 1992;1116:129–36.

Muramoto K, Kagawa D, Sato T, Ogawa T, Nishida Y, Kamiya H. Functional and structural characterization of multiple galectins from the skin mucus of conger eel, *Conger myriaster*. Comp Biochem Physiol B. 1999;123:33–45.

Murdy EO, Musick JA. Field guide to fishes of the Chesapeake Bay. JHU Press; 2013. p. 360.

Muus BJ, Dahlström P. Süßwasserfische. München: BLV Verlagsgesellschaft; 1968. p. 224.

Muus BJ, Nielsen JG. Die Meeresfische Europas in Nordsee, Ostsee und Atlantik. Stuttgart. 1999;Kosmos-Naturführer:336.

Myers RF. Micronesian reef fishes. 2nd ed. Guam: Coral Graphics; 1991. p. 298.

Nakayama N, Matsunuma M, Endo H. A preliminary review and in situ observations of the spookfish genus *Harriotta* (Holocephali: Rhinochimaeridae). Ichthyol Res. 2019;67(1):82–91.

NatureServe. Ameiurus nebulosus. The IUCN Red List of Threatened Species; 2013. https://doi.org/10.2305/IUCN.UK.2013-1.RLTS.T202676A2746713.en

Nelson JS. Fishes of the world. 3rd ed. New York: Wiley; 1994.

New Zealand Water Ways Restoration. First pest fish survey of the Motukaraka Drainage Scheme, Mercer. Report KOI-MDS-49 I NZ Waterways Restoration Ltd I PO Box 413 Orewa; 2019. p. 43.

Nigrelli RF. Dutchman's baccy juice or growth promoting and growth-inhibiting substances of marine origin. Trans N Y Acad Sci. 1958;II(20):248–62.

Nigrelli RF. Antimicrobial substances from marine organisms. introduction: the role of antibiosis in the sea. Trans N Y Acad Sci. 1962;II(24):496–7.

Okamoto M, Tsutsui S, Tasumi S, Suetake S, Kikuchi K, Suzuki Y. Tandem repeat L-rhamnose-binding lectin from the skin mucus of ponyfish, *Leiognathus nuchalis*. Biochem Biophys Res Commun. 2005;333:463–9.

Page LM, Burr BM. A field guide to freshwater fishes of North America north of Mexico. Boston: Houghton Mifflin Harcourt; 2011. p. 663.

Patel MR, Wells S. Lionfish envenomation of the hand. J Hand Surg Am. 1993;18:523–5.

Paulin C, Roberts C. The rockpool fishes of New Zealand (Te ika aaria o Aotearoa). Museum of New Zealand (Te Papa Tongarewa); 1992. p. 177.

Paulin C, Stewart A, Roberts C, McMillan P. New Zealand fish: a complete guide. National Museum of New Zealand Miscellaneous Series No. 19; 1989. p. 279.

Pawlak M, Stankowski S, Schwarz G. Melittin induced voltage dependent conductance in DOPC lipid bilayers. Biochim Biophys Acta. 1991;1062:94–102.

Pawlowsky E. Zur Kenntnis der Giftdrusen von Scorpaena porcus und Trachi nus draco. Trav Soc Imp Natur d St. Petersbourg. 1906;37, Fasc. 1, Nos. 78. (Original text in Russian.)

Pawlowsky EN. Sur la structure des glands a` venin de certains poissons et en particulier de celles de Plotosus. Compt Rènd Soc Biol. 1913;74:1033–6.

Pawlowsky EN. Pisons and poison-producing organs in the animal kingdom (in Russian). Russ J Trop Med. 1929;7:4–11.

Pawson MG, Ellis JR. Stock identity of elasmobranchs in the Northeast Atlantic in relation to assessment and management. J Northw Atl Fish Sci. 2005;35:173–93.

Pearson TH, Rosenberg R. Macrobenthic succession in relation to organic enrichment and pollution of the marine environment. Oceanogr Mar Biol Ann Rev. 1978;16:229–311.

Pequeño G. Peces de Chile. Lista sistematica revisada y comentada. Rev Biol Mar, Valparaiso. 1989;24(2):1–132.

Phisalix M. Animaux venimeux et venins, vol. 2 volumes. Paris: Masson et Cie; 1922.

Poh CH, Yuen R, Khoo HE, Chung MCM, Gwee MCE, Gopalakrishnakone P. Purification and partial characterization of Stonustoxin (lethal factor) from Synanceia horrida venom. Comp Biochem Physiol. 1991;99:793–8.

Ramos AD, Conceição K, Silva PI, Richardson M, Lima C, Lopes-Ferreira M. Specialization of the sting venom and skin mucus of *Cathorops spixii* reveals functional diversification of the toxins. Toxicon. 2012;59(6):651–65.

Randall JE, Cea A. Shore fishes of Easter Island. University of Hawai'i Press; 2011. p. 164.

Randall JE, Allen GR, Steene RC. Fishes of the Great Barrier Reef and Coral Sea. Second Edition. Revised and expanded edition. Crawford House Publishing Pty Ltd. Bathurst, NSW, Australia; 1997. p. 557.

Ravi V. Report on the rare distribution of leopard whip ray, *Himantura undulate* (Bleeker, 1852) from Nagapattinam, Tamil Nadu coast. Geobios-Jodhpur. 2006;33:219–20.

Ravi V. Stingray and electric ray (Chondrichthyes: Rajiformes) diversity along Parangipettai and Nagapattinam coasts, Tamil Nadu. J Aquat Biol. 2007;22(1):55–8.

Reardon MB, Bray DJ. *Hydrolagus homonycteris. Fishes of Australia*; 2023. Retrieved 27 July 2023.

Reed JG. Food. In: Pryor JC, editor. Naval hygiene. Philadelphia: P. Blackiston's Son and co.; 1900a. p. 153–5.

Reed JG. Marine animal life dangerous to man. In: Pryor JC, editor. Naval hygiene. Philadelphia: P. Blackiston's Son and co.; 1900b. p. 309–18.

Reed JG. Notes on the poison organs in fishes. Science. 1906;24:293.

Reed JG. The poison glands of Noturus and Schilbeodes. Am Nat. 1907;41:553–66.

Reed JG, Lloyd TJ. The nature of the spines in catfishes. Trans Amer Fish Soc. 1916;45:202–5.

Riede K. Global register of migratory species - from global to regional scales. Final Report of the R&D-Projekt 808 05 081. Federal Agency for Nature Conservation, Bonn, Germany; 2004. p. 329.

Rigby CL, Chin A, Derrick D. *Bathytoshia brevicaudata*. The IUCN Red List of Threatened Species; 2021. https://doi.org/10.2305/IUCN.UK.2021-2.RLTS.T104039923A104039985.en

Ritter P. Beträge zur kenntriss der stacheden von trygon und acanthias. Thesis, Rostock; 1900.

Roberts C, Stewart AL, Struthers CD, Barker J, Kortet S, Freeborn M. The fishes of New Zealand, vol. 2. Wellington: Te Papa Press; 2015. p. 39.

Robins CR, Ray GC. A field guide to Atlantic coast fishes of North America. Boston: Houghton Mifflin Company; 1986. p. 354.

Roedel PM, Riply WE. California sharks and rays. California Department of Natural Resources, Division of Fish and game, Fish Bulletin; 1950, p. 75.

Rojo A. Osteological atlas of the brown bullhead (Ameiurus nebulosus) from Nova Scotia waters: A morphological and biometric study. Curatorial Report Number 100, Nova Scotia Museum, Halifax; 2013.

Romero P. An etymological dictionary of taxonomy. Madrid, unpublished; 2002.

Rosen N. Studies on the plectognaths. The integument. Arkiv Zool Stockholm. 1913a;8:1–29.

Rosen N. Studies on plectognaths. The body muscles. Arkiv Zool Stockholm. 1913b;8:1–14.

Roth BJ, Geller SM. Deep soft-tissue necrosis of the foot and ankle caused by catfish envenomation: a case report. J Am Podiatr Med Assoc. 2010;100:493–6.

Russell F. Studies on the mechanism of death from stingray venom: a report of two fatal cases. Am J Med Sci. 1958;235:566–84.

Russell FE. Marine foams and venomous and poisonous marine animals. In: Advances to marine biology, vol. 3. London: Academic Press; 1965a. p. 235.

Russell FE. Marine toxins and venomous and poisonous marine animals. In: Russell FS, editor. Advances in marine biology, vol. 3. New York: Academic Press; 1965b. p. 255–384.

Russell BC. The food and feeding habits of rocky reef fish of North-Eastern New Zealand. NZ J Mar Freshwat Res. 1983;17(2):121–45.

Russell FE. The venomous and poisonous marine invertebrates of the Indian Ocean. Enfield: Science Publishers Inc.; 1996. p. 1–14.

Saminathan R, Babuji S, Sethupathy S, Viswanathan P, Balasubaramanian T, Gopalakrishanakone P. Clinicotoxinological characterization of the acute effects of the venom of the marine snail. Conus loroisii Acta Trop. 2006;97:75–87.

Sanches JG. Catálogo dos principais peixes marinhos da República de Guiné-Bissau. Publ Avuls Inst Nac Invest Pescas. 1991;16:429.

Satora L, Kuciel M, Gawlikowski T. Catfish stings and the venom apparatus of the African catfish Clarias gariepinus (Burchell, 1822), and stinging catfish Heteropneustes fossilis (Bloch, 1794). Annls Agric Environ Med. 2008;15:163–6.

Saunders PR, Taylor PB. Venom of the lionfish Pterois volitans. Am J Phys. 1959;197:437–40.

Schilling B, Couperus AS, Bos OG. Fotogids mesopelagische vissen, Noordoost Atlantische Oceaan. Versie 1.0. [Photo guide mesopelagic fish, Northeast Atlantic

Ocean. Version 1.0]. Waginengen Marine Research; 2019.

Schwartz FJ. Tail spine characteristics of stingrays (order Myliobatiformes) found in the Northeast Atlantic, Mediterranean, and Black seas. Electron J Ichthyol. 2005;1(1):1–9.

Scott WB, Crossman EJ. Freshwater fishes of Canada. Bull Fish Res Board Can. 1973;184:xi+1-966.

Selby J. Steve Irwin's final words: cameraman present at death opens up about deadly stingray attack for the first time. The Independent; 10 March 2014. Retrieved 31 Jul 2023.

Shao KT. A checklist of fishes recorded in Taiwan and their distribution around Taiwan. Unpublished database, version of April 1997; 1997.

Shao K-T, Chen J-P. Twelve new records of fishes from Taiwan. J Taiwan Mus. 1988;41(1):113–25.

Singletary EM, Rochman AS, Bodmer JCA, Holstege CP. Envenomations. Med Clin North Am. 2005;89:1195–224.

Sire JY. Development and fine structure of the bony scutes in *Corydoras arcuatus* (Siluriformes, Callichthyidae). J Morphol. 1993;215(3):225–44.

Slaughter RJ, Beasley DM, Lambie BS, Schep LJ. New Zealand's venomous creatures. NZ Med J. 2009;122(1290):83–97.

Smith JLB. The sea fishes of southern Africa. Cape Town: Central News Agency, ltd; 1950. p. 1–180. 8 pls., 3.

Smith MM. Ostraciidae. In: Smith MM, Heemstra PC, editors. Smiths' sea fishes. Berlin: Springer-Verlag; 1986. p. 890–3.

Smith CL. National Audubon Society field guide to tropical marine fishes of the Caribbean, the Gulf of Mexico, Florida, The Bahamas, and Bermuda. New York: Alfred A. Knopf, Inc; 1997. p. 720.

Smith-Vaniz WF, Satapoomin U, Allen GR. *Meiacanthus urostigma*, a new fangblenny from the northeastern Indian Ocean, with discussion and examples of mimicry in species of Meiacanthus (Teleostei: Blenniidae: Nemophini). Aqua J Ichthyol Aquat Biol. 2001;5:25–43.

Sommer C, Schneider W, Poutiers JM. FAO species identification field guide for fishery purposes. The living marine resources of Somalia. FAO, Rome; 1996. p. 376.

Spanier E. Dangerous marine organisms in the coastal waters. In: Nezer Y, Epstein Y, editors. The way out, the skill to survive. Tel Aviv: Israel Ministry of Defense Publishing Office; 1987. p. 235–74. (Hebrew).

Stewart J, Hughes JM. Life-history traits of the southern hemisphere eastern red scorpionfish, *Scorpaena cardinalis* (Scorpaenidae: Scorpaeninae). Mar Freshw Res. 2010;61:1290–7.

Taiwan Data Base. *Scorpaena onaria*. Taiwan Fish Database. Retrieved 13 July 2023.

Tang Y. Beitrag zur kenntnis der morphologie des giftapparates bei den japanischen fischen, nebst bemerkungen über dessen giftigkeit. II Uber den giftapparatbei *Pterois lunulata* Temminck et Sclegel. Yokohama Med Bull. 1953;4:178–84.

Tang Y. Beitrag zur kenntnis der morphologie des giftapparates bei den japanischen fischen, nebst bemerkungen über dessen giftigkeit. V. Uber den giftapparatbei *Hypodytes rubipinnis* (Temminck et Sclegel). Yokohama Med Bull. 1954;5:42–8.

Tani I. Toxicological studies on Japanese puffers (in Japanese). Teikoku Tosho Kabushiki Kaisha. 1945;2(3):103.

Tartar D, Limova M, North J. Clinical and histopathologic findings in cutaneous stingray wounds: a case report. Dermatol Online J. 2013;19(8):19261.

Tasumi S, Yang WJ, Usami T, Tsutsui S, Ohira T, Kawazoe I, Wilder MN, Aida K, Suzuki Y. Characteristics and primary structure of a galectin in the skin mucus of the Japanese eel, Anguilla japonica. Dev Comp Immunol. 2004;28:325–35.

Tauranga City Council. Aquatic management contract survey & reporting for Tauranga City Council. Tauriko Lakes and Papamoa Stormwater System; 2019. p. 57.

The Age (Australia). Croc Hunter Irwin killed by stingray. The Age (Australia); 4 September 2006. Retrieved 31 Jul 2023.

The Daily Telegraph (UK). Crocodile Hunter's final stunt with sea snake. The Daily Telegraph (UK); 30 December 2006. Retrieved 31 Jul 2016.

Theiss SM, Huveneers C, Ebert DA. *Hydrolagus homonycteris*. The IUCN Red List of Threatened Species; 2015. https://doi.org/10.2305/IUCN.UK.2015-4.RLTS.T195433A68629582.en

Tinker SW. Fishes of Hawaii, a handbook of the marine fishes of Hawaii and the Central Pacific Ocean. Honolulu: Hawaiian Service Inc.; 1978. p. 568.

Toyoshima T. Serological study of toxin of the fish Plotosus anguillaris Lace'pe'de. (in Japanese). J Japan Protoz Soc. 1918;6:45–270.

Trickett R, Whitaker IS, Boyce DE. Sting-ray injuries to the hand: case report, literature review and a suggested algorithm for management. J Plast Reconstr Aesthet Surg. 2009;62:270–3.

Tsutsui S, Tasumi S, Suetake H, Suzuki Y. Lectins homologous to those of monocotyledonous plants in the skin mucus and intestine of pufferfish, *Fugu rubripes*. J Biol Chem. 2003;278:20882–9.

Tsutsui S, Iwamoto K, Nakamura O, Watanabe T. Yeast binding C-type lectin with opsonic activity from conger eel (*Conger myriaster*) skin mucus. Mol Immunol. 2007;44:691–702.

University of Wisconsin Sea Grant Institute. The brown bullhead. Archived from the original on April 21, 2017; 2023. Retrieved July 23, 2023.

Vetrano SJ, Lebowitz JB, Marcus S. Lionfish envenomation. J Emerg Med. 2002;23:379–82.

Walker TI, Francis MP, Reardon MB. *Callorhinchus milii*. The IUCN Red List of Threatened Species; 2015. https://doi.org/10.2305/IUCN.UK.2015-4.RLTS.T41743A68610951.en

Walovich KA, Ebert DA, Kemper JM. *Hydrolagus erithacus* sp. nov. (Chimaeriformes: Chimaeridae), a new species of chimaerid from the southeastern Atlantic and southwestern Indian oceans. Zootaxa. 2017;4226(4):509–20.

Wheeler A. Das grosse Buch der Fische. Stuttgart: Eugen Ulmer GmbH & Co.; 1977. p. 356.

White WT, Yearsley GK, Last PR. Clarification of the status of *Squalus tasmaniensis* and a diagnosis of *Squalus acanthias* from Australasia, including a key to the Indo-Australian species of *Squalus*. CSIRO Marine and Atmospheric Research Paper, 14; 2007, p. 109–115.

Whyte SK. The innate immune response of finfish – a review of current knowledge. Fish Shellfish Immunol. 2007;23:1127–51.

Wiener S. Observations on the venom of the stone fish (Synanceja trachynis). Med J Aust. 1959;1(19):620–7.

Wikipedia. 2016. https://en.wikipedia.org/wiki/Steve_Irwin. Retrieved 31 Jul 2023.

Wright JJ. Diversity, phylogenetic distribution, and origins of venomous catfishes. BMC Evol Biol. 2009;9(1):282.

Wright JJ. Adaptive significance of venom glands in the tadpole madtom Noturus gyrinus (Siluriformes: Ictaluridae). J Exp Biol. 2012;215:1816–23. https://doi.org/10.1242/jeb.068361.

Common Name Index

L. A. Jawad, *Dangerous Fishes of New Zealand*, https://doi.org/10.1007/978-3-031-69698-5

Scientific Name Index

L. A. Jawad, *Dangerous Fishes of New Zealand*, https://doi.org/10.1007/978-3-031-69698-5

General Index

L. A. Jawad, *Dangerous Fishes of New Zealand*, https://doi.org/10.1007/978-3-031-69698-5